# Advances in Genome Science: Probing Intracellular Regulation

## Volume 2

## Editor

**Christian Neri**

*INSERM, Laboratory of Neuronal Cell Biology & Pathology Psychiatry and Neuroscience Center*

*U894 2 ter rue d'Alesia*

*75014 Paris*

*France*

# CONTENTS



***Contd…..***

# PREFACE

Genome science has become essential in advancing knowledge in the wide field of biological and clinical research. The e-book 'Advances in Genome Science' is intended to provide the expert and non-expert community with a multi-disciplinary and accessible view of some of the latest developments in the field, allowing readers to capture the essentiality and diversity of genomics. The second volume of the e-book reach out to studies at the intersection of genomics and cell molecular biology, which together with Volume 1, may provide a valuable reference and handbook for researchers and clinicians, as well as students and medical professionals interested in the impact of genome science on our understanding of biological processes.

Christian Neri
INSERM, Laboratory of Neuronal Cell Biology &
Pathology Psychiatry and Neuroscience Center
U894 2 ter rue d'Alesia
75014 Paris
France
E-mail: christian.neri@inserm.fr

# LIST OF CONTRIBUTORS

**Ignacio Rego-Pérez**
Biomedical Research Institute of Corunna, INIBIC-University Hospital La Coruña La Coruña, Spain

**Natividad Oreiro-Villar**
Biomedical Research Institute of Corunna, INIBIC-University Hospital La Coruña La Coruña, Spain

**Carlos Fernández-López**
Biomedical Research Institute of Corunna, INIBIC-University Hospital La Coruña La Coruña, Spain

**Jose-Luis Fernández-Sueiro**
Biomedical Research Institute of Corunna, INIBIC-University Hospital La Coruña La Coruña, Spain

**Francisco J. Blanco**
Biomedical Research Institute of Corunna, INIBIC-University Hospital La Coruña La Coruña, Spain

**Fabio Coppedè**
Faculty of Medicine, University of Pisa, *Via* Roma 67, 56126 Pisa, Italy

**Natália Freitas**
Unit of Medical Microbiology, Center of Malaria and other Tropical Diseases, Institute of Hygiene and Tropical Medicine, New University of Lisbon Street Junqueira, 100 1349-008 Lisbon, Portugal

**Celso Cunha**
Unit of Medical Microbiology, Center of Malaria and other Tropical Diseases, Institute of Hygiene and Tropical Medicine, New University of Lisbon Street Junqueira, 100 1349-008 Lisbon, Portugal

**Alessandra di Masi**
Interdepartmental Laboratory of Electron Microscopy, University Roma Tre, *Via* della Vasca Navale 79, I-00146 Roma, Italy

**Francesco Berardinelli**
Department of Biology, University Roma Tre, Viale Guglielmo Marconi 446, I-00146 Roma, Italy

**Domenica Cilli**
Interdepartmental Laboratory of Electron Microscopy, University Roma Tre, *Via* della Vasca Navale 79, I-00146 Roma, Italy

**Antonio Antoccia**
Department of Biology, University Roma Tre, Viale Guglielmo Marconi 446, I-00146 Roma, Italy

**Claudia Ghigna**
Institute of Molecular Genetics, National Research Council (IGM-CNR), Pavia, Italy

**Cristina Valacca**
Institute of Molecular Genetics, National Research Council (IGM-CNR), Pavia, Italy

**Giuseppe Biamonti**
Institute of Molecular Genetics, National Research Council (IGM-CNR), Pavia, Italy

**Li J. Xiao**
Department of Biological Sciences, Louisiana State University, Baton Rouge, LA 70803, USA

**Anne Grove**
Department of Biological Sciences, Louisiana State University, Baton Rouge, LA 70803, USA

**Reinhard Hehl**
Technical University of Braunschweig, Institute of Genetics, Spielmannstr. 7, D-38106 Brunswick, Germany

**Julio. C. Bolívar**
Technical University of Braunschweig, Institute of Genetics, Spielmannstr. 7, D-38106 Brunswick, Germany

**Jeannette Koschmann**
Technical University of Braunschweig, Institute of Genetics, Spielmannstr. 7, D-38106 Brunswick, Germany

**Yuri Brill**
Technical University of Braunschweig, Institute of Genetics, Spielmannstr. 7, D-38106 Brunswick, Germany

**Lorenz Bülow**
Technical University of Braunschweig, Institute of Genetics, Spielmannstr. 7, D-38106 Brunswick, Germany

**Ronny V.L. Joosen**
Laboratory of Plant Physiology, Wageningen University, Droevendaalsesteeg 1, NL-6708 PB Wageningen, The Netherlands

**Wilco Ligterink**
Laboratory of Plant Physiology, Wageningen University, Droevendaalsesteeg 1, NL-6708 PB Wageningen, The Netherlands

**Henk W.M. Hilhorst**
Laboratory of Plant Physiology, Wageningen University, Droevendaalsesteeg 1, NL-6708 PB Wageningen, The Netherlands

**Beatriz Llamusí**
Foundation Research University Teaching Hospital /INCLIVA, Valencia, Spain

**Ariadna Bargiela**
Department of Genetics, University of Valencia, Doctor Moliner, 50, E46100 Burjasot, Valencia, Spain

**Ruben Artero**
Department of Genetics, University of Valencia, Doctor Moliner, 50, E46100 Burjasot, Valencia, Spain

**Mark E. Samuels**
Centre de Recherche du CHU Ste-Justine, University of Montreal, Montreal, Canada

**Jianbing Mu**
LMVR, NIAID, NIH, 12735 Twinbrook Parkway, Room 3E-32C, Rockville, MD 20852, USA

**Karl B. Seydel**
Michigan State University, East Lansing, MI, USA

**Adam Bates**
LMVR, NIAID, NIH, 12735 Twinbrook Parkway, Room 3E-32C, Rockville, MD 20852, USA

**Junhui Duan**
Laboratory of Malaria Immunology and Vaccinology, National Institute of Allergy and Infectious Diseases, National Institutes of Health, Bethesda, MD, USA

**Lubin Jiang**
Institut Pasteur of Shanghai, Chinese Academy of Sciences, Shanghai, China

**Xin-zhuan Su**
LMVR, NIAID, NIH, 12735 Twinbrook Parkway, Room 3E-32C, Rockville, MD 20852, USA

*Advances in Genome Science, Vol. 2*, 2013, 3-27 

# CHAPTER 1

# Pharmacogenetics of Rheumatoid Arthritis and Spondyloarthropathies

**Ignacio Rego-Pérez[*,1], Natividad Oreiro-Villar[1], Carlos Fernández-López[1], Jose-Luis Fernández-Sueiro[1] and Francisco J. Blanco[1,2]**

*[1]Rheumatology Division, Genomic Group, INIBIC-Hospital Universitario A Coruña, Spain; [2]Department of Medicine, Universidad Santiago Compostela, Spain*

**Abstract:** Rheumatoid arthritis (RA), ankylosing spondylitis (AS) and psoriatic arthritis (PsA) are chronic and systemic diseases with an important inflammatory component and differing manifestations, all causing a significant decline in the quality of life of patients. RA produces bone and cartilage destruction, as well as a wide variety of extra-articular manifestations; AS mainly affects joints in the spine and sacroilium in the pelvis, and can cause eventual fusion of the spine; and PsA is an inflammatory arthritis associated with psoriasis.

These diseases are characterized by an unknown aetiology as well as a strong genetic contribution, supported by evidence from family and twin studies, linkage studies and population-based association studies. Proinflammatory cytokines, such as tumour necrosis factor-α (TNF-α) and several interleukins (IL-1, IL-6, IL-12) mediate the pathogeneses of these diseases. The human leukocyte antigen (HLA) region has been consistently associated with autoimmune pathologies, with HLA-DR mainly associated with RA and HLA-B27 with both AS and PsA.

These chronic inflammatory diseases are highly heterogeneous in all their aspects, including drug efficacy and toxicity in individual patients. A substantial proportion of patients do not respond efficiently to classical therapies and/or experience toxicity limiting further treatment, both of which lead to progression of inflammation. This high inter-individual variability in drug response and toxicity gives rise to the need to individualize and optimize therapy with anti-rheumatic agents, such as DMARDs [*i.e.*, Methotrexate (MTX) and biological agents (*i.e.*, anti-TNF agents)]. Studies in the field of pharmacogenetics are necessary to identify potential markers associated with clinical response. To date, many polymorphisms in genes associated not only with the diseases, but also with drug-metabolizing enzymes, drug targets and transporters, have been analysed, showing in some cases conflicting results. The aim of this review is to describe the state of the art of pharmacogenetics in three important chronic inflammatory diseases, RA, AS and PsA.

**Keywords:** Genetics, pharmacogenetics, inflammatory disease, rheumatoid arthritis, psoriatic arthritis, ankylosing spondylitis.

***Address correspondence to Ignacio Rego-Pérez:** INIBIC-Hospital Universitario A Coruña, Edificio Anexo Hospital Materno Infantil, A Coruña, Spain; Tel: 34-981-176399; Fax: 34-981-176398; E-mail: Ignacio.rego.perez@sergas.es

**Christian Neri (Ed)**

## INTRODUCTION

Rheumatoid arthritis (RA), ankylosing spondylitis (AS) and psoriatic arthritis (PsA) are chronic and systemic diseases with an important inflammatory component and differing manifestations, all causing a significant decline in the quality of life of patients. RA is a highly heterogeneous disease affecting 0.5-1% of the population worldwide [1], characterized by chronic inflammation of the synovial joints, which leads to bone and cartilage destruction, as well as several extra-articular manifestations. A wide variety of mediators, both inflammatory and non-inflammatory, contribute to the pathogenesis of RA, including Interleukin-1β (IL-1β), tumor necrosis factor-α (TNF-α), the metalloproteinases (MMPs), and B lymphocytes or macrophages. The heterogeneity of this disease is reflected by difficulty in making an early diagnosis and predicting individual responses to therapy, as well as its clinical manifestations. Studies of serological factors, such as the rheumatoid factor (RF), which is unique to RA, and the anti-citrullinated protein antibodies (ACPA), which are highly specific for RA [2], have led to sub-classifications of the phenotype of RA. For example, patients with ACPA-positive RA have a more aggressive clinical course of the disease compared to patients lacking ACPA [3].

AS is a hyper-responsive joint disease associated with substantial bone and cartilage apposition, mainly affecting joints in the spine and sacroilium in the pelvis and causing eventual fusion of the spine. Low back pain is the earliest clinical manifestation for AS, indicating inflammation in the above-mentioned joints that can be identified by magnetic resonance imaging (MRI) [4].

PsA was considered a subtype of RA that occurred in combination with skin psoriasis (PS) [5]; up to 30% of PS patients develop PsA. PsA patients suffer significant joint damage and disability over time and, like RA, PsA is an erosive disease that leads to resorption of cortical bone. This pathology also shows the formation of bony spurs along the insertion sites of the entheses (entesiophytes) [6].

Sibling and twin pair studies demonstrated that genetic factors have a substantial impact on the severity of these diseases and, for RA, the genetic contribution ranges from 50% to 60% [7], while for AS the genetic factors contribute over 90% to overall susceptibility, the rest being due to random environmental effects [8]. Studies show that PsA also has a strong and complex genetic component.

The efficacy of new drugs used to treat these diseases is variable, especially those that block TNF. The main focus of pharmacogenetics is the study of not only drug-metabolizing enzymes, but also polymorphisms in drug targets and transporters [9]. With this background in mind, the purpose of this review is to describe the current status of the clinical influences of the most important genetic polymorphisms, as well as their influence on the response to treatment for the major inflammatory joint diseases, RA, AS and PsA.

## RHEUMATOID ARTHRITIS

### Susceptibility Genes in RA

Although the pathogenesis of RA is poorly understood, progress has been made in identifying genetic factors that contribute to the disease. To date, the most important genetic factor in RA is found in the HLA locus, the contribution of which to genetic variation in RA has been estimated to be 37% [10], particularly the HLA molecules that carry the "shared epitope" (SE), consisting of HLA-DRB1 alleles that encode a conserved sequence of amino acids at positions 70 to 74 in the DRβ1 chain [11]. The presence and dosage of alleles encoding the SE have been associated with more rapid development of erosions, presence of rheumatic nodules, and increased need for surgery [12]. More recent studies calculate the contribution of HLA alleles to the genetic variance as about 40% for ACPA-positive RA, but only 2% for ACPA-negative RA [13]. This indicates that, from the genetic perspective, ACPA-positive and ACPA-negative represent two forms of RA.

Since the performance of genome-wide association studies (GWAS) led to powerful case control studies, an increased number of non-HLA genes have recently been associated with RA. The results of these studies revealed more than 30 gene regions associated with RA [2, 14, 15] (Fig. **1**).

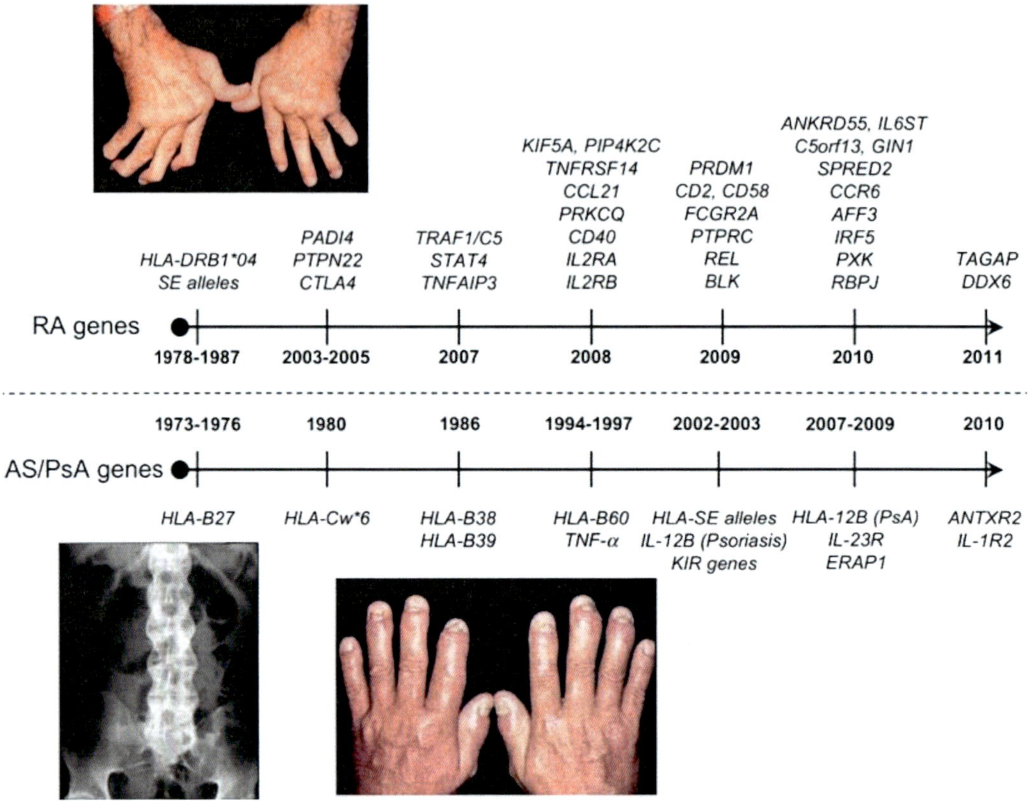


**Figure 1:** Different non-HLA gene regions associated mainly to RA have been described in the last years as a consequence of the powerful case control studies performed by means of GWAS.

The first non-HLA gene clearly associated with RA was peptidylarginine deiminase type 4 (PADI4) in a Japanese population in 2003 [16]. This gene encodes the enzyme that changes arginine into citrulline, the target of ACPA [2]. This was followed by the discovery of the protein tyrosine phosphatase non-receptor type 22 gene (PTPN22) in 2004 [17] and cytotoxic T-lymphocyte antigen 4 (CTLA4) in 2005 [18], found during a candidate gene study.

However, despite the rapidly expanding number of single nucleotide polymorphisms (SNPs) found to be associated with RA from GWAS studies, the odds ratios (ORs) of the associations with non-HLA loci are considerably lower than those of HLA [13].

## Pharmacogenetics of RA

Therapy for RA is actually based on strict monitoring of disease activity and tight control treatment to prevent progression of joint damage and functional disability [19]. In current clinical practice, newly diagnosed RA patients are treated with traditional disease modifying anti-rheumatic drugs (DMARDs), such as methotrexate (MTX) [20]. In cases with low response to DMARDs or those who experience undesirable side effects or toxicity, a combination of MTX and biologicals is recommended [21]. However, 40-60% of RA patients fail to achieve a satisfactory response and 15-30% develop adverse drug events [21, 22], indicating a substantial number of unidentified SNPs influencing the treatment outcome of DMARDs and biological agents. This review will focus on the responses to MTX and TNF inhibitors.

### *Pharmacogenetics of methotrexate in RA*

MTX is one of the most widely used drugs for treatment of RA and, although its exact mechanism of action is unclear, there are many enzymes important to its immunosuppressive effects [23]. The principal pharmacological effect of MTX is thought to be antagonism to folate. MTX enters the cell by the transporter-enzyme reduced folate carrier-1 (RFC1) and is converted intracellularly to MTX polyglutamates by the enzyme folypolyglutamate synthetase (FPGS). This situation can be reversed by gamma-glutamyl hydrolase (GGH). MTX can efflux the cell by ATP-binding cassette transporters (ABC) [24].

Polyglutamated MTX inhibits several enzymes of the folate pathway directly, such as thymidylate synthase (TYMS), dehydrofolate reductase (DHFR) and the conversion of 5-aminoimidazole-4-carboxamide ribonucleotide (AICAR) to formyl-AICAR, facilitated by the enzyme AICAR transformylase (ATIC). The accumulation of AICAR inhibits other enzymes, such as adenosine monophosphate deaminase (AMPD1), leading to the release of adenosine, a potential anti-inflammatory agent [25]. Polyglutamated-MTX also indirectly inhibits methylenetetrahydrofolate reductase (MTHFR), a key enzyme in the folate pathway [23].

To date, several SNPs in genes encoding enzymes involved in metabolic pathways and transport enzymes of MTX have been described (Table **1**), however, their association with efficacy and/or toxicity has shown mixed results.

**Table 1: Pharmacogenetics of methotrexate (MTX) in Rheumatoid Arthritis (RA)**

| Gene Symbol | Polymorphism | Effect of Polymorphism/Function | Pharmacogenetics |
|---|---|---|---|
| MTHFR | 677C>T | Thermolabile variant of MTHFR enzyme with decreased enzyme activity | Increased gastrointestinal side effects and increased hepatic toxicity [26]; TT associated with poor response [29]; CC associated with good response [30, 31]; No effects on toxicity or efficacy [27, 28] |
| | 1298A>C | Decreased MTHFR activity | C allele associated with Increased MTX efficacy [32, 33] and toxicity [29, 34, 35]; AA genotype associated with decrease in DAS score [30] |
| MTHFD1 | 1958G>A | Affects enzymatic activity | Inefficacy to MTX [35] |
| SHMT1 | 1420C>T | Affects enzymatic activity | CC genotype associated with increased response to MTX [29] and with alopecia and central nervous system side effects [36] |
| TYMS | 3´UTR 6bp deletion | Decreased mRNA stability and expression | Increased MTX efficacy [27] |
| DHFR | -473G>A | Direct target of MTX. Involved in the synthesis of purines | No effects on efficacy or toxicity [30] |
| | 3589G>A | Direct target of MTX. Involved in the synthesis of purines | No effects on efficacy and toxicity [30] |
| ATIC | 347C>G | AICAR accumulation and increased adenosine | GG genotype associated with improved DAS score and increased toxicity [30] |
| RFC1 | 80G>A | Increased MTX entry into cell | AA associates with increased response [37] |
| ABCB1 | C3435T | Involved in MTX transport | TT associated with efficacy [38, 39] and toxicity [40]; TT associated with poor response to treatment [41] |
| ABCC2 | 1249G>A | Involved in MTX transport | Increased gastrointestinal toxicity in African Americans [42] |
| | 1058G>A | Involved in MTX transport | Increased hepatotoxicity in African Americans [42] |

**(Table 1) contd.....**

| Gene Symbol | Polymorphism | Effect of Polymorphism/Function | Pharmacogenetics |
|---|---|---|---|
| FPGS | 114G>A | MTX polyglutamation | No associations with efficacy and toxicity [43] |
| | 1994A>G | MTX polyglutamation | No associations with efficacy and toxicity [43] |
| GGH | 16C>T | MTX polyglutamation | Non-TT associated with liver dysfunction in Japanese [44] |
| | 452C>T | MTX polyglutamation | No significant influence in efficacy and toxicity [43] |
| | -401C>T | MTX polyglutamation | No significant influence in efficacy and toxicity [43] |

MTHFR: methylenetetrahydrofolate reductase; MTHFD1: methylenetetrahydrofolate dehydrogenase; SHMT1: methylenetetrahydrofolate; TYMS: thymidylate synthase; DHFR: dehydrofolate reductase; ATIC: aminoimidazole carboxamide ribonucleotide transformylase; RFC1 = reduced folate carrier 1; ABCB1; adenosine triphospate-binding cassette B1; ABCC2: adenosine triphosphate-binding cassette C2; FPGS: folypolyglutamate synthetase; GGH: gammaglutamyl hydrolase; bp: basepair

Best-studied polymorphisms influencing the metabolic pathways of MTX include SNPs at positions 677C>T and 1298A>C within the gene encoding for MTHFR. The association of these two SNPs with both treatment outcome and toxicity has been reported. Both 677T and 1298C variants associate with decreased enzyme activity [26] and, eventually, could influence the gastrointestinal tract in RA patients receiving MTX therapy [26]. However, contradictory results are shown in the literature; some studies found no association with toxicity and efficacy [27, 28], while other reports show that carriers of the MTHFR 677TT genotype are less likely to respond to MTX therapy than those with alternate genotypes [29], and patients genotyped for MTHFR 677CC were associated with a better response [30, 31]. For RA patients carrying allele C at position 1298, some studies exhibit increased MTX efficacy [32, 33] and increased risk of toxicity [29, 34, 35], although one report found an association between the MTHFR 1298AA genotype and a decrease in the DAS score [30].

In addition to MTHFR, there are other enzymes indirectly influenced by polyglutamated MTX, such as methylenetetrahydrofolate dehydrogenase (MTHFD1) and methylenetetrahydrofolate (SHMT1) [23]. The 1958G>A polymorphism in the gene encoding for MTHFD1 was correlated with inefficacy of MTX treatment [35]. For SHMT1, the SNP 1420C>T affects enzymatic activity and an association between the 1420CC genotype and increased efficacy has been reported [29]. Some authors report no association with toxicity [29, 36], while others report an association of the 1420CC genotype with alopecia and central nervous system side effects [36].

TYMS, DHFR and ATIC are direct targets of MTX. One study reported an association of one polymorphism of TYMS, a 6 bp deletion at nucleotide 1494 in the 3′-UTR of the gene, with increased MTX efficacy [27]. TYMS is an enzyme that converts deoxyuridine monophosphate (dUMP) to deoxythymidine monophosphate (dTMP). DHFR is an enzyme that catalyzes the reduction of dihydrofolic acid (DHF) to tetrahydrofolic acid (THF), which is required for the *de novo* synthesis of purines. Two SNPs of DHFR have been described, -473G>A and 3589G>A, both having no effect on efficacy or toxicity [30]. For ATIC, one SNP has been described, 347C>G, reported to be associated with improvement in the DAS score, but the 347GG genotype is associated with toxicity [30].

Associations of polymorphisms in the transport enzyme genes RFC1, ABCB1 and ABCC2 have been reported, with contradictory results. Individuals carrying the RFC1 80AA genotype have a greater response to MTX than patients with the wild type genotype 80GG [37]. In contrast, studies of ABCB1 3435C>T reported that genotype 3435TT was associated with efficacy [38, 39], toxicity [40] and worse response to MTX [41]. In addition, for ABCC2, 1249G>A was associated with gastrointestinal toxicity and 1058G>A with hepatotoxicity in African Americans [42].

FPGS and GGH are enzymes involved in MTX polyglutamation. Two SNPs have been described in the gene encoding FPGS, 114G>A and 1994A>G, although no association with efficacy and toxicity [43] was found. The non-TT genotype at 16C>T of GGH was associated with a high risk of liver dysfunction in Japanese patients [44]. Two other SNPs for GGH have been reported, 452C>T and -401C>T, both with no significant influence on efficacy or toxicity [43].

### *Pharmacogenetics of Biological Agents in RA*

The introduction of biological agents has dramatically improved the treatment of RA. Controlled trials with the three anti-TNF agents currently approved for use in daily clinical practice for treatment of RA, Etanercept (Enbrel), Infliximab (Remicade) and Adalimumab (Humira), have been found to reduce the symptoms and signs of the disease, and even slow down radiographic progression [45]. The therapeutic effect of these biological agents is achieved by blocking the interaction of TNF-α with the accessory TNF cell surface receptors. However, there is consistent evidence that 25-30% of RA patients do not respond to these therapies [46], which means that, given their expense, the early identification of patients who respond positively to these drugs would greatly help to establish a cost-effective treatment profile for the use of these molecules [47].

Most studies of genetic predictors of anti-TNF response have focused on candidate genes that play a role in susceptibility to RA, like the HLA-DRB1 SE [48]. Other studies have investigated TNF itself and its signalling pathway [49-51], as well as other cytokines [52]. Contradictory results have been reported about some SNPs of these genes (Table **2**).

The genetic polymorphism in the promoter region of TNF (TNF -308 G>A) has been the most widely studied SNP. Again, contradictory results with response to Infliximab, Etanercept or Adalimumab have been shown. A meta-analysis carried out by O´Rielly, which included a total of 692 RA patients from nine studies that met the inclusion criteria, concluded that the less common allele -308A is associated with a poor response to any TNF-α inhibitor [53]. Other SNPs in this gene that have been analyzed include -238G>A and 857C>T. The haplotypes -238G,-308G, and -857C were associated with inefficacy to treatment with Adalimumab [54], and the allele T of the SNP -857C>T was reported to be associated with good response to Etanercept [55].

SNPs in genes encoding the two TNF-α receptors (TNFRSF1A and TNFRSF1B) have been associated with clinical response in several reports. The SNP TNFRSF1B 196TT has been correlated with higher response to anti-TNF therapy over 24 weeks [56], and the allele G of SNP 676T>G has been correlated with inefficacy to Infliximab and Etanercept [57], while other studies found no significant associations [58]. None of the three SNPs described for the gene TNFRSF1A at positions -609, -580 and -383 had an effect on efficacy to etanercept [48].

**Table 2: Pharmacogenetics of Anti-TNF Agents in Rheumatoid Arthritis (RA), Ankylosing Spondylitis (AS) and Psoriatic Arthritis (PsA)**

| Gene Symbol | Polymorphism | Effect of Polymorphism/Function | Pharmacogenetics |
|---|---|---|---|
| TNF-α | -308G>A | TNF-α production and regulation | A allele associated with poor response to treatment [53] |
| | -238G>A | TNF-α production and regulation | No effect on response to treatment [48]; haplotype -308G/-238G/-857C associated with inefficacy to Adalimumab [54] |
| | -857C>T | TNF-α production and regulation | T allele associated with good response to treatment [55] |
| TNFRSF1B | 196T>G | TNF-α soluble receptor type 1 | TT associated with increased response to treatment after 24 weeks [56] |
| | 676T>G | | G allele associated with inefficacy to Infliximab and Etanercept [57]; no significant association with response to treatment [58] |
| TNFRSF1A | -609 | TNF-α soluble receptor type 2 | No effect on efficacy of treatment [48] |
| | -580 | | |
| | -383 | | |
| IL-10 | -1082G>A | Influence IL-10 production and regulation | GG associated with increased response to etanercept in combination with TNF-α -308GG [52] |
| | -819T>C | | Patients with lower inflammatory haplotype are better etanercept responders [52] |
| | -592A>C | | |
| IL-1β | 3954C>T | Altered production of IL-1β | No association with efficacy of treatment [59, 97] |
| HLA | Specific SE alleles (HLA-DR) | Increased susceptibility to and severity of RA | Specific SE alleles associated with increased response to etanercept [48]; no association with response to treatment [60-62] |
| EYA4 | rs17301249 (G>C) | Stimulates the expression of interferon-β in response to undigested DNA of apoptotic cells (Okabe *et al.* 2009) | The minor allele C associated with increased response to treatment [63] |
| PDZD2 | rs1532269 (G>C) | Influence the secretion of insulin (Tsang *et al.* 2010) | The minor allele C associated with poor response to treatment [63] |

TNF-α: Tumor necrosis factor alpha; TNFRSF1B: TNF-α soluble receptor type 1; TNFRSF1A: TNF-α soluble receptor type 2; IL.10: Interleukin-10; IL-1β: Interleukin-1 beta; HLA; Human leukocyte antigen; EYA4: Eyes absent homolog 4; PDZD2: PDZ domain-containing protein 2; SE: Shared epitope.

Genetic polymorphisms in the gene encoding for IL-10 also described are -1082G>A, -819T>C and -592A>C. An interesting association between the haplotype IL10 -1082GG and TNF -308GG (which reduces inflammatory response) is reported to produce a good response to Etanercept [52]. The genetic polymorphism 3954C>T in the gene encoding for IL-1β did not show any association with the efficacy of Infliximab [59].

Since the HLA region has been widely associated with susceptibility to RA, several studies have been performed to assess the influence of the SE with response to anti-TNF therapy. Only one report showed that RA patients with two copies of the HLA-DR SE alleles had an improved response to Etanercept [48], while other studies did not find any association with response to treatment [60-62].

A recent GWAS carried out by Plant and co-workers in 566 anti-TNF-treated RA patients identified the association of seven genetic loci with response to anti-TNF therapy [63]. The strongest effect was at the EYA4 gene (rs17301249) with the minor allele C conferring improved response to treatment. In contrast, the minor allele C of the PDZD2 gene (rs1532269) was associated with a reduced treatment response. The EYA4 gene has been found to stimulate the expression of interferon-β (IFN-β) in response to undigested DNA of apoptotic cells [64]; and in fact, cross-talk between type I IFN and TNF has recently been investigated in patients with RA [65]. Interestingly, the PDZD2 gene has been reported to influence the secretion of insulin in an animal model [66], and an elevated insulin level is a feature of severe disease in early RA [67].

Other biological agents are used for the treatment of RA, but little or no pharmacogenetic data are available. These treatments are Anakinra, an IL-1 blocker; Tozilizumab, an IL-6 receptor inhibitor; Certolizumab, a new TNF blocker; Abatacept, an inhibitor of the co-stimulation of T cells; and Rituximab, which destroys B cells that have CD-20 on their surfaces.

## ANKYLOSING SPONDYLITIS

### Susceptibility Genes in AS

The recurrence risk ratio for AS in siblings of probands with AS ranges from 50 to 80 and the heritability from twin studies is estimated to be over 90% [8]. A list of the genes described below is shown in Table **3**.

**Table 3: Susceptibility Genes in Ankylosing Spondylitis (AS)**

| Gene Symbol | Polymorphism/Gene Region | Incidence/Comment |
|---|---|---|
| HLA | HLA-B27 | Increased susceptibility to AS. The most significant association with AS [68] |
| | HLA-B60 | The second most notable association with AS [71] |
| IL-23R | rs11209026 (G>A) | The minor allele A associated with decreased risk of AS [73, 78], probably due to a reduction in cellular response to IL-23 [75] |
| ERAP1 | Haplotypes including both SNPs rs30187 (C>T) and/or rs27044 (C>G) | Downregulation of the signalling of inflammatory cytokines [81] |
| IL-1R2 | rs2310173 (G>T) | Impedes the binding of IL-1 to IL-R1. Modest incidence in AS [74] |
| ANTXR2 | rs4333130 (T>C) | Binds to collagen IV and lamimin. Modest effect on AS [74] |

HLA: Human leukocyte antigen; IL-23R: Interleukin-23 receptor; ERAP1: Endoplasmic reticulum aminopeptidase 1; IL-1R2 : Interleukin-1 receptor 2 ; ANTXR2 : Anthraxin receptor 2.

As found for RA, the major histocompatibility complex (MHC) region, which is characterized by extreme linkage disequilibrium, is associated with susceptibility to AS. Specifically, HLA-B27 forms the most significant association with AS, noted since the early 1970s [68]. HLA-B27 is present in over 90% of patients with AS; however, less than 5% of HLA-B27-positive individuals develop AS. Over 69 subtypes of HLA-B27 are currently recognised [69]; these subtypes evolved from the parent and predominant B27 allele B*2705. In Western European Caucasians, HLA-B*2705 comprises about 90% of the population, and HLA-B*2702 about 19% [70].

HLA-B27 is not the only susceptibility factor within the MHC region. The entire genetic contribution of the MHC region to AS is estimated to be 33%, of which HLA-B27 accounts for about 16% [68]. However, among the several association studies performed, the most notable is HLA-B60 [71]. Other associations have not been widely replicated [72].

GWAS have led to the identification of other non-HLA related genes, the two most important associations being the Interleukin 23 receptor (IL-23R) and endoplasmic reticulum aminopeptidase (ERAP1).

In several studies, IL-23R is consistently associated as a susceptibility locus for AS. The most significant association is with the SNP rs11209026 [73, 74]. This SNP was associated as part of an extended haplotype and the association has been subsequently confirmed in studies from Canada, Spain, Portugal and Hungary [75-77].

ERAP1 was identified as a candidate gene for AS in a GWAS performed in 1500 Caucasian controls and 1000 AS patients [78]. ERAP 1 is the strongest non-HLA gene associated with AS, with a genetic contribution of about 26%. This association has also been found in other Caucasian populations [79]. The two major functions of this gene are (i) to trim peptide antigens to optimal length for binding to MHC class I molecules [80], and (ii) to cleave cell surface receptors for the inflammatory cytokines, IL-1 (IL-1R2), IL-6 (IL-6Rα) and TNF (TNFR1), thereby downregulating their signalling [81].

Other candidate genes with more modest evidence for association with AS include the IL-1 gene cluster family [82], particularly IL-1R2 [74], a decoy receptor impeding the binding of IL-1 to IL-R1 [83], and anthraxin receptor 2 (ANTXR2) that binds to collagen type IV and laminin. These genes, however, are not differentially expressed in AS, and their association may reflect linkage with another nearby gene region. Other SNPs have been found, but do not achieve genome-wide significance and/or have not been replicated in other cohorts

## PSORIATIC ARTHRITIS

### Susceptibility Genes in PsA

The prevalence of psoriasis is much higher among first-degree of probands with PsA than in the general population. The genetic factors underlying susceptibility to PsA are closely intertwined with those of cutaneous psoriasis; almost all patients diagnosed with PsA have either a personal or family history of psoriasis. While the heritability of PsA is higher than that of psoriasis, but not as high as that of AS [84], the genetics of PsA is not well defined, probably due to phenotypic heterogeneity and case ascertainment, although associations with HLA class I alleles have consistently been reported [70]. A list of the most representative genes is shown in Table **4**.

**Table 4: Susceptibility Genes in Psoriatic Arthritis (PsA)**

| Gene Symbol | Polymorphism/Gene Region | Incidence/Comment |
|---|---|---|
| HLA | Cw*6 | Associated with Psoriasis and PsA [85] |
| | Cw*0602 | Associated with earlier age of onset [86] |
| | Homozygosity for HLA-Cw | Increased risk of developing PsA [86] |
| | B7, B13, B16 | Disease-susceptibility associations (Rahman *et al.* 2005) |
| | B27 | Associated with axial PsA [87] |
| | B38, B39 | Associated with peripheral polyarthritis [87] |
| | SE alleles | Associated with radiographic erosions [88] |
| | HLA-Cw*6 + HLA-DRB1*07 | Less severe course of arthritis [89] |
| IL-23R | rs11209026 (G>A) | Same as AS [90] |
| IL-12B | rs3212227 (A>C) | Replicated polymorphism associated with PsA and psoriasis [92, 93] |
| | rs6887695 (G>C) | Replicated polymorphism associated with PsA and psoriasis [92, 93] |
| TNF-α | -238G>A | Meta-analysis confirmed association with PsA [91,94] |

HLA: Human leukocyte antigen; SE: Shared epitope; IL-23R: Interleukin-23 receptor; IL-12B: Interleukin-12 B; TNF-α: Tumor necrosis factor alpha.

HLA studies in PsA have revealed disease-susceptibility associations with HLA-B7, -B13, -B16 and -B27 [85]. However, the strongest association is with HLA-Cw*6, which is associated with psoriasis and PsA [85]. In addition, HLA-Cw*0602 associates with an earlier age of onset in PsA, and homozygosity for the HLA-Cw group is significantly associated with increased risk for development of PsA [86]. Other HLA alleles also appear to be important in disease expression, such as HLA-B27, which is associated with axial PsA, and –B38 and –B39, both associated with peripheral polyarthritis [87]. Also, the RA SE was found to be associated with radiographic erosions, while patients with PsA carrying the HLA-Cw*6 and HLA-DRB1*07 alleles have a less severe course of arthritis [88, 89].

Association studies with PsA have primarily involved candidate gene studies because, unlike psoriasis, to date an adequately powered GWAS on PsA has not been performed, although multiple efforts are underway. Therefore, a meta-analysis of genome-wide scans and further studies must be performed.

Replicated association studies identified other non-HLA genes in the PsA-like IL-1 gene cluster: Killer-cell immunoglobulin-like receptor (KIR) genes and IL-23R [86, 90, 91]. Specifically, the association between the SNP rs11209026 in the IL-23R gene and PsA have been confirmed [90], as well as two SNPs at the IL-12B gene (rs3212227 and rs6887695) [92, 93]. Genetic association studies with the TNF-α gene initially showed conflicting results, however, a subsequent meta-analysis confirmed an association between TNF-α -238 polymorphism and PsA [91, 94].

## PHARMACOGENETICS OF AS AND PsA

There is accumulating evidence that anti-TNF therapy is highly effective in AS and PsA, being even more effective than in RA patients. Thus, TNF blockers are considered the first-line immunosuppressive agents in those patients with active AS and PsA for whom treatment with non-steroidal anti-inflammatory drugs and DMARDs for AS and PsA was not successful. However, there is much individual variation in drug efficacy and adverse drug events. There is also evidence that PsA patients treated with MTX suffer increased toxicity when compared with RA patients.

Currently, pharmacogenetic studies in AS and PsA are lacking. After an extensive PubMed search, only one interesting review by Rahman and co-workers described possible effects of several SNPs in candidate genes on the response to anti-TNF treatment in PsA [95]. Most of these SNPs are the same as those for RA patients treated with the same therapy, such as IL-1, IL-10 and TNF-α (Table **2**), and are the genetic polymorphisms of the promoter site of the TNF-α gene, which is the major association to response to anti-TNF therapy. A double-blind, randomized and controlled trial, demonstrated the therapeutic efficacy of an IL-12/23 monoclonal antibody in psoriasis [96]; however, no pharmacogenetic information is available regarding therapeutic response and IL-23 variants in psoriasis.

In summary, convincing pharmacogenetic targets for both AS and PsA are currently lacking.

## CONFLICT OF INTEREST

FJB has received Grants (Clinical Trials, conferences, advisor and publications) from: Abbvie, Amgen, Bioiberica, Bristol Mayer, Celgene, Celltrion, Cellerix, Grunenthal, Gebro Pharma, Lilly, MSD, Merck Serono, Pfizer, Pierre-Fabra, Roche, Sanofi, Servier, UCB. IRP, NOV, JCFL and JLFS declare they do not have any conflict of interest.

## ACKNOWLEDGEMENTS

This work is dedicated to the memory of the author J.L. Fernández-Sueiro, M.D. This study was supported by grants from Fundacion Espanola de Reumatologia (programa GEN-SER) and from Fondo Investigacion Sanitaria (CIBERCB06/01/0040)-Spain, Fondo Investigacion Sanitaria-PI 08/2028, Ministerio Ciencia e Innovacion PLE2009-0144, with a contribution of funds from FEDER (European Community). IR-P was supported by Contrato Miguel Servet-Fondo Investigacion Sanitaria (CP12/03192).

## DISCLOSURE

This is an updated report of an earlier review article by Ignacio Rego-Pérez, Mercedes Fernández-Moreno and Francisco J. Blanco. Curr. Genomics, 2008, 9: 381-393.

## REFERENCES

[1] Silman, A. J.; Pearson, J. E., Epidemiology and genetics of rheumatoid arthritis. *Arthritis Res* **2002**, 4 Suppl 3, S265-72.

[2] Bax, M.; van Heemst, J.; Huizinga, T. W.; Toes, R. E., Genetics of rheumatoid arthritis: what have we learned? *Immunogenetics* **2011**.

[3] van Gaalen, F. A.; Linn-Rasker, S. P.; van Venrooij, W. J.; de Jong, B. A.; Breedveld, F. C.; Verweij, C. L.; Toes, R. E.; Huizinga, T. W., Autoantibodies to cyclic citrullinated peptides predict progression to rheumatoid arthritis in patients with undifferentiated arthritis: a prospective cohort study. *Arthritis Rheum* **2004**, 50, 709-15.

[4] Marzo-Ortega, H.; McGonagle, D.; O'Connor, P.; Hensor, E. M.; Bennett, A. N.; Green, M. J.; Emery, P., Baseline and 1-year magnetic resonance imaging of the sacroiliac joint and lumbar spine in very early inflammatory back pain. Relationship between symptoms, HLA-B27 and disease extent and persistence. In *Ann Rheum Dis*, England, **2009**; Vol. 68, pp 1721-7.

[5] Schett, G.; Coates, L. C.; Ash, Z. R.; Finzel, S.; Conaghan, P. G., Structural damage in rheumatoid arthritis, psoriatic arthritis, and ankylosing spondylitis: traditional views, novel insights gained from TNF blockade, and concepts for the future. In *Arthritis Res Ther*, England, **2011**; Vol. 13 Suppl 1, p S4.
[6] Finzel, S.; Englbrecht, M.; Engelke, K.; Stach, C.; Schett, G., A comparative study of periarticular bone lesions in rheumatoid arthritis and psoriatic arthritis. In *Ann Rheum Dis*, England, **2011**; Vol. 70, pp 122-7.
[7] MacGregor, A. J.; Snieder, H.; Rigby, A. S.; Koskenvuo, M.; Kaprio, J.; Aho, K.; Silman, A. J., Characterizing the quantitative genetic contribution to rheumatoid arthritis using data from twins. *Arthritis Rheum* **2000**, 43, 30-7.
[8] Brown, M. A.; Kennedy, L. G.; MacGregor, A. J.; Darke, C.; Duncan, E.; Shatford, J. L.; Taylor, A.; Calin, A.; Wordsworth, P., Susceptibility to ankylosing spondylitis in twins: the role of genes, HLA, and the environment. *Arthritis Rheum* **1997**, 40, 1823-8.
[9] Evans, W. E.; Relling, M. V., Pharmacogenomics: translating functional genomics into rational therapeutics. In *Science*, United States, **1999**; Vol. 286, pp 487-91.
[10] Deighton, C. M.; Walker, D. J.; Griffiths, I. D.; Roberts, D. F., The contribution of HLA to rheumatoid arthritis. *Clin Genet* **1989**, 36, 178-82.
[11] Gregersen, P. K.; Silver, J.; Winchester, R. J., The shared epitope hypothesis. An approach to understanding the molecular genetics of susceptibility to rheumatoid arthritis. *Arthritis Rheum* **1987**, 30, 1205-13.
[12] Reveille, J. D., Genetic studies in the rheumatic diseases: present status and implications for the future. In *J Rheumatol Suppl*, Canada, **2005**; Vol. 72, pp 10-3.
[13] de Vries, R., Genetics of rheumatoid arthritis: time for a change! *Curr Opin Rheumatol* **2011**, 23, 227-32.
[14] Gregersen, P. K., Susceptibility genes for rheumatoid arthritis - a rapidly expanding harvest. *Bull NYU Hosp Jt Dis* **2010**, 68, 179-82.
[15] Stahl, E. A.; Raychaudhuri, S.; Remmers, E. F.; Xie, G.; Eyre, S.; Thomson, B. P.; Li, Y.; Kurreeman, F. A.; Zhernakova, A.; Hinks, A.; Guiducci, C.; Chen, R.; Alfredsson, L.; Amos, C. I.; Ardlie, K. G.; Barton, A.; Bowes, J.; Brouwer, E.; Burtt, N. P.; Catanese, J. J.; Coblyn, J.; Coenen, M. J.; Costenbader, K. H.; Criswell, L. A.; Crusius, J. B.; Cui, J.; de Bakker, P. I.; De Jager, P. L.; Ding, B.; Emery, P.; Flynn, E.; Harrison, P.; Hocking, L. J.; Huizinga, T. W.; Kastner, D. L.; Ke, X.; Lee, A. T.; Liu, X.; Martin, P.; Morgan, A. W.; Padyukov, L.; Posthumus, M. D.; Radstake, T. R.; Reid, D. M.; Seielstad, M.; Seldin, M. F.; Shadick, N. A.; Steer, S.; Tak, P. P.; Thomson, W.; van der Helm-van Mil, A. H.; van der Horst-Bruinsma, I. E.; van der Schoot, C. E.; van Riel, P. L.; Weinblatt, M. E.; Wilson, A. G.; Wolbink, G. J.; Wordsworth, B. P.; Wijmenga, C.; Karlson, E. W.; Toes, R. E.; de Vries, N.; Begovich, A. B.; Worthington, J.; Siminovitch, K. A.; Gregersen, P. K.; Klareskog, L.; Plenge, R. M., Genome-wide association study meta-analysis identifies seven new rheumatoid arthritis risk loci. In *Nat Genet*, United States, **2010**; Vol. 42, pp 508-14.
[16] Suzuki, A.; Yamada, R.; Chang, X.; Tokuhiro, S.; Sawada, T.; Suzuki, M.; Nagasaki, M.; Nakayama-Hamada, M.; Kawaida, R.; Ono, M.; Ohtsuki, M.; Furukawa, H.; Yoshino, S.; Yukioka, M.; Tohma, S.; Matsubara, T.; Wakitani, S.; Teshima, R.; Nishioka, Y.; Sekine, A.; Iida, A.; Takahashi, A.; Tsunoda, T.; Nakamura, Y.; Yamamoto, K., Functional haplotypes of PADI4, encoding citrullinating enzyme peptidylarginine deiminase 4, are

associated with rheumatoid arthritis. In *Nat Genet*, United States, **2003**; Vol. 34, pp 395-402.

[17] Begovich, A. B.; Carlton, V. E.; Honigberg, L. A.; Schrodi, S. J.; Chokkalingam, A. P.; Alexander, H. C.; Ardlie, K. G.; Huang, Q.; Smith, A. M.; Spoerke, J. M.; Conn, M. T.; Chang, M.; Chang, S. Y.; Saiki, R. K.; Catanese, J. J.; Leong, D. U.; Garcia, V. E.; McAllister, L. B.; Jeffery, D. A.; Lee, A. T.; Batliwalla, F.; Remmers, E.; Criswell, L. A.; Seldin, M. F.; Kastner, D. L.; Amos, C. I.; Sninsky, J. J.; Gregersen, P. K., A missense single-nucleotide polymorphism in a gene encoding a protein tyrosine phosphatase (PTPN22) is associated with rheumatoid arthritis. In *Am J Hum Genet*, United States, **2004**; Vol. 75, pp 330-7.

[18] Plenge, R. M.; Padyukov, L.; Remmers, E. F.; Purcell, S.; Lee, A. T.; Karlson, E. W.; Wolfe, F.; Kastner, D. L.; Alfredsson, L.; Altshuler, D.; Gregersen, P. K.; Klareskog, L.; Rioux, J. D., Replication of putative candidate-gene associations with rheumatoid arthritis in >4,000 samples from North America and Sweden: association of susceptibility with PTPN22, CTLA4, and PADI4. In *Am J Hum Genet*, United States, **2005**; Vol. 77, pp 1044-60.

[19] Vencovsky, J.; Huizinga, T. W., Rheumatoid arthritis: the goal rather than the health-care provider is key. In *Lancet*, England, **2006**; Vol. 367, pp 450-2.

[20] Pincus, T.; Yazici, Y.; Sokka, T.; Aletaha, D.; Smolen, J. S., Methotrexate as the "anchor drug" for the treatment of early rheumatoid arthritis. *Clin Exp Rheumatol* **2003**, 21, S179-85.

[21] Klareskog, L.; van der Heijde, D.; de Jager, J. P.; Gough, A.; Kalden, J.; Malaise, M.; Martin Mola, E.; Pavelka, K.; Sany, J.; Settas, L.; Wajdula, J.; Pedersen, R.; Fatenejad, S.; Sanda, M., Therapeutic effect of the combination of etanercept and methotrexate compared with each treatment alone in patients with rheumatoid arthritis: double-blind randomised controlled trial. In *Lancet*, England, **2004**; Vol. 363, pp 675-81.

[22] Breedveld, F. C.; Weisman, M. H.; Kavanaugh, A. F.; Cohen, S. B.; Pavelka, K.; van Vollenhoven, R.; Sharp, J.; Perez, J. L.; Spencer-Green, G. T., The PREMIER study: A multicenter, randomized, double-blind clinical trial of combination therapy with adalimumab plus methotrexate *versus* methotrexate alone or adalimumab alone in patients with early, aggressive rheumatoid arthritis who had not had previous methotrexate treatment. *Arthritis Rheum* **2006**, 54, 26-37.

[23] Chan, E. S.; Cronstein, B. N., Molecular action of methotrexate in inflammatory diseases. *Arthritis Res* **2002**, 4, 266-73.

[24] Norris, M. D.; De Graaf, D.; Haber, M.; Kavallaris, M.; Madafiglio, J.; Gilbert, J.; Kwan, E.; Stewart, B. W.; Mechetner, E. B.; Gudkov, A. V.; Roninson, I. B., Involvement of MDR1 P-glycoprotein in multifactorial resistance to methotrexate. In *Int J Cancer*, United States, **1996**; Vol. 65, pp 613-9.

[25] Cronstein, B. N., Low-dose methotrexate: a mainstay in the treatment of rheumatoid arthritis. In *Pharmacol Rev*, United States, **2005**; Vol. 57, pp 163-72.

[26] Haagsma, C. J.; Blom, H. J.; van Riel, P. L.; van't Hof, M. A.; Giesendorf, B. A.; van Oppenraaij-Emmerzaal, D.; van de Putte, L. B., Influence of sulphasalazine, methotrexate, and the combination of both on plasma homocysteine concentrations in patients with rheumatoid arthritis. *Ann Rheum Dis* **1999**, 58, 79-84.

[27] Kumagai, K.; Hiyama, K.; Oyama, T.; Maeda, H.; Kohno, N., Polymorphisms in the thymidylate synthase and methylenetetrahydrofolate reductase genes and sensitivity to the

low-dose methotrexate therapy in patients with rheumatoid arthritis. *Int J Mol Med* **2003**, 11, 593-600.

[28] Berkun, Y.; Levartovsky, D.; Rubinow, A.; Orbach, H.; Aamar, S.; Grenader, T.; Abou Atta, I.; Mevorach, D.; Friedman, G.; Ben-Yehuda, A., Methotrexate related adverse effects in patients with rheumatoid arthritis are associated with the A1298C polymorphism of the MTHFR gene. In *Ann Rheum Dis*, England, **2004**; Vol. 63, pp 1227-31.

[29] Dervieux, T.; Greenstein, N.; Kremer, J., Pharmacogenomic and metabolic biomarkers in the folate pathway and their association with methotrexate effects during dosage escalation in rheumatoid arthritis. *Arthritis Rheum* **2006**, 54, 3095-103.

[30] Wessels, J. A.; de Vries-Bouwstra, J. K.; Heijmans, B. T.; Slagboom, P. E.; Goekoop-Ruiterman, Y. P.; Allaart, C. F.; Kerstens, P. J.; van Zeben, D.; Breedveld, F. C.; Dijkmans, B. A.; Huizinga, T. W.; Guchelaar, H. J., Efficacy and toxicity of methotrexate in early rheumatoid arthritis are associated with single-nucleotide polymorphisms in genes coding for folate pathway enzymes. *Arthritis Rheum* **2006**, 54, 1087-95.

[31] Kurzawski, M.; Pawlik, A.; Safranow, K.; Herczynska, M.; Drozdzik, M., 677C>T and 1298A>C MTHFR polymorphisms affect methotrexate treatment outcome in rheumatoid arthritis. *Pharmacogenomics* **2007**, 8, 1551-9.

[32] Urano, W.; Taniguchi, A.; Yamanaka, H.; Tanaka, E.; Nakajima, H.; Matsuda, Y.; Akama, H.; Kitamura, Y.; Kamatani, N., Polymorphisms in the methylenetetrahydrofolate reductase gene were associated with both the efficacy and the toxicity of methotrexate used for the treatment of rheumatoid arthritis, as evidenced by single locus and haplotype analyses. *Pharmacogenetics* **2002**, 12, 183-90.

[33] Taniguchi, A.; Urano, W.; Tanaka, E.; Furihata, S.; Kamitsuji, S.; Inoue, E.; Yamanaka, M.; Yamanaka, H.; Kamatani, N., Validation of the associations between single nucleotide polymorphisms or haplotypes and responses to disease-modifying antirheumatic drugs in patients with rheumatoid arthritis: a proposal for prospective pharmacogenomic study in clinical practice. In *Pharmacogenet Genomics*, United States, **2007**; Vol. 17, pp 383-90.

[34] Herrlinger, K. R.; Cummings, J. R.; Barnardo, M. C.; Schwab, M.; Ahmad, T.; Jewell, D. P., The pharmacogenetics of methotrexate in inflammatory bowel disease. In *Pharmacogenet Genomics*, United States, **2005**; Vol. 15, pp 705-11.

[35] Wessels, J. A.; van der Kooij, S. M.; le Cessie, S.; Kievit, W.; Barerra, P.; Allaart, C. F.; Huizinga, T. W.; Guchelaar, H. J., A clinical pharmacogenetic model to predict the efficacy of methotrexate monotherapy in recent-onset rheumatoid arthritis. *Arthritis Rheum* **2007**, 56, 1765-75.

[36] Weisman, M. H.; Furst, D. E.; Park, G. S.; Kremer, J. M.; Smith, K. M.; Wallace, D. J.; Caldwell, J. R.; Dervieux, T., Risk genotypes in folate-dependent enzymes and their association with methotrexate-related side effects in rheumatoid arthritis. *Arthritis Rheum* **2006**, 54, 607-12.

[37] Dervieux, T.; Furst, D.; Lein, D. O.; Capps, R.; Smith, K.; Walsh, M.; Kremer, J., Polyglutamation of methotrexate with common polymorphisms in reduced folate carrier, aminoimidazole carboxamide ribonucleotide transformylase, and thymidylate synthase are associated with methotrexate effects in rheumatoid arthritis. *Arthritis Rheum* **2004**, 50, 2766-74.

[38] Pawlik, A.; Wrzesniewska, J.; Fiedorowicz-Fabrycy, I.; Gawronska-Szklarz, B., The MDR1 3435 polymorphism in patients with rheumatoid arthritis. *Int J Clin Pharmacol Ther* **2004**, 42, 496-503.

[39] Drozdzik, M.; Rudas, T.; Pawlik, A.; Kurzawski, M.; Czerny, B.; Gornik, W.; Herczynska, M., The effect of 3435C>T MDR1 gene polymorphism on rheumatoid arthritis treatment with disease-modifying antirheumatic drugs. *Eur J Clin Pharmacol* **2006**, 62, 933-7.

[40] Bohanec Grabar, P.; Logar, D.; Lestan, B.; Dolzan, V., Genetic determinants of methotrexate toxicity in rheumatoid arthritis patients: a study of polymorphisms affecting methotrexate transport and folate metabolism. *Eur J Clin Pharmacol* **2008**, 64, 1057-68.

[41] Takatori, R.; Takahashi, K. A.; Tokunaga, D.; Hojo, T.; Fujioka, M.; Asano, T.; Hirata, T.; Kawahito, Y.; Satomi, Y.; Nishino, H.; Tanaka, T.; Hirota, Y.; Kubo, T., ABCB1 C3435T polymorphism influences methotrexate sensitivity in rheumatoid arthritis patients. In *Clin Exp Rheumatol*, Italy, **2006**; Vol. 24, pp 546-54.

[42] Ranganathan, P.; Culverhouse, R.; Marsh, S.; Mody, A.; Scott-Horton, T. J.; Brasington, R.; Joseph, A.; Reddy, V.; Eisen, S.; McLeod, H. L., Methotrexate (MTX) pathway gene polymorphisms and their effects on MTX toxicity in Caucasian and African American patients with rheumatoid arthritis. In *J Rheumatol*, Canada, **2008**; Vol. 35, pp 572-9.

[43] van der Straaten, R. J.; Wessels, J. A.; de Vries-Bouwstra, J. K.; Goekoop-Ruiterman, Y. P.; Allaart, C. F.; Bogaartz, J.; Tiller, M.; Huizinga, T. W.; Guchelaar, H. J., Exploratory analysis of four polymorphisms in human GGH and FPGS genes and their effect in methotrexate-treated rheumatoid arthritis patients. *Pharmacogenomics* **2007**, 8, 141-50.

[44] Yanagimachi, M.; Naruto, T.; Hara, T.; Kikuchi, M.; Hara, R.; Miyamae, T.; Imagawa, T.; Mori, M.; Kaneko, T.; Morita, S.; Goto, H.; Yokota, S., Influence of polymorphisms within the methotrexate pathway genes on the toxicity and efficacy of methotrexate in patients with juvenile idiopathic arthritis. *Br J Clin Pharmacol* **2011**, 71, 237-43.

[45] Genovese, M. C.; Bathon, J. M.; Martin, R. W.; Fleischmann, R. M.; Tesser, J. R.; Schiff, M. H.; Keystone, E. C.; Wasko, M. C.; Moreland, L. W.; Weaver, A. L.; Markenson, J.; Cannon, G. W.; Spencer-Green, G.; Finck, B. K., Etanercept *versus* methotrexate in patients with early rheumatoid arthritis: two-year radiographic and clinical outcomes. *Arthritis Rheum* **2002**, 46, 1443-50.

[46] Furst, D. E.; Breedveld, F. C.; Kalden, J. R.; Smolen, J. S.; Burmester, G. R.; Bijlsma, J. W.; Dougados, M.; Emery, P.; Keystone, E. C.; Klareskog, L.; Mease, P. J., Updated consensus statement on biological agents, specifically tumour necrosis factor {alpha} (TNF{alpha}) blocking agents and interleukin-1 receptor antagonist (IL-1ra), for the treatment of rheumatic diseases, 2005. In *Ann Rheum Dis*, England, **2005**; Vol. 64 Suppl 4, pp iv2-14.

[47] Ferraccioli, G., The possible clinical application of pharmacogenetics in rheumatology. In *J Rheumatol*, Canada, **2003**; Vol. 30, pp 2517-20.

[48] Criswell, L. A.; Lum, R. F.; Turner, K. N.; Woehl, B.; Zhu, Y.; Wang, J.; Tiwari, H. K.; Edberg, J. C.; Kimberly, R. P.; Moreland, L. W.; Seldin, M. F.; Bridges, S. L., Jr., The influence of genetic variation in the HLA-DRB1 and LTA-TNF regions on the response to treatment of early rheumatoid arthritis with methotrexate or etanercept. *Arthritis Rheum* **2004**, 50, 2750-6.

[49] Mugnier, B.; Balandraud, N.; Darque, A.; Roudier, C.; Roudier, J.; Reviron, D., Polymorphism at position -308 of the tumor necrosis factor alpha gene influences outcome of infliximab therapy in rheumatoid arthritis. *Arthritis Rheum* **2003**, 48, 1849-52.

[50] Cuchacovich, M.; Ferreira, L.; Aliste, M.; Soto, L.; Cuenca, J.; Cruzat, A.; Gatica, H.; Schiattino, I.; Perez, C.; Aguirre, A.; Salazar-Onfray, F.; Aguillon, J. C., Tumour necrosis factor-alpha (TNF-alpha) levels and influence of -308 TNF-alpha promoter polymorphism

on the responsiveness to infliximab in patients with rheumatoid arthritis. In *Scand J Rheumatol*, Norway, **2004**; Vol. 33, pp 228-32.

[51] Fabris, M.; Di Poi, E.; D'Elia, A.; Damante, G.; Sinigaglia, L.; Ferraccioli, G., Tumor necrosis factor-alpha gene polymorphism in severe and mild-moderate rheumatoid arthritis. *J Rheumatol* **2002**, 29, 29-33.

[52] Padyukov, L.; Lampa, J.; Heimburger, M.; Ernestam, S.; Cederholm, T.; Lundkvist, I.; Andersson, P.; Hermansson, Y.; Harju, A.; Klareskog, L.; Bratt, J., Genetic markers for the efficacy of tumour necrosis factor blocking therapy in rheumatoid arthritis. *Ann Rheum Dis* **2003**, 62, 526-9.

[53] O'Rielly, D. D.; Roslin, N. M.; Beyene, J.; Pope, A.; Rahman, P., TNF-alpha-308 G/A polymorphism and responsiveness to TNF-alpha blockade therapy in moderate to severe rheumatoid arthritis: a systematic review and meta-analysis. In *Pharmacogenomics J*, United States, **2009**; Vol. 9, pp 161-7.

[54] Miceli-Richard, C.; Comets, E.; Verstuyft, C.; Tamouza, R.; Loiseau, P.; Ravaud, P.; Kupper, H.; Becquemont, L.; Charron, D.; Mariette, X., A single tumour necrosis factor haplotype influences the response to adalimumab in rheumatoid arthritis. In *Ann Rheum Dis*, England, **2008**; Vol. 67, pp 478-84.

[55] Kang, C. P.; Lee, K. W.; Yoo, D. H.; Kang, C.; Bae, S. C., The influence of a polymorphism at position -857 of the tumour necrosis factor alpha gene on clinical response to etanercept therapy in rheumatoid arthritis. In *Rheumatology (Oxford)*, England, **2005**; Vol. 44, pp 547-52.

[56] Fabris, M.; Tolusso, B.; Di Poi, E.; Assaloni, R.; Sinigaglia, L.; Ferraccioli, G., Tumor necrosis factor-alpha receptor II polymorphism in patients from southern Europe with mild-moderate and severe rheumatoid arthritis. *J Rheumatol* **2002**, 29, 1847-50.

[57] Ongaro, A.; De Mattei, M.; Pellati, A.; Caruso, A.; Ferretti, S.; Masieri, F. F.; Fotinidi, M.; Farina, I.; Trotta, F.; Padovan, M., Can tumor necrosis factor receptor II gene 676T>G polymorphism predict the response grading to anti-TNFalpha therapy in rheumatoid arthritis? *Rheumatol Int* **2008**, 28, 901-8.

[58] Toonen, E. J.; Coenen, M. J.; Kievit, W.; Fransen, J.; Eijsbouts, A. M.; Scheffer, H.; Radstake, T. R.; Creemers, M. C.; de Rooij, D. J.; van Riel, P. L.; Franke, B.; Barrera, P., The tumour necrosis factor receptor superfamily member 1b 676T>G polymorphism in relation to response to infliximab and adalimumab treatment and disease severity in rheumatoid arthritis. In *Ann Rheum Dis*, England, **2008**; Vol. 67, pp 1174-7.

[59] Marotte, H.; Pallot-Prades, B.; Grange, L.; Tebib, J.; Gaudin, P.; Alexandre, C.; Blond, J. L.; Cazalis, M. A.; Mougin, B.; Miossec, P., The shared epitope is a marker of severity associated with selection for, but not with response to, infliximab in a large rheumatoid arthritis population. In *Ann Rheum Dis*, England, **2006**; Vol. 65, pp 342-7.

[60] Martinez, A.; Salido, M.; Bonilla, G.; Pascual-Salcedo, D.; Fernandez-Arquero, M.; de Miguel, S.; Balsa, A.; de la Concha, E. G.; Fernandez-Gutierrez, B., Association of the major histocompatibility complex with response to infliximab therapy in rheumatoid arthritis patients. *Arthritis Rheum* **2004**, 50, 1077-82.

[61] Pinto, J. A.; Rego, I.; Rodriguez-Gomez, M.; Canete, J. D.; Fernandez-Lopez, C.; Freire, M.; Fernandez-Sueiro, J. L.; Sanmarti, R.; Blanco, F. J., Polymorphisms in genes encoding tumor necrosis factor-alpha and HLA-DRB1 are not associated with response to infliximab in patients with rheumatoid arthritis. *J Rheumatol* **2008**, 35, 177-8.

[62] Potter, C.; Hyrich, K. L.; Tracey, A.; Lunt, M.; Plant, D.; Symmons, D. P.; Thomson, W.; Worthington, J.; Emery, P.; Morgan, A. W.; Wilson, A. G.; Isaacs, J.; Barton, A., Association of rheumatoid factor and anti-cyclic citrullinated peptide positivity, but not carriage of shared epitope or PTPN22 susceptibility variants, with anti-tumour necrosis factor response in rheumatoid arthritis. In *Ann Rheum Dis*, England, **2009**; Vol. 68, pp 69-74.

[63] Plant, D.; Bowes, J.; Potter, C.; Hyrich, K. L.; Morgan, A. W.; Wilson, A. G.; Isaacs, J. D.; Barton, A., Genome-wide association study of genetic predictors of anti-tumor necrosis factor treatment efficacy in rheumatoid arthritis identifies associations with polymorphisms at seven loci. *Arthritis Rheum* **2011**, 63, 645-53.

[64] Okabe, Y.; Sano, T.; Nagata, S., Regulation of the innate immune response by threonine-phosphatase of Eyes absent. In *Nature*, England, **2009**; Vol. 460, pp 520-4.

[65] van Baarsen, L. G.; Wijbrandts, C. A.; Rustenburg, F.; Cantaert, T.; van der Pouw Kraan, T. C.; Baeten, D. L.; Dijkmans, B. A.; Tak, P. P.; Verweij, C. L., Regulation of IFN response gene activity during infliximab treatment in rheumatoid arthritis is associated with clinical response to treatment. In *Arthritis Res Ther*, England, **2010**; Vol. 12, p R11.

[66] Tsang, S. W.; Shao, D.; Cheah, K. S.; Okuse, K.; Leung, P. S.; Yao, K. M., Increased basal insulin secretion in Pdzd2-deficient mice. In *Mol Cell Endocrinol*, 2009 Elsevier Ireland Ltd: Ireland, **2010**; Vol. 315, pp 263-70.

[67] Shahin, D.; Eltoraby, E.; Mesbah, A.; Houssen, M., Insulin resistance in early untreated rheumatoid arthritis patients. In *Clin Biochem*, 2010. Published by Elsevier Inc: United States, **2010**; Vol. 43, pp 661-5.

[68] Brown, M. A.; Crane, A. M.; Wordsworth, B. P., Genetic aspects of susceptibility, severity, and clinical expression in ankylosing spondylitis. *Curr Opin Rheumatol* **2002**, 14, 354-60.

[69] Reveille, J. D.; Maganti, R. M., Subtypes of HLA-B27: history and implications in the pathogenesis of ankylosing spondylitis. *Adv Exp Med Biol* **2009**, 649, 159-76.

[70] Reveille, J. D., The genetic basis of spondyloarthritis. In *Ann Rheum Dis*, England, **2011**; Vol. 70 Suppl 1, pp i44-50.

[71] Wei, J. C.; Tsai, W. C.; Lin, H. S.; Tsai, C. Y.; Chou, C. T., HLA-B60 and B61 are strongly associated with ankylosing spondylitis in HLA-B27-negative Taiwan Chinese patients. In *Rheumatology (Oxford)*, England, **2004**; Vol. 43, pp 839-42.

[72] Ward, M. M.; Hendrey, M. R.; Malley, J. D.; Learch, T. J.; Davis, J. C., Jr.; Reveille, J. D.; Weisman, M. H., Clinical and immunogenetic prognostic factors for radiographic severity in ankylosing spondylitis. *Arthritis Rheum* **2009**, 61, 859-66.

[73] Duerr, R. H.; Taylor, K. D.; Brant, S. R.; Rioux, J. D.; Silverberg, M. S.; Daly, M. J.; Steinhart, A. H.; Abraham, C.; Regueiro, M.; Griffiths, A.; Dassopoulos, T.; Bitton, A.; Yang, H.; Targan, S.; Datta, L. W.; Kistner, E. O.; Schumm, L. P.; Lee, A. T.; Gregersen, P. K.; Barmada, M. M.; Rotter, J. I.; Nicolae, D. L.; Cho, J. H., A genome-wide association study identifies IL23R as an inflammatory bowel disease gene. In *Science*, United States, **2006**; Vol. 314, pp 1461-3.

[74] Reveille, J. D.; Sims, A. M.; Danoy, P.; Evans, D. M.; Leo, P.; Pointon, J. J.; Jin, R.; Zhou, X.; Bradbury, L. A.; Appleton, L. H.; Davis, J. C.; Diekman, L.; Doan, T.; Dowling, A.; Duan, R.; Duncan, E. L.; Farrar, C.; Hadler, J.; Harvey, D.; Karaderi, T.; Mogg, R.; Pomeroy, E.; Pryce, K.; Taylor, J.; Savage, L.; Deloukas, P.; Kumanduri, V.; Peltonen, L.; Ring, S. M.; Whittaker, P.; Glazov, E.; Thomas, G. P.; Maksymowych, W. P.; Inman, R. D.; Ward, M. M.; Stone, M. A.; Weisman, M. H.; Wordsworth, B. P.; Brown, M. A.,

Genome-wide association study of ankylosing spondylitis identifies non-MHC susceptibility loci. In *Nat Genet*, United States, **2010**; Vol. 42, pp 123-7.

[75] Rueda, B.; Orozco, G.; Raya, E.; Fernandez-Sueiro, J. L.; Mulero, J.; Blanco, F. J.; Vilches, C.; Gonzalez-Gay, M. A.; Martin, J., The IL23R Arg381Gln non-synonymous polymorphism confers susceptibility to ankylosing spondylitis. In *Ann Rheum Dis*, England, **2008**; Vol. 67, pp 1451-4.

[76] Pimentel-Santos, F. M.; Ligeiro, D.; Matos, M.; Mourao, A. F.; Sousa, E.; Pinto, P.; Ribeiro, A.; Sousa, M.; Barcelos, A.; Godinho, F.; Cruz, M.; Fonseca, J. E.; Guedes-Pinto, H.; Trindade, H.; Evans, D. M.; Brown, M. A.; Branco, J. C., Association of IL23R and ERAP1 genes with ankylosing spondylitis in a Portuguese population. In *Clin Exp Rheumatol*, Italy, **2009**; Vol. 27, pp 800-6.

[77] Karaderi, T.; Harvey, D.; Farrar, C.; Appleton, L. H.; Stone, M. A.; Sturrock, R. D.; Brown, M. A.; Wordsworth, P.; Pointon, J. J., Association between the interleukin 23 receptor and ankylosing spondylitis is confirmed by a new UK case-control study and meta-analysis of published series. In *Rheumatology (Oxford)*, England, **2009**; Vol. 48, pp 386-9.

[78] Burton, P. R.; Clayton, D. G.; Cardon, L. R.; Craddock, N.; Deloukas, P.; Duncanson, A.; Kwiatkowski, D. P.; McCarthy, M. I.; Ouwehand, W. H.; Samani, N. J.; Todd, J. A.; Donnelly, P.; Barrett, J. C.; Davison, D.; Easton, D.; Evans, D. M.; Leung, H. T.; Marchini, J. L.; Morris, A. P.; Spencer, C. C.; Tobin, M. D.; Attwood, A. P.; Boorman, J. P.; Cant, B.; Everson, U.; Hussey, J. M.; Jolley, J. D.; Knight, A. S.; Koch, K.; Meech, E.; Nutland, S.; Prowse, C. V.; Stevens, H. E.; Taylor, N. C.; Walters, G. R.; Walker, N. M.; Watkins, N. A.; Winzer, T.; Jones, R. W.; McArdle, W. L.; Ring, S. M.; Strachan, D. P.; Pembrey, M.; Breen, G.; St Clair, D.; Caesar, S.; Gordon-Smith, K.; Jones, L.; Fraser, C.; Green, E. K.; Grozeva, D.; Hamshere, M. L.; Holmans, P. A.; Jones, I. R.; Kirov, G.; Moskivina, V.; Nikolov, I.; O'Donovan, M. C.; Owen, M. J.; Collier, D. A.; Elkin, A.; Farmer, A.; Williamson, R.; McGuffin, P.; Young, A. H.; Ferrier, I. N.; Ball, S. G.; Balmforth, A. J.; Barrett, J. H.; Bishop, T. D.; Iles, M. M.; Maqbool, A.; Yuldasheva, N.; Hall, A. S.; Braund, P. S.; Dixon, R. J.; Mangino, M.; Stevens, S.; Thompson, J. R.; Bredin, F.; Tremelling, M.; Parkes, M.; Drummond, H.; Lees, C. W.; Nimmo, E. R.; Satsangi, J.; Fisher, S. A.; Forbes, A.; Lewis, C. M.; Onnie, C. M.; Prescott, N. J.; Sanderson, J.; Matthew, C. G.; Barbour, J.; Mohiuddin, M. K.; Todhunter, C. E.; Mansfield, J. C.; Ahmad, T.; Cummings, F. R.; Jewell, D. P.; Webster, J.; Brown, M. J.; Lathrop, M. G.; Connell, J.; Dominiczak, A.; Marcano, C. A.; Burke, B.; Dobson, R.; Gungadoo, J.; Lee, K. L.; Munroe, P. B.; Newhouse, S. J.; Onipinla, A.; Wallace, C.; Xue, M.; Caulfield, M.; Farrall, M.; Barton, A.; Bruce, I. N.; Donovan, H.; Eyre, S.; Gilbert, P. D.; Hilder, S. L.; Hinks, A. M.; John, S. L.; Potter, C.; Silman, A. J.; Symmons, D. P.; Thomson, W.; Worthington, J.; Dunger, D. B.; Widmer, B.; Frayling, T. M.; Freathy, R. M.; Lango, H.; Perry, J. R.; Shields, B. M.; Weedon, M. N.; Hattersley, A. T.; Hitman, G. A.; Walker, M.; Elliott, K. S.; Groves, C. J.; Lindgren, C. M.; Rayner, N. W.; Timpson, N. J.; Zeggini, E.; Newport, M.; Sirugo, G.; Lyons, E.; Vannberg, F.; Hill, A. V.; Bradbury, L. A.; Farrar, C.; Pointon, J. J.; Wordsworth, P.; Brown, M. A.; Franklyn, J. A.; Heward, J. M.; Simmonds, M. J.; Gough, S. C.; Seal, S.; Stratton, M. R.; Rahman, N.; Ban, M.; Goris, A.; Sawcer, S. J.; Compston, A.; Conway, D.; Jallow, M.; Rockett, K. A.; Bumpstead, S. J.; Chaney, A.; Downes, K.; Ghori, M. J.; Gwilliam, R.; Hunt, S. E.; Inouye, M.; Keniry, A.; King, E.; McGinnis, R.; Potter, S.; Ravindrarajah, R.; Whittaker, P.; Widden, C.; Withers, D.; Cardin, N. J.; Ferreira, T.; Pereira-Gale, J.; Hallgrimsdo'ttir, I. B.; Howie, B. N.; Su, Z.;

Teo, Y. Y.; Vukcevic, D.; Bentley, D.; Mitchell, S. L.; Newby, P. R.; Brand, O. J.; Carr-Smith, J.; Pearce, S. H.; Reveille, J. D.; Zhou, X.; Sims, A. M.; Dowling, A.; Taylor, J.; Doan, T.; Davis, J. C.; Savage, L.; Ward, M. M.; Learch, T. L.; Weisman, M. H.; Brown, M., Association scan of 14,500 nonsynonymous SNPs in four diseases identifies autoimmunity variants. In *Nat Genet*, United States, **2007**; Vol. 39, pp 1329-37.

[79] Maksymowych, W. P.; Inman, R. D.; Gladman, D. D.; Reeve, J. P.; Pope, A.; Rahman, P., Association of a specific ERAP1/ARTS1 haplotype with disease susceptibility in ankylosing spondylitis. *Arthritis Rheum* **2009**, 60, 1317-23.

[80] Saveanu, L.; Carroll, O.; Lindo, V.; Del Val, M.; Lopez, D.; Lepelletier, Y.; Greer, F.; Schomburg, L.; Fruci, D.; Niedermann, G.; van Endert, P. M., Concerted peptide trimming by human ERAP1 and ERAP2 aminopeptidase complexes in the endoplasmic reticulum. In *Nat Immunol*, United States, **2005**; Vol. 6, pp 689-97.

[81] Cui, X.; Hawari, F.; Alsaaty, S.; Lawrence, M.; Combs, C. A.; Geng, W.; Rouhani, F. N.; Miskinis, D.; Levine, S. J., Identification of ARTS-1 as a novel TNFR1-binding protein that promotes TNFR1 ectodomain shedding. *J Clin Invest* **2002**, 110, 515-26.

[82] Timms, A. E.; Crane, A. M.; Sims, A. M.; Cordell, H. J.; Bradbury, L. A.; Abbott, A.; Coyne, M. R.; Beynon, O.; Herzberg, I.; Duff, G. W.; Calin, A.; Cardon, L. R.; Wordsworth, B. P.; Brown, M. A., The interleukin 1 gene cluster contains a major susceptibility locus for ankylosing spondylitis. In *Am J Hum Genet*, United States, **2004**; Vol. 75, pp 587-95.

[83] Brown, M. A., Genetics of ankylosing spondylitis. *Curr Opin Rheumatol* **2010**, 22, 126-32.

[84] Chandran, V.; Rahman, P., Update on the genetics of spondyloarthritis--ankylosing spondylitis and psoriatic arthritis. In *Best Pract Res Clin Rheumatol*, 2010 Elsevier Ltd: Netherlands, **2010**; Vol. 24, pp 579-88.

[85] Rahman, P.; Elder, J. T., Genetic epidemiology of psoriasis and psoriatic arthritis. In *Ann Rheum Dis*, England, **2005**; Vol. 64 Suppl 2, pp ii37-9; discussion ii40-1.

[86] Nelson, G. W.; Martin, M. P.; Gladman, D.; Wade, J.; Trowsdale, J.; Carrington, M., Cutting edge: heterozygote advantage in autoimmune disease: hierarchy of protection/susceptibility conferred by HLA and killer Ig-like receptor combinations in psoriatic arthritis. In *J Immunol*, United States, **2004**; Vol. 173, pp 4273-6.

[87] Gladman, D. D.; Anhorn, K. A.; Schachter, R. K.; Mervart, H., HLA antigens in psoriatic arthritis. *J Rheumatol* **1986**, 13, 586-92.

[88] Korendowych, E.; Dixey, J.; Cox, B.; Jones, S.; McHugh, N., The Influence of the HLA-DRB1 rheumatoid arthritis shared epitope on the clinical characteristics and radiological outcome of psoriatic arthritis. In *J Rheumatol*, Canada, **2003**; Vol. 30, pp 96-101.

[89] Ho, P. Y.; Barton, A.; Worthington, J.; Thomson, W.; Silman, A. J.; Bruce, I. N., HLA-Cw6 and HLA-DRB1*07 together are associated with less severe joint disease in psoriatic arthritis. In *Ann Rheum Dis*, England, **2007**; Vol. 66, pp 807-11.

[90] Rahman, P.; Inman, R. D.; Maksymowych, W. P.; Reeve, J. P.; Peddle, L.; Gladman, D. D., Association of interleukin 23 receptor variants with psoriatic arthritis. In *J Rheumatol*, Canada, **2009**; Vol. 36, pp 137-40.

[91] Rahman, P.; Siannis, F.; Butt, C.; Farewell, V.; Peddle, L.; Pellett, F.; Gladman, D., TNFalpha polymorphisms and risk of psoriatic arthritis. In *Ann Rheum Dis*, England, **2006**; Vol. 65, pp 919-23.

[92] Filer, C.; Ho, P.; Smith, R. L.; Griffiths, C.; Young, H. S.; Worthington, J.; Bruce, I. N.; Barton, A., Investigation of association of the IL12B and IL23R genes with psoriatic arthritis. *Arthritis Rheum* **2008**, 58, 3705-9.

[93] Liu, Y.; Helms, C.; Liao, W.; Zaba, L. C.; Duan, S.; Gardner, J.; Wise, C.; Miner, A.; Malloy, M. J.; Pullinger, C. R.; Kane, J. P.; Saccone, S.; Worthington, J.; Bruce, I.; Kwok, P. Y.; Menter, A.; Krueger, J.; Barton, A.; Saccone, N. L.; Bowcock, A. M., A genome-wide association study of psoriasis and psoriatic arthritis identifies new disease loci. *PLoS Genet* **2008**, 4, e1000041.

[94] Reich, K.; Huffmeier, U.; Konig, I. R.; Lascorz, J.; Lohmann, J.; Wendler, J.; Traupe, H.; Mossner, R.; Reis, A.; Burkhardt, H., TNF polymorphisms in psoriasis: association of psoriatic arthritis with the promoter polymorphism TNF*-857 independent of the PSORS1 risk allele. *Arthritis Rheum* **2007**, 56, 2056-64.

[95] Rahman, P.; O'Rielly, D. D., Psoriatic arthritis: genetic susceptibility and pharmacogenetics. *Pharmacogenomics* **2008**, 9, 195-205.

[96] Krueger, G. G.; Langley, R. G.; Leonardi, C.; Yeilding, N.; Guzzo, C.; Wang, Y.; Dooley, L. T.; Lebwohl, M., A human interleukin-12/23 monoclonal antibody for the treatment of psoriasis. In *N Engl J Med*, 2007 Massachusetts Medical Society.: United States, **2007**; Vol. 356, pp 580-92.

[97] Tolusso, B.; Pietrapertosa, D.; Morelli, A.; De Santis, M.; Gremese, E.; Farina, G.; Carniello, S. G.; Del Frate, M.; Ferraccioli, G., IL-1B and IL-1RN gene polymorphisms in rheumatoid arthritis: relationship with protein plasma levels and response to therapy. *Pharmacogenomics* **2006**, 7, 683-95.

CHAPTER 2

# One-Carbon Metabolism and Alzheimer's Disease: The DNA Methylation Hypothesis

**Fabio Coppedè***

*Faculty of Medicine, University of Pisa, Via Roma 67, 56126 Pisa, Italy; Program of Genetics and Epigenetics, Pisa University Hospital, Via S. Giuseppe 22, 56126 Pisa, Italy*

**Abstract:** Increasing evidence supports a role for epigenetic modifications in AD pathogenesis. Folate metabolism, also known as one-carbon metabolism, is required for the production of S-adenosylmethionine (SAM), which is the major DNA methylating agent. AD individuals are characterized by decreased serum folate values, as well as increased plasma homocysteine (Hcy) levels, and there is indication of impaired DNA methylation potential (SAM/SAH levels) in AD tissues. Some polymorphisms of genes participating in one-carbon metabolism have been associated with AD risk and/or with folate, Hcy or vitamin B12 levels in AD individuals. However, it is not yet clear wheter impairments in one-carbon metabolism are causative of AD or rather a consequence of the disease. Studies in rodents and in neuroblastoma cell cultures have shown that dietary restrictions of folate and other B vitamins result in epigenetic modifications of AD related genes as well as altered production of enzymes involved in DNA methylation, and that SAM treatment is able to restore both methylation and expression levels of those genes. Exposure to lead (Pb) resulted in epigenetic modification of AD related genes in rodents, monkeys and cell cultures. Moreover the amyloid-beta peptide itself possesses epigenetic properties. A few studies have been performed in human post-mortem specimens, suggesting that some genes might be epigenetically modified within the disease pathogenesis. This chapter focuses on the current knowledge on DNA methylation in AD tissues, cell cultures and animal models.

**Keywords:** Alzheimer's disease, folate, homocysteine, DNA methylation, gene polymorphisms, epigenetics.

## INTRODUCTION

### Alzheimer's Disease

Alzheimer's disease (AD) is an age related neurodegenerative disorder and the most common form of dementia in the elderly. The disease is clinically

***Address correspondence to Fabio Coppedè:** Faculty of Medicine, University of Pisa. Principal Investigator at Pisa University Hospital, Program of Genetics and Epigenetics, *Via* S. Giuseppe 22, 56126 Pisa, Italy; Tel: +39 050 2211028; Fax: +39 050 2211034; E-mail: f.coppede@geog.unipi.it

Christian Neri (Ed)

characterized by a progressive neurodegeneration in selected brain regions, including the temporal and parietal lobes and restricted regions within the frontal cortex and the cingulate gyrus, leading to memory loss accompanied by changes of behaviour and personality severe enough to affect daily life. Globally, more than 26 million people have been diagnosed with AD and it has been estimated that following the global aging of the world's population this number is expected to exceed 100 million people by 2050 [2]. Therefore, AD is becoming a serious health concern in both developed and developing countries, particularly because there is actually no available cure.

The occurrence of extracellular amyloid deposits or senile plaques (SP) and the presence of neurofibrillary tangles (NFT) composed of intraneuronal aggregates of hyperphosphorylated tau protein are the neuropathological hallmarks of the disease [3]. The primary component of the extracellular amyloid deposits in AD brains is an approximately 40-residue long peptide, known as amyloid β (Aβ) peptide, resulting from the proteolytic processing of its precursor, the amyloid precursor protein (APP), mediated by the subsequential action of β-secretase (β-site APP cleaving enzyme 1, BACE1) and γ-secretase (a protein complex composed by presenilins and other proteins).

Rare mutations in *APP*, presenilin-1 (*PSEN1*) and presenilin-2 (*PSEN2*) genes cause early-onset (< 65 years) familial forms of the disease, but mutations of these gene are responsible for less than 1% of the total AD cases, leaving part of the genetics of familial AD yet to be identified [4]. However, the vast majority of AD are sporadic forms, diagnosed in people aged over 65 years, and referred as to late-onset forms (LOAD), likely resulting from complex interactions among genetic, epigenetic, environmental and stochastic factors superimposed on an age-related cognitive decline [5]. Hundreds of genes have been investigated in genetic association studies as possible AD susceptibility or modifier genes. The AlzGene database [6] is a continuously updated database containing meta-analyses of published studies, including genome-wide association studies (GWAS), in order to provide a list of the genetic variants that most likely affect disease risk. According to this database the gene coding for the apolipoprotein E ε4 allele (*APOE ε4*) is the major genetic risk factor for LOAD forms, with an odds ratio = 3.68. About 40 additional genes have been associated with disease risk, all of them playing a minor role with odds ratios ranging from 0.8 to 1.3, suggesting a multi-factorial nature of the disease and complex gene-gene and gene-environment interactions yet to be clarified [6]. Several environmental agents have been proposed as possible LOAD risk factors, often with conflicting or inconclusive results [5]. Among them metals, pesticides, brain injuries and the resulting inflammation and oxidative stress, have been suggested to increase LOAD risk; by contrast, Mediterranean diet, the consumption of fruit, vegetables and fish, dietary antioxidants, a regular physical activity and brain stimulation until late in life are considered to be among LOAD protective factors [5]. However, despite active research in the field the aetiology of sporadic AD is still uncertain.

Folate metabolism, also known as one-carbon metabolism, is required for the production of S-adenosylmethionine (SAM), which is the major DNA methylating agent [7]. Impaired one-carbon metabolism is often observed in AD subjects and increasing evidence suggests a possible contribution of epigenetic processes in LOAD pathogenesis [8]. This manuscript updates the current knowledge on one-carbon metabolism in AD, with focus on methylation and epigenetic modifications.

## ONE-CARBON METABOLISM: AN OVERVIEW

Folate (vitamin B9) is an essential nutrient required for one-carbon biosynthetic and epigenetic processes. Folates are entirely derived from dietary sources, mainly from the consumption of green vegetables, fruits, cereals, and meat. Folic acid is the synthetic form added to foods and found in dietary supplements. After intestinal absorption, folate metabolism requires reduction and methylation into the liver to form 5-methyltetrahydrofolate (5-MTHF), release into the blood and cellular uptake; then it can be used as a donor of single carbons for the synthesis of DNA and RNA precursors or for the conversion of homocysteine (Hcy) to methionine, which is then used to form SAM. Folic acid is converted to a natural biological form of the vitamin as it passes through the intestinal wall, with enzymatic reduction and methylation resulting in the circulating form of the vitamin, 5-MTHF [7, 9].

Folates do not cross biological membranes by diffusion alone, but require several transport systems to enter the cells, the best characterized being the reduced folate carrier (RFC1). Methylenetetrahydrofolate reductase (MTHFR) is the first enzyme in the DNA methylation pathway since it reduces 5,10-methylentetrahydrofolate (5,10-MTHF) to 5-MTHF. Subsequently, methionine synthase (MTR) transfers a methyl group from 5-MTHF to Hcy forming methionine and tetrahydrofolate (THF). Methionine is then converted to SAM in a reaction catalyzed by methionine adenosyltransferase (MAT). Most of the SAM generated is used in transmethylation reactions, whereby SAM is converted to S-adenosylhomocysteine (SAH) by transferring the methyl group to diverse biological acceptors, including proteins and DNA. Vitamin B12 (or cobalamin) is a cofactor of MTR, and methionine synthase reductase (MTRR) is required for the maintenance of MTR in its active state. If not converted into methionine, Hcy can be condensed with serine to form cystathionine in a reaction catalyzed by cystathionine β-synthase (CBS), which requires vitamin B6 as a cofactor. Cystathionine can be then utilized to form the antioxidant compound glutathione (GSH). Another important function of folates is in the *de novo* synthesis of DNA and RNA precursors, required during nucleic acid synthesis and for DNA repair processes. Thymidylate synthase (TYMS) converts deoxyuridine monophosphate (dUMP) and 5,10-MTHF to deoxythymine monophospate (dTMP) and dihydrofolate (DHF) in the *de novo* synthesis of pyrimidines (Fig. **1**). Methylenetetrahydrofolate dehydrogenase (MTHFD1) consists of three activities: 5,10-MTHF dehydrogenase, 5,10-methenyltetrahydrofolate cyclohydrolase and 10-formyltetrahydrofolate synthetase, respectively, and catalyzes three sequential reactions in the inter-conversion of THF derivatives. Once generated by the 10-formyltetrahydrofolate synthetase activity, 10-formyltetrahydrofolate can be utilized for the production of purines by the phosphoribosylglycineamide transformylase (GART) enzyme. Therefore, depending on cellular demands folate derivatives can be used for the synthesis of SAM or for the synthesis of nucleic acid precursors, and the one-carbon metabolic pathway is tightly regulated by intracellular levels of metabolites and cofactors [7, 9]. A diagram illustrating folate metabolism is shown in Fig. **1**.

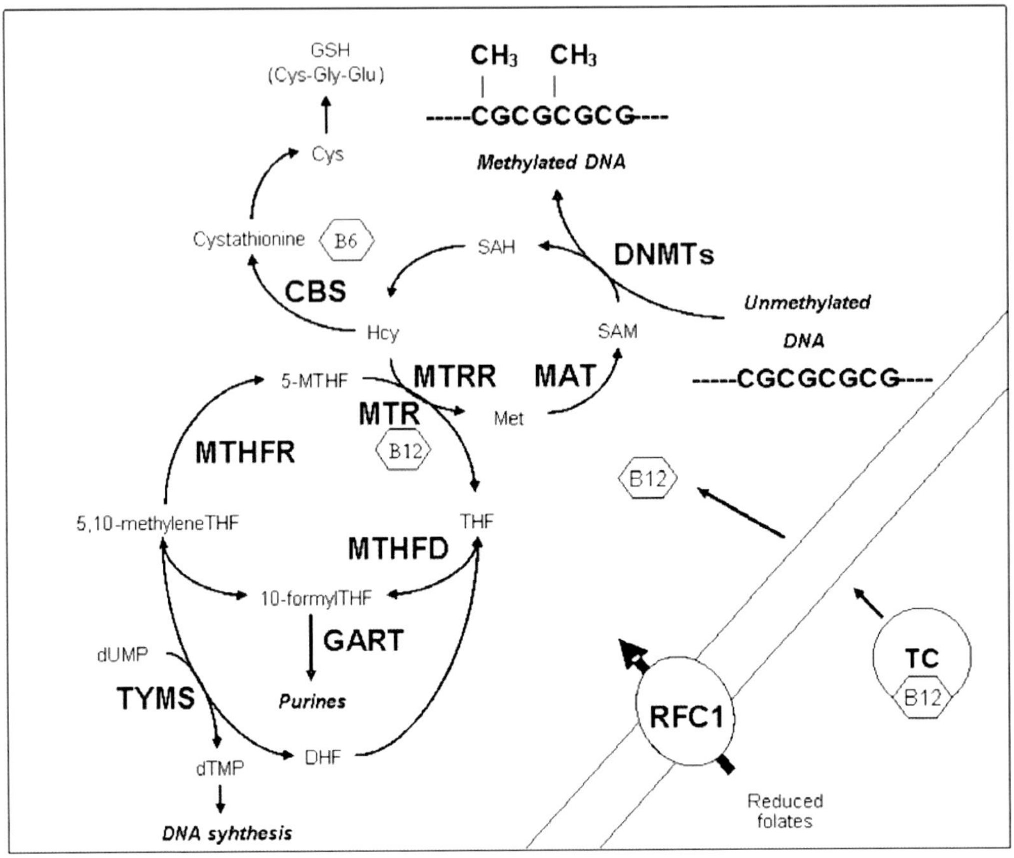


**Figure 1:** Overview of the folate metabolic pathway adapted from Coppedè *et al.* [205]: *Metabolites:* Cys = cysteine; dTMP = deoxythymidine monophosphate; dUMP = deoxyuridine monophosphate; DHF = dihydrofolate; 10-formyl-THF = 10-formyl-tetrahydrofolate; GSH = glutathione; Hcy = homocysteine; Met = methionine; 5-MTHF = 5-methyltetrahydrofolate; 5,10-MTHF = 5,10-methylentetrahydrofolate; SAH = S-adenosylhomocysteine; SAM = S-adenosylmethionine; THF = tetrahydrofolate. *Enzymes:* CBS = cystathionine β-synthase; DNMTs = DNA methyltransferases; GART = phosphoribosylglycineamide transformylase; MAT = methionine adenosyltransferase; MTHFD = methylenetetrahydrofolate dehydrogenase; MTHFR = methylenetetrahydrofolate reductase; MTR = methionine synthase; MTRR = methionine synthase reductase; RFC1 = reduced folate carrier; TC = transcobalamin; TYMS = thymidylate synthase. *Cofactors:* B6 = vitamin B6; B12 = vitamin B12.

## ONE-CARBON METABOLISM IN ALZHEIMER'S DISEASE

### Homocysteine, Folate and Other B Vitamins

In 1990, Regland and colleagues first reported elevated Hcy levels in patients with primary degenerative dementia [10]. Since then, several researchers have investigated the levels of Hcy, folate, and other B group vitamins in AD subjects and healthy matched controls [15-36]. Table **1** lists most of the retrospective studies focusing on the comparison between plasma Hcy levels in AD patients and healthy matched controls. Overall, the majority of those studies observed increased plasma Hcy values in AD subjects with respect to controls. Several prospective studies have been also performed following a group of healthy subjects for several months/years and recording the incidence of AD/dementia (reviewed in [37]). Again, the evidence from most of them converges to suggest that a moderate elevation in Hcy levels is a potential risk factor for AD. However, there are also conflicting results, and it remains controversial whether hyperhomocysteinemia (HHcy) is really an AD risk factor or merely a disease biomarker [37]. Data from recent meta-analyses are also conflicting. Wald and colleagues [38] performed a meta-analysis of data from eight cohort studies (involving 8,669 participants; median duration of study: 5 years) observing a statistically significant association between serum Hcy and the incidence of dementia: the odds ratio for a 5 μmol/L increase in serum Hcy was 1.35 (1.02-1.79) or 1.50 (1.13-2.00) adjusted for regression dilution bias. The odds ratio for a 3 μmol/L decrease in serum Hcy was 0.78 (0.66-0.93) [38]. Ho and co-workers performed a meta-analysis of 12 retrospective case-control studies that compared the Hcy levels between patients suffering from AD and controls, observing significantly higher Hcy levels in AD subjects with respect to controls [38]. They also performed a meta-analysis of four prospective studies involving a total of 2,631 participants, followed for a 24 months period, and failed to find a causal relationship between HHcy and the risk of developing dementia [39], suggesting that a more in-depth understanding of the underlying mechanisms linking Hcy to cognitive decline is required. The screening of 732 elderly subjects to determine association between Hcy and brain atrophy revealed that those with higher Hcy levels showed greater frontal, parietal, and occipital white matter atrophy in the entire cohort, irrespective of diagnosis, age, or sex. This association was also found when considering mild cognitive impairment individuals separately [40]. There is also some "although controversial" indication that Hcy levels are increased in the cerebrospinal fluid (CSF) of AD patients with respect to controls [41-43]. A recent study including 70 AD patients, 33 patients with another type of dementia, and 30 age-matched control subjects, revealed that folates in CSF were significantly different between groups, but not Hcy. In addition, the average folate in CSF was lower in AD patients compared with controls, and in AD CSF there was a significant inverse correlation between Hcy and folate [44].

**Table 1: Homocysteine Levels in AD Patients and Matched Healthy Controls**

| Reference | AD/Controls | Total Hcy[1] | Total Hcy[1] | *P*-Value |
|---|---|---|---|---|
| **Hcy Levels Increased in AD** | | **AD patients** | **Controls** | |
| [11] | 52/49 | 18.3 μM | 12.3 μM | <0.01 |
| [12] | 164/108 | 15.3 μM | 13.2 μM | <0.05 |
| [13] | 74/74 | 20.9 μM | 11.8 μM | <0.001 |
| [14] | 83/71 | 14.7 μM | 10.7 μM | <0.001 |
| [15] | 27/25 | 21.0 μM | 16.0 μM | <0.001 |
| [16] | 50/57 | 12.6 μM | 9.6 μM | <0.001 |
| [17] | 25/25 | 23.5 μM | 19.0 μM | <0.001 |
| [18] | 137/42 | 21.4 μM | 15.5 μM | <0.01 |
| [19] | 22/24 | 22.3 μM | 10.7 μM | <0.01 |
| [20] | 75/155 | 20.6 μM | 16.4μM | <0.05 |
| [21] | 11/207 | 24.5 μM | 14.9 μM | <0.001 |
| [22] | 23/21 | 13.3 μM | 8.3 μM | <0.01 |
| [23] | 105/102 | 16.0 μM | 11.9 μM | <0.001 |
| [24] | 43/50 | 18.3 μM | 15.3 μM | <0.05 |
| [25] | 71/44 | 19.1 μM | 12.9 μM | <0.001 |
| [26] | 51/40 | 14.2 μM | 10.3 μM | <0.001 |
| [27] | 106/104 | 9.03 μM | 8.2 μM | <0.05 |
| [28] | 60/60 | 14.1 μM | 12.7 μM | <0.01 |
| [29] | 94/74 | 22.7 μM | 14.5 μM | <0.05 |
| **No Significant Difference Detected** | | **AD patients** | **Controls** | |
| [30] | 14/17 | 19.4 μM | 17.9 μM | NS |
| [31] | 137/277 | 14.7 μM | 12.8 μM | NS |
| [32] | 60/60 | 12.3 μM | 11.5 μM | NS |
| [33] | 74/55 | 16.8 μM | 14.6 μM | NS |
| [34] | 121/795 | 17.0 μM | 16.4 μM | NS |
| [35] | 92/67 | 12.9 μM | 12.1 μM | NS |

[1]Mean or median values as derived from the original manuscript.
NS = Not significant.

HHcy is a surrogate marker for B vitamin deficiency (folate, B12, B6). Several retrospective studies (Table **2**) observed significantly decreased serum folate values in AD subjects with respect to controls, and an inverse correlation between plasma Hcy and serum folate. Less data are available concerning vitamins B12 (Table **3**) and B6 (Table **4**). Some authors observed significantly decreased levels of vitamin B12 in plasma of AD subjects with respect to controls, but results are conflicting (Table **3**). Moore and co-workers [45] recently reviewed 43 studies investigating the association of vitamin B12 and cognitive impairment or dementia observing that vitamin B12 levels in the subclinical low-normal range (<250 ρmol/L) are associated with AD, vascular dementia, and Parkinson's disease, and that vitamin B12 deficiency (<150 ρmol/L) is associated with cognitive impairment. They also observed that vitamin B12 supplements (1 mg daily) improved cognition only in patients with pre-existing vitamin B12 deficiency (serum vitamin B12 levels <150 ρmol/L or serum Hcy levels >19.9 µmol/L) [45]. Data on vitamin B6 levels in blood of AD patients are still scarce (Table **4**). The concept of improving the AD patient's clinical outcome by lowering of Hcy with B vitamins seems to be attractive. However, systematic review and meta-analysis of randomized trials revealed that supplementation of vitamins B12, B6, and folic acid alone or in combination does not appear to improve cognitive function in individuals with or without existing cognitive impairment [46, 47]. Those study, however do not exclude an involvement of folate and other B vitamins in AD. For example, a recent study was performed on asymptomatic and partially symptomatic Alzheimer's disease individuals, a series of cognitive states wherein subjects have substantial AD pathology, but are non-demented and have normal or only partially impaired cognitive function. Restriction of the population to subjects with near normal cognition on the cognitive state score indicated an elevated association with blood folate levels, suggesting that folate status may influence the expression of clinical symptoms of AD and aid in the delay or prevention of dementia [48]. In summary, whether HHcy and impaired one-carbon nutrients are causative agents or merely bioindicators of dementia is still largely questioned, but a causal role cannot be ruled out [37].

**Table 2: Folate Levels in AD Patients and Healthy Matched Controls**

| Reference | AD/Controls | Folate[1] | Folate[1] | *P*-Value |
|---|---|---|---|---|
| **Folate Levels Decreased in AD** | | **AD Patients** | **Controls** | |
| [12] | 164/108 | 17.6 nmol/l | 22.9 nmol/l | <0.05 |
| [13] | 74/74 | 5.7 ng/ml | 8.5 ng/ml | <0.001 |
| [14] | 83/71 | 4.5 μg/l | 5.1 μg/l | <0.01 |
| [15] | 27/25 | 14.7 nmol/l | nmol/l | <0.001 |
| [16] | 50/57 | 9.0 μg/l | 10.5 μg/l | <0.05 |
| [19] | 22/24 | 10.0 nmol/l | 13.9 nmol/l | <0.01 |
| [26] | 51/40 | 9.4 ng/ml | 12.4 ng/ml | <0.001 |
| [27] | 106/104 | 5.25 μg/l | 6.7 μg/l | <0.001 |
| [29] | 94/74 | 5.7 ng/ml | 7.8 ng/ml | <0.01 |
| [36] | 30/30 | 4.9 ng/ml | 7.5 ng/ml | <0.01 |
| **No Significant Difference Detected** | | **AD Patients** | **Controls** | |
| [11] | 52/49 | 3.5 ng/ml | 3.8 ng/ml | NS |
| [18] | 137/42 | 5.1 ng/ml | 6.2 ng/ml | NS |
| [20] | 75/155 | 4.3 nmol/l | 4.8 nmol/l | NS |
| [22] | 23/21 | 7.0 nmol/l | 8.7 nmol/l | NS |
| [28] | 60/60 | 15.6 nmol/l | 14.05 nmol/l | NS |
| [33] | 74/55 | 13.6 nmol/l | 16.9 nmol/l | NS |
| [34] | 121/795 | 9.5 ng/ml | 10.0 ng/ml | NS |

[1]Mean or median values as derived from the original manuscript.
NS = Not significant.

## Polymorphisms in One-Carbon Metabolizing Genes, Interaction with Hcy and B-Vitamins, and Risk of Alzheimer's Disease

### *Methylenetetrahydrofolate Reductase*

Polymorphisms of genes participating in one-carbon metabolism have been investigated as candidate AD risk factors (Table **5**). Methylenetetrahydrofolate reductase is the flavoprotein that catalyzes the conversion of 5,10-methylentetrahydrofolate (5,10-MTHF) to 5-methylTHF (Fig. **1**). The *MTHFR*

**Table 3: Vitamin B12 Levels in AD Patients and Healthy Matched Controls**

| Reference | AD/Controls | Vitamin B12[1] | Vitamin B12[1] | *P*-Value |
|---|---|---|---|---|
| **Vit B12 levels Decreased in AD** | | **AD Patients** | **Controls** | |
| [13] | 74/74 | 491 pg/ml | 780 pg/ml | <0.001 |
| [15] | 27/25 | 254 pmol/l | 322 pmol/l | <0.01 |
| [19] | 22/24 | 392 pmol/l | 438 pmol/l | <0.05 |
| [26] | 51/40 | 281 pg/ml | 390 pg/ml | <0.001 |
| **No Significant Difference Detected** | | **AD Patients** | **Controls** | |
| [11] | 52/49 | 284 pg/ml | 284 pg/ml | NS |
| [12] | 164/108 | 236 pmol/l | 253 pmol/l | NS |
| [14] | 83/71 | 351 pg/ml | 379 pg/ml | NS |
| [18] | 137/42 | 376 pg/ml | 498 pg/ml | NS |
| [20] | 75/155 | 323 pmol/l | 351 pmol/l | NS |
| [22] | 23/21 | 483 pg/ml | 494 pg/ml | NS |
| [28] | 60/60 | 272 pmol/l | 248 pmol/l | NS |
| [29] | 94/74 | 387 pg/ml | 451 pg/ml | NS |
| [33] | 74/55 | 281 pmol/l | 278 pmol/l | NS |
| [34] | 121/795 | 276 pg/ml | 269 pg/ml | NS |

[1]Mean or median values as derived from the original manuscript.
NS = Not significant.

**Table 4: Vitamin B6 Levels in AD Patients and Healthy Matched Controls**

| Reference | AD/Controls | Vitamin B6[1] | Vitamin B6[1] | *p*-Value |
|---|---|---|---|---|
| **No Significant Difference Detected** | | **AD Patients** | **Controls** | |
| [14] | 83/71 | 5.6 ng/ml | 6.4 ng/ml | NS |
| [19] | 22/24 | 52.0 nmol/l | 57.5 nmol/l | NS |

[1]Mean or median values as derived from the original manuscript.
NS = Not significant.

gene has been largely studied in AD association studies. Particularly, two common *MTHFR* polymorphisms, namely 677C>T (Ala222Val) and 1298A>C (Glu429Ala), are known to reduce MTHFR activity [49]. Numerous studies have shown that the *MTHFR* 677T allele is associated with increased total plasma Hcy levels (tHcy) and/or decreased serum folate levels, mainly in 677TT homozygous subjects [50-52]. Several authors investigated the *MTHFR* 677C>T polymorphism as a candidate AD risk factor [12, 23, 24, 27, 29, 53-62] and meta-analyses have been performed [6, 63, 64], the most recent including a total of 3,299 cases and 4,363 controls [64]. Overall, the *MTHFR* 677T allele was significantly associated with increased AD risk in combined populations [OR =1.13 (1.05-1.21)]. Also the combined (CT + TT) genotype was associated with increased AD risk [OR =1.18 (1.07-1.31)]. After stratification into ethnic groups, the *MTHFR* 677C>T polymorphisms was significantly associated with AD risk only in Asians, but not in Caucasians [46]. However, we recently performed the largest Italian case-control study, including 378 AD cases and 308 controls, observing that both the *MTHFR* 677T allele [OR =1.29 (1.04-1.60)] and the combined (CT + TT) genotype [OR =1.5 (1.08-2.09)] resulted associated with increased AD risk [29]. Some authors observed that *MTHFR* 677TT homozygous AD subjects had higher plasma tHcy values and/or decreased folate values compared to carriers of the *MTHFR* 677CT or 677CC genotypes [13, 23, 56, 58, 61]. We also recently observed interaction between the *MTHFR* 677C>T polymorphism and both folate and vitamin B12 levels in AD subjects [29]. Others observed interaction between the *MTHFR* 677T allele and the *APOE* genotype in modifying AD risk [27, 58, 59, 62].

**Table 5: Polymorphisms in Folate/Homocysteine Metabolizing Genes and AD Risk**

| AD Cases/Controls or (Range)[1] | Gene | Polymorphism | Observation | Refs. |
|---|---|---|---|---|
| (3000-4000) | *MTHFR* | 677C>T (Ala222Val) | Associated with ↑ AD risk in Asians Conflicting results in Caucasians | [6, 29, 63, 64] |
| (50-400) | *MTHFR* | 677C>T (Ala222Val) | Associated with ↑plasma Hcy and/or ↓folate values in AD patients | [13, 23, 29, 56, 58, 61] |
| 94-74 | *MTHFR* | 677C>T (Ala222Val) | Associated with ↑ vitamin B12 levels in AD | [29] |
| (50-400) | *MTHFR* | 677C>T (Ala222Val) | Associated with ↑AD risk in combination with the *APOE* genotype | [27, 58, 59, 62] |

*Table 5 contd…..*

| AD Cases/Controls or (Range)[1] | Gene | Polymorphism | Observation | Refs. |
|---|---|---|---|---|
| (80-250) | *MTHFR* | 677C>T (Ala222Val) | Associated with ↑AD risk in combination with *interleukin* gene polymorphisms | [67, 68] |
| (2000-5000) | *MTHFR* | 1298A>C (Glu429Ala) | Conflicting results in genetic association studies | [6, 65-67] |
| 129/178 | *MTHFR* | 677C/1298C/1793G (haplotype) | Associated with ↓AD risk | [72] |
| 223/323 | *MTHFR* | -713G>A (promoter region) | Not associated with AD risk | [73] |
| 223/323 | *MTHFR* | -393C>A (promoter region) | Not associated with AD risk | [73] |
| (1000-3000) | *MTHFD1L* | Rs11754661 | Associated with ↑AD risk in 3 out of 4 studies | [74-77] |
| (29-386) | *MTHFD1* | 1958G>A (Arg653Gln) | Inconclusive results | [82, 83] |
| 386/375 | *RFC1* | 80G>A (Arg27His) | Associated with ↑AD risk | [27] |
| (150-380) | *MTR* | 2756A>G (Asp 919Gly) | Inconclusive results, possible association with Hcy levels or dementia severity | [6, 29, 79, 80, 82, 97, 98] |
| 378/308 | *MTRR* | 66A>G (Asp 919Gly) | Associated with ↑AD risk | [29] |
| 94/74 | *MTRR* | 66A>G (Asp 919Gly) | Associated with ↑vitamin B12 levels in healthy controls | [29] |
| (70-200) | *TC* | 776C>G (Pro259Arg) | Conflicting results in genetic association studies | [58, 97, 111, 112] |
| (100-400) | *CBS* | 844ins68 (insertion) | Conflicting results in genetic association studies | [23, 83, 121] |
| 378/308 | *TYMS* | 28bp repeat (promoter region) | Not associated with AD risk | [29] |
| 376/308 | *DNMT3B* | -149C>T (promoter region) | Not associated with AD risk | [126] |
| 376/308 | *DNMT3B* | -579G>T (promoter region) | Not associated with AD risk | [126] |

[1]When only 1 reference is quoted the exact number of AD cases/controls is shown. When more than 1 reference is quoted the range of samples in case-control studies or meta-analyses is given into brackets (min-max).

The *MTHFR* 1298A>C polymorphism has been studied less extensively than the 677C>T in AD association studies, and results are still inconclusive [6, 24, 55, 58, 65]. In 2007, Bertram and colleagues reported the first meta-analysis concerning the *MTHFR* 1298A>C polymorphism and AD risk, observing that it was significantly associated with decreased AD risk [6], subsequent studies gave conflicting results [65-67]. A more recent genome-wide association study performed in French Caucasians [66] revealed a significant association between the *MTHFR* 1298A>C polymorphism and decreased AD risk [OR = 0.92 (0.85-1.00)]. However, the *MTHFR* 1298C allele was associated with increased AD risk in India, and in the same study the *MTHFR* 677T allele was associated with increased AD risk in combination with interleukin-6 (*IL-6*) polymorphism [67]. A previous study reported interaction between *MTHFR* 677C>T and interleukin-8 (*IL-8*) polymorphism and AD risk [68]. *MTHFR* 677C>T and 1298A>C polymorphisms are in strong linkage disequilibrium (LD), particularly the 677T allele has been nearly always observed in *cis* with the 1298C allele. A study suggested that the 677T variant arose later than the 1298C variant on a chromosome harbouring 1298A [69]. LD is not complete; however frequencies below 0.005 have generally been reported for the rare 677T-1298C haplotype [70]. A biological explanation for the LD existing between the two different *MTHFR* polymorphisms has been suggested [71]. MTHFR works as a dimer and monomers associate head to tail, but the stability of the dimer depends on both folate availability and on what aminoacid is present at position 222 and what at position 429, resulting from the *MTHFR* 677/1298 genotype. Based on this model it was proposed that the combined presence of both variant alleles in homozygosis would impair significantly the stability and the activity of the dimer protein [71]. Wakutani and co-workers [72] investigated *MTHFR* haplotypes generated by the combinations of three polymorphisms, 677C>T (Ala222Val), 1298A>C (Glu429Ala), and 1793A>G (Arg594Gln), in AD subjects and controls, suggesting that the haplotype 677C/1298C/1793G could be protective against the development of AD [72]. Therefore, differences observed among ethnic groups or when considering different polymorphisms could be due to the fact that *MTHFR* haplotypes, rather than each single polymorphism, are likely to affect disease risk, and those complex interactions duly exist among one-carbon nutrients (folate, Hcy, and other B-vitamins) and the stability and activity of this protein [1].

Polymorphisms in the regulatory region of the *MTHFR* gene (-713G>A and -393C>A, upstream of the start codon) were not associated with AD risk [73].

### *Methylenetetrahydrofolate Dehydrogenase (NADP+ Dependent) 1-Like*

Methylenetetrahydrofolate dehydrogenase (NADP+ dependent) 1-like (MTHFD1L) is involved in THF synthesis in the mitochondrion, catalyzing the reversible synthesis of 10-formyl-THF to formate and THF, an important step in homocysteine conversion to methionine. Unlike MTHFD1, however, MTHFD1L is a monofunctional enzyme, containing only the 10-formyl-THF synthetase activity. In 2010 Naj and colleagues [74] published data of a GWAS of 931 LOAD cases and 1104 controls. Moreover, significant results were replicated in an additional 1,338 cases and 2,003 controls. Their analysis revealed that the SNP rs11754661 in the gene *MTHFD1L* was significantly associated with LOAD, with an OR of about 2. Statistically significant associations in multiple SNPs (rs803424; rs2073067; rs2072064), all in linkage disequilibrium with rs11754661, were also observed [74]. Subsequently, others failed to replicate the association between rs11754661 and risk of AD in an independent study of this genetic variant in 2467 individuals from Spain [75]. However, two recent independent studies in Chinese Han populations have replicated the association between rs11754661 and LOAD risk, with ORs of 1.8 [76] and 1.7 [77], respectively. Moreover, association with LOAD risk was also observed for rs2073067, only in *APOEε4* carriers [77]. Overall, albeit some conflicting results, three out of four association studies support a role for *MTHFD1L* genetic variants in LOAD pathogenesis. However, the biological implication of the associations of rs11754661 and rs2073067 with LOAD remains unclear. Neither of these SNPs is suggested to alter MTHFD1L activity directly because they are both intronic. Therefore, it is expected that they may be in LD with an as yet unidentified functional one [77]. An additional *MTHFD1L* polymorphism (rs6922269) is included in the AlzGene database [6], but it showed no significant association with AD risk in three publications [78-80].

### *Methylenetetrahydrofolate Dehydrogenase*

Unlike the mitochondrial MTHFD1L, methylenetetrahydrofolate dehydrogenase (MTHFD1) possesses three distinct enzymatic activities, 5,10-MTHF dehydrogenase, 5,10-methenyltetrahydrofolate cyclohydrolase and 10-formyltetrahydrofolate synthetase, and catalyzes three sequential reactions in the inter-conversion of THF derivatives, which are substrates for methionine, thymidylate, and *de novo* purine syntheses (Fig. **1**). A G to A substitution at nucleotide 1958 within *MTHFD1* (*MTHFD1* 1958G>A) causes the arginine to glutamine substitution of the amino acid residue 653 in the synthetase domain of MTHFD1 protein, likely impairing protein stability and synthetase activity [81]. A small case-control study, including 29 AD subjects and 50 controls, revealed that the frequency of the 1958A allele was similar between AD cases and controls. However, AD cases had lower methionine/homocysteine levels than controls in the subjects with the *MTHFD1* 1958AA genotype [82]. A subsequent study on 386 Chinese Han sporadic AD patients and 375 matched controls revealed no association between the *MTHFD1* 1958G>A polymorphism and plasma levels of cysteine and Hcy in the Chinese subjects [83]. Moreover, the authors failed to find association of the *MTHFD1* 1958G>A polymorphism with AD risk in the total samples. When stratified by age at onset, it was found that A allele and AA genotype frequencies in early onset AD cases were higher than in controls and the differences were close to significant [83], suggesting that the A allele might be a weak risk factor for early onset AD. Confirmation in larger case-control cohorts is required. Overall, results on the contribution of the *MTHFD1* 1958G>A polymorphism to AD risk are still inconclusive [82, 83].

### *Reduced Folate Carrier*

The first report of a *reduced folate carrier* (*RFC1* or *SLC19A1*) gene polymorphism was in 2000 by Chango and co-workers [84] that described a high frequency 80G>A single nucleotide polymorphism resulting in replacement of an arginine by histidine (Arg27His). Authors found a moderate, but significant, increase in tHcy levels in doubly homozygous *RFC1* 80GG/ *MTHFR* 677TT subjects as compared to *RFC1* 80GG/ *MTHFR* 677CC or CT subjects. In addition, individuals who were *RFC1* 80AA/ *MTHFR* 677CT had higher plasma folate levels than those who were *RFC1* 80GG/ *MTHFR* 677CT [84]. Further studies provided conflicting results, therefore the effect of the *RFC1* 80G>A polymorphism on plasma folate and Hcy levels is still debated [85-88]. Bi and co-workers [27] investigated *RFC1* 80G>A and *MTHFR* 677C>T polymorphisms in

386 AD patients and 375 controls. Significant associations of the *RFC1* 80G allele and GG genotype with AD risk was found [OR = 1.6 (1.0-2.4) and OR= 2.2 (1.1-4.7), respectively]. However, no interaction between the two studied polymorphisms was found, nor the *RFC1* 80G variant was associated with plasma folate and Hcy levels [27]. The study is pending replication in other populations.

### *Methionine Synthase*

A common *methionine synthase* (*MTR*) 2756A>G (Asp919Gly) polymorphism is known, and there is indication from large scale population studies that it can have an effect on Hcy levels [89]. However, results are still conflicting and the contribution of the *MTR* 2756A>G polymorphism to Hcy concentrations has not been yet fully clarified. Some studies reported increased Hcy levels in the presence of the wild type (*MTR* 2756A) allele [90, 91], whereas others observed increased Hcy levels in the presence of the mutant (*MTR* 2756G) allele [92, 93]. There is also indication that the heterozygous genotype *MTR* 2756AG is associated with increased Hcy concentrations in Down syndrome individuals [94]. These apparent discrepancies might be explained by evidence suggesting that the interaction among different polymorphisms may totally modify their individual effect, and that the same genotype combinations could have different effects on tHcy levels in different individuals, depending on age, health status and interactions with nutritional and lifestyle factors [95]. In 2003 Beyer and co-workers observed association between the *MTR* 2756AA genotype and increased AD risk [96]. Subsequent studies gave conflicting results [29, 79, 80, 82, 97, 98], and a meta-analysis of relevant papers available at the AlzGene database [6] does not confirm association of this polymorphism with AD risk [OR = 0.90 (0.7-1.1)]. However, albeit Bosco and co-workers did not observe a difference in genotype frequencies between AD and controls, association of the *MTR* 2756AA genotype with dementia severity of sporadic AD was observed [97]. More recently, we observed similar *MTR* 2756A>G allele and genotype frequencies in LOAD patients and controls, but the polymorphism was significantly associated with Hcy levels in AD subjects [29]. Therefore, we cannot yet exclude a contribution of this variant to the AD phenotype.

### *Methionine Synthase Reductase*

Vitamin B12, in the form of methylcobalamin, serves as a coenzyme for MTR during the remethylation of Hcy to methionine (Fig. **1**). Methionine synthase reductase maintains adequate levels of activated cobalamin, the enzyme-bound cofactor for MTR. A common *MTRR* 66A>G variant, leading to isoleucine to methionine substitution (Ile22Met), was identified in homocystinuric patients with severe MTRR deficiency [99], and conflicting results were reported concerning its association with circulating Hcy values [100-102]. We recently observed that the *MTRR* 66G allele, particularly in homozygosis, might modestly increase LOAD risk in an Italian cohort of 378 LOAD patients and 308 healthy controls [OR = 1.25 (1.01-1.54) for the G allele, and OR = 1,57 (1.01-2.46) for the GG genotype, respectively] [29]. Our search of PubMed as well as other public available databases, such as the AlzGene database [6], revealed no previous genetic association studies aimed at addressing the contribution of this polymorphisms to LOAD risk. Therefore, this association should be taken with caution until confirmed in other populations. No correlation with plasma Hcy or serum folate levels was observed, but we observed that *MTRR* 66GG individuals showed a mean increase in vitamin B12 levels in all the studied groups, and the difference reached significance in healthy controls [29]. This is not surprising since vitamin B12 uptake and reactivation occurs through specific protein interactions in the MTR-MTRR complex [103].

### *Transcobalamin*

In circulation, vitamin B12 is bound to two plasma proteins: transcobalamin or haptocorrin. Transcobalamin (TC) is the transport protein required for cellular uptake of vitamin B12. Specific membrane receptors recognize the transcobalamin-vitamin B12 complex, whereas free vitamin B12 or haptocorrin-bound vitamin B12 is not taken up by the cell [104, 105]. Several studies have related holo-transcobalamin (holo-TC) levels to AD risk [106-108]. A common *TC* 776C>G polymorphism results in the replacement of proline with arginine (Pro259Arg) and negatively affects vitamin B12 metabolism, thus increasing plasma Hcy levels [109]. Conflicting results have been obtained when investigating the *TC* 776C>G polymorphism as a candidate AD risk factor.

Zetterberg and co-workers reported that this polymorphism influences holo-TC concentration in the CSF from AD patients [110], and suggested that it could be a modifiable AD genetic risk factor [111]. McCaddon and co-workers observed that serum holo-TC levels were significantly higher in *TC* 776CC individuals and that proportionately fewer *TC* 776CC homozygotes appear to develop AD at any given age [112]. Others failed to find association between the *TC* 677C>G polymorphism and sporadic AD risk [58, 97].

### *Cystathionine β-Synthase*

Human cystathionine β-synthase (CBS) is a hemoprotein which catalyzes the condensation of Hcy and serine to form cystathionine, which is then used to form GSH (Fig. **1**). Insufficiency in CBS activity may lead to hyperhomocysteinemia and a gross deficiency in CBS activity is associated with homocystinuria, an inborn recessive metabolic disorder [113, 114]. The *CBS* gene is known to have a large number of mutations, including missense and nonsense ones, as well as some insertion, deletion and splice site variants, some of which are polymorphic in nature [113]. The identification of an 844ins68 insertion in the *CBS* gene was first reported in a patient affected by homocysteinuria due to CBS deficiency [115]. Subsequent studies have revealed that this insertion is not a disease causing mutation but rather a common polymorphism whose frequency is largely different among human populations, with the variant allele being prevalent in African, European and North American populations [116-118]. Several studies report that the *CBS* 844ins68 polymorphism alone has not a relevant effect on tHcy concentrations [119, 120]. Beyer and co-workers genotyped 206 AD patients and 186 age-matched controls, observing that the 844ins68 mutation was associated with AD risk in subjects aged 75 years or more at onset [121]. By contrast, Zhang and co-workers observed no difference in the distribution of the *CBS* 844ins68 allele between 105 AD patients and 102 matched controls [23]. Moreover, no association between the polymorphism and plasma Hcy levels was observed [23]. More recently, Bi and co-workers screened 386 AD patients and 375 matched controls, and no significant difference of allele and genotype frequencies of the *CBS* 844ins68 polymorphism between AD cases and controls was detected, before and after stratification by *APOE* e4-carrying status, age, age at onset and gender

[83]. Therefore, the contribution of this polymorphism to AD risk is still controversial.

### *Thymidylate Synthase*

Thymidylate synthase (TYMS) uses 5,10-MTHF for the conversion of deoxyuridine monophosphate (dUMP) to deoxythymine monophospate (dTMP) in the *de novo* synthesis of DNA precursors. Therefore, polymorphisms affecting TYMS activity might shift the pools of 5,10-MTHF from DNA synthesis toward DNA methylation (Fig. **1**). Several polymorphisms are known in the *TYMS* gene, including a double (2R) or triple (3R) 28 bp tandem repeat sequence in the promoter enhancer region and a 6 bp deletion/insertion (6bp−/6bp+) at position 1494 (1494del6) in the 3'-untranslated region (3'-UTR). The first influences *TYMS* mRNA production and the latter is thought to influence mRNA expression and/or stability [122, 123]. We recently investigated the contribution of the *TYMS* 28bp repeat polymorphism to LOAD risk. The *TYMS* polymorphism was not associated with LOAD risk or disease age at onset. We also failed to observe correlation among the studied polymorphism and plasma Hcy or serum folate and vitamin B12 levels [29].

### *DNA Methyltransferase 3b*

DNA methyltransferase 3b (DNMT3b) is an enzyme involved in *de novo* methylation of the genome during embryogenesis, expressed in progenitor cells during neurogenesis (see the next sections of this chapter). *DNMT3B* promoter polymorphisms are believed to affect gene expression levels and have been associated with various kinds of cancer [124, 125]. We evaluated two functional *DNMT3B* promoter polymorphisms, namely -149 C>T (rs2424913) and -579 G>T (rs1569686), as candidate LOAD risk factors. Our analysis of 376 Italian LOAD patients and 308 matched controls revealed no difference in allele frequencies between the case an the control group. Also the genotype distributions of both polymorphisms were closely similar between groups, and no significant effect on disease age at onset was observed. We also failed to observe correlation among the studied polymorphisms and plasma Hcy or serum folate and vitamin B12 levels [126].

## The Methylation Potential: S-Adenosylmethionine, S-Adenosylhomocysteine and Methionine Adenosyltransferase in AD Tissues

DNA methylation is closely dependent on the DNA methylation potential, which is referred to as the ratio between SAM and SAH levels. Several studies have been performed to measure SAM and SAH levels, as well as methionine adenosyltransferase activity, in plasma, CSF and brain regions of AD subjects (reviewed in [1]). Table **6** lists some of those studies that, even if "somewhat controversial", indicate impaired methylation potential in the AD examined tissues [28, 127-133]. It has been also suggested that increased SAH concentrations in the brains of AD patients inhibit methyltransferases and this was related to cognitive impairment [133]. A recent study showed that Hcy plasma levels in the highest quartile were more frequent in AD patients than in controls. In addition, AD patients had significantly lower CSF levels of the methyl group donor SAM. Accordingly, the SAM/SAH ratio, *i.e.* the methylation capacity, was significantly lower in the CSF of the AD patients. Further, explorative analysis of all subjects showed that CSF SAM levels were lower in carriers of the *APOE ε4* allele compared with non-carriers [28]. Studies in AD animal models also

**Table 6: S-Adenosylmethionine (SAM), S-Adenosylhomocysteine (SAH) and Methionine Adenosyltransferase (MAT) Activity in AD Patients and Controls**

| AD Cases/Controls | Observation | Ref. |
|---|---|---|
| 9/13 | ↓ SAM levels in AD CSF | [127] |
| 11/14 | ↓ SAM levels in AD brains<br>↓ SAH levels in AD brains | [128] |
| 5/5 | Normal MAT activity in AD brains | [129] |
| 9/10 | ↓ MAT activity in AD erythrocytes | [129] |
| 30/28 | No difference in CSF SAM levels between AD cases and controls<br>No difference in CSF SAH levels between AD cases and controls | [130] |
| 26/29 | ↑ plasma SAH levels in AD patients<br>↑ plasma SAM levels in AD patients | [131] |
| 25/25 | ↓ adenosine levels in the plasma of AD patients | [132] |
| 34/43 | ↑ SAH levels in the prefrontal cortex of AD patients<br>↑ SAH in AD brain inhibits methyltransferases | [133] |
| 60/60 | ↓ CSF SAM levels in AD patients<br>↓ CSF SAM/SAH ratio in AD patients<br>↓ CSF SAM levels in *APOE ε4* carriers | [28] |

revealed altered SAM levels in the brain. However, the analysis of SAM and SAH levels in the aging brain of APP/PS1 Alzheimer mice during aging revealed that SAM levels decreased in the brains of wild-type mice, whereas SAH levels diminished in both wild type and APP/PS1 mice. In contrast to wild-type mice, SAM levels in APP/PS1 mice are not decreased during aging, probably related to less demand due to neurodegeneration [134].

## Linking One-Carbon Metabolism to AD: The DNA Methylation Hypothesis

Several hypotheses have been formulated to explain the observed increased plasma Hcy levels (Table **1**) and decreased serum folate values (Table **2**) in AD subjects in an attempt to link them to AD pathogenesis. For instance, folate deficiency fosters a decline in SAM, decreasing DNA methylation during aging and AD [128, 135]. Folate deficiency and resultant SAM depletion lead to increased levels of Hcy, which in turn potentiate Aβ peptide toxicity [136]. Hcy is a critical branch point metabolite that can influence cellular levels of SAM and SAH, which regulate the activity of methyltransferases during DNA methylation and posttranslational modification of proteins [137]. Studies in rodents showed that Hcy accumulation reduces cellular levels of SAM, stimulates glutamate excitotoxicity and increases oxidative damage [138]. Hcy has been also associated to vascular disease in AD, with attention focused on vascular changes related to AD as a consequence of Aβ peptide toxicity and its deposition [139]. Several studies suggest a correlation between plasma Hcy concentrations and plasma Aβ levels [140, 141]. Moreover, there is indication that elevated Hcy causes tau hyperphosphorilation, NFT formation and SP formation *via* inhibition of methyltransferases and reduced methylation of protein phosphatase 2A [142, 143]. However, one of the most exciting hypothesis linking one-carbon metabolism to AD risk suggests that impaired folate/Hcy metabolism and subsequent reduction of SAM levels might result in epigenetic modifications of the promoters of AD-related genes leading to increased Aβ peptide production [144, 145]. One of the most studied epigenetic modifications is the change of methylation patterns of CpG rich regions in the promoters of specific genes, resulting in gene silencing (hypermethylation) or overexpression (hypomethylation). In the next sections are discussed evidences from cell cultures, animal models and humans, linking one-carbon metabolism to epigenetic modifications of AD-related genes (Table 7). Before describing those studies in details a brief description of DNA methylation/demethylation processes and their role in CNS formation, aging, learning and memory is worth.

**Table 7: Epigenetic Modifications of AD-Related Genes**

| Experimental Model | Observation | Refs. |
|---|---|---|
| Human neuroblastoma SK-N-SH or SK-N-BE cells | Folate and vitamin B12 deprivation induced epigenetic modifications in the promoter of *PSEN1*, resulting in up-regulation of gene expression and increased production of presenilin 1, BACE1, and APP | [145] |
| | B vitamin deficiency induced decreased *de novo* DNA methylation activity | [167] |
| | SAM administration resulted in down-regulation of *PSEN1* expression | [144] |
| | SAM administration was able to modulate the expression of 7 out of 588 genes of the nervous system | [169] |
| Human neuroblastoma SH-SY5Y cells | Treatment with anisomycic (a MAPKs activator) resulted in *APP*, *PSEN1* and *BACE1* demethylation with increased protein expression and Aβ overproduction | [171] |
| | Exposure to lead (Pb) increased the expression of BACE1 and APP proteins and downregulated that of proteins involved in DNA methylation | [180] |
| Rat pheochromocytoma (PC12) cells | Exposure to Pb acetate resulted in *APP* demethylation and impaired DNMT1 expression | [181] |
| Cells overexpressing Aβ (APP-sw expressing cells) | Hypermethylation of *DDR2* and hypomethylation of *CTIF* and *NXT2* genes | [174] |
| BV-2 mouse microglial cells | SAH administration increased the production of Aβ peptide likely through induction of hypomethylation of *APP* and *PSEN1* gene promoters | [166] |
| | Methylation of two CpG sites at position +298 and +351 in the *BACE1* 5'-UTR resulted in gene sylencing | [175] |
| Murine cerebral endothelial cells | Aβ reduces global DNA demethylation whilst increasing DNA methylation of the gene encoding neprilysin | [172] |
| Rodents | B vitamin deprivation altered the methylation potential and induced hypomethylation of the promoter of *PSEN1*, resulting in increased gene expression | [162, 163] |
| | B vitamin deficiency induced decreased *de novo* DNA methylation activity | [167] |
| Rodents and monkeys | Early life exposure to Pb resulted in inhibition of DNA-methyltransferase, hypomethylation of the promoter of *APP* and delayed up-regulation of gene expression later in life | [176, 177] |
| Post-mortem human brains | AD brains showed a marked reduction of markers of DNA methylation | [182, 183] |

*Table 7 contd.....*

| Experimental Model | Observation | Refs. |
|---|---|---|
| | The analysis of DNA methylation in selected regions of *MAPT*, *APP*, and *PSEN1* genes revealed no difference between control samples and AD samples | [184] |
| | The analysis of DNA methylation of *APP* promoter showed no methylation in cortex regions and cerebellum of familial AD cases | [186] |
| | Only two (*SORBS3* and *S100A2*) out of 50 analyzed loci showed AD-related methylation changes, which appeared to reflect an acceleration of age-related changes | [186] |
| | AD brains showed an increased epigenetic drift with unusual methylation patterns, and inter-individual variations concerning *PSEN1*, *APOE*, *MTHFR* and *DNMT1* genes | [187] |
| | A genome-wide DNA methylation analysis of 918 genes showed only a weak (2.9%) difference in DNA methylation between AD and control brains. The *TMEM59* gene was hypomethylated in AD brains | [188] |
| | The promoter of nucleolar rRNA genes (rDNA) was hypermethylated in brain regions of AD and MCI individuals | [189] |
| Blood DNA | Higher levels of LINE-1 methylation in AD subjects with respect to controls | [190] |

## DNA Methylation

DNA methylation mediates gene repression by serving as a maintainable mark that can be used to reconstruct silent chromatin following each round of replication. During development, germline DNA methylation is erased in the blastocyst, and a bimodal pattern is established anew at the time of implantation when the entire genome gets methylated while CpG islands are protected. This brings about global repression and allows housekeeping genes to be expressed in all cells of the body. Postimplantation development is characterized by stage- and tissue-specific changes in methylation that ultimately mold the epigenetic patterns that define each individual cell type [146]. DNA methylatranferases (DNMTs) are the key enzymes for DNA methylation (Fig. **1**) and catalyze the transfer of a methyl group from SAM to cytosine, thus forming 5-methyl-cytosine. There are multiple families of DNMTs in mammals. DNMT1 is primarily involved in the maintenance of DNA methylation patterns during development and cell division, whereas DNMT3a and DNMT3b are the *de novo* methyltransferases and establish DNA methylation patterns during early development. It was also proposed that DNMT3a and DNMT3b may be involved in the maintenance of DNA methylation patterns by correcting errors left by DNMT1 [147]. DNMT3L induces *de novo* DNA methylation by recruitment or activation of DNMT3a, whilst DNMT2 is primarily involved in the methylation of transfer RNA molecules [148, 149]. Alongside with their role in *de novo* DNA methylation, DNMT3a/b have been also involved in dynamic demethylation processes, thus allowing a transient regulation of the transcription of several genes [150]. The analysis of Dnmt3a and Dnmt3b proteins during the development of the mouse central nervous system (CNS) revealed that whereas Dnmt3b is important for the early phase of neurogenesis, Dnmt3a likely plays a dual role in regulating neurogenesis prenatally and CNS maturation and function postnatally. Particularly, Dnmt3b is specifically expressed in progenitor cells during neurogenesis, suggesting an important role in the initial steps of progenitor cell differentiation. Dnmt3a is expressed in postmitotic young neurons following the Dnmt3b expression and may be required for the establishment of tissue-specific methylation patterns of the genome [151, 152]. Dnmt1 and Dnmt3a are expressed in postmitotic neurons and required for synaptic plasticity, learning and memory through their overlapping roles in maintaining DNA methylation and modulating neuronal gene expression in adult CNS rodent neurons [153]. Moreover, Dnmt3a is expressed in postnatal neural stem cells and methylates intergenic regions and gene bodies flanking proximal promoters of several genes, many of which encode regulators of neurogenesis. Surprisingly, Dnmt3a-dependent nonproximal promoter methylation of these genes promotes gene expression and may be used for maintaining active chromatin states of genes critical for development [154]. Epigenetic changes in gene expression that do not alter the underlying DNA sequence-may be the answer for how can memories last a lifetime. Indeed, increasing evidence supports a role for epigenetic mechanisms including DNA methylation and acetylation or methylation of the histone proteins that package DNA in the context of animal learning (for an updated review see [155]). Moreover, epigenetic mechanisms have now emerged as key contributors to the alterations of genome structure and function that accompany aging, and epigenetic defects might represent a major factor in the pathophysiology of aging and aging-related diseases, including cancer and neurodegeneration [156]. Little is still known concerning DNA demethylation in the adult mammalian brain. Guo and co-workers [157] have recently demonstrated that the 5-methylcytosine hydroxylase TET, by converting 5-methylcytosine to 5-hydroxymethylcytosines (5hmCs), promotes DNA demethylation in mammalian cells through a process that requires the DNA base excision repair pathway. Demethylation of 5hmCs is promoted by the AID (activation-induced deaminase)/APOBEC (apolipoprotein B mRNA-editing enzyme complex) family of cytidine deaminases. Furthermore, Tet1 and Apobec1 are involved in neuronal activity-induced, region-specific, active DNA demethylation and subsequent gene expression in the dentate gyrus of the adult mouse brain *in vivo* [157]. Various studies have shown that 5-hmC is present in high levels in the brain. Its lower affinity to methyl-binding proteins as compared to 5-mC suggests that it might have a different role in the regulation of gene expression, while it is also implicated in the DNA demethylation process. Aging is believed to affect epigenetic marking of brain DNA with 5-methylcytosine and possibly *via* the 5-methylcytosine to 5hmC conversion by TET enzymes. A recent study investigated the impact of aging on hippocampal DNA 5-hydroxymethylation in mice revealing that global hippocampal 5hmC content increased during aging and was not related to oxidative stress or to an increased expression of the three TET enzymes (TET 1-3). Global 5hmC content was increased during aging in the absence of 5-methylcytosine decrease, suggesting that 5hmC could act as an epigenetic marker and not only as an intermediary in DNA demethylation [158]. The analysis of epigenetic modifications of mitochondrial DNA (mtDNA) in brain samples of mice revealed that aging decreased mtDNA 5hmC but not 5-methylcytosine levels in the frontal cortex but not in the cerebellum. Transcript levels of selected mtDNA-encoded genes increased during aging in the frontal cortex only. Moreover, in the frontal cortex, aging decreased mtDNMT1 mRNA levels without affecting TET1-TET3 mRNAs. In the cerebellum, TET2 and TET3 mRNA content was increased but mtDNMT1 mRNA was unaffected [159]. Further studies are indeed required to clarify the contribution of 5hmC in the aging brain and in AD pathology [160].

## EPIGENETIC MODIFICATIONS OF AD-RELATED GENES

### Studies in Cell Cultures and Disease Animal Models

#### *Dietary Manipulations (B Vitamins, SAM, Vitamin E)*

As described above, DNA methylation represents one of the most important epigenetic processes alongside with histone tail modifications and mechanisms involving small RNA molecules. Methylation of CpG sequences might induce chromatin conformational modifications and inhibit the access of the transcriptional machinery to gene promoter regions, thus altering gene expression levels. Therefore, promoter hypermethylation is commonly associated with gene silencing and promoter demethylation with gene expression, though several exceptions to this rule have been reported, and the methylation of other gene sequences apart from the promoter (gene-body methylation) deserves further investigation in terms of its biological significance [161]. Several studies have been performed using cell cultures and animal models in order to investigate the methylation levels of AD related genes under different environmental conditions (Table 7). Particularly, *APP* and *PSEN1* promoter methylation was analyzed in human neuroblastoma SK-N-SH or SK-N-BE cell lines under conditions of folate and vitamin B12 deprivation from the media. The study revealed demethylation of a CpG site in the promoter of *PSEN1*, resulting in increased production of presenilin1, BACE1 and APP proteins [145]. By contrast, SAM administration to human neuroblastoma SK-N-SH cells resulted in down-regulation of *PSEN1* gene expression and reduced Aβ peptide production [144]. Similarly, a combination of dietary folate, vitamin B12 and vitamin B6 deprivation (B vitamin deprivation) resulted in hyperhomocysteinemia, increased brain SAH levels, depletion of brain SAM, hypomethylation of specific CpG moieties in the 5'-flanking region of *PSEN1* and enhancement of presenilin 1 and BACE1 expression and Aβ deposition in mice [162, 163]. Dietary deficiency of folate and vitamin E, in condition of oxidative stress (the diet contained iron as a pro-oxidant), increased presenilin 1 expression, BACE1 activity, and Aβ levels in normal adult mice. These increases were particularly evident in mice lacking apolipoprotein E. On the contrary, dietary SAM supplementation attenuated these deleterious consequences [164]. A similar experiment was performed in mice expressing the human *APOE* gene. Mice expressing human apolipoprotein ε4, apolipoprotein ε3, or apolipoprotein ε2, were subjected to a diet lacking folate and vitamin E, and containing iron as a pro-oxidant. The study revealed that presenilin 1 and gamma-secretase were over-expressed in ε3 mice to the same extent as in ε4 mice even under a complete diet, and were not alleviated by SAM supplementation. Aβ increased only in ε4 mice maintained under the complete diet, and was alleviated by SAM supplementation [165]. Lin and co-workers showed that SAH increases the production of Aβ in BV-2 mouse microglial cells possibly by an increased expression of APP and induction of hypomethylation of *APP* and *PSEN1* gene promoters [166]. A recent study demonstrated that B vitamin deficiency induces DNMT3a and DNMT3b protein down-regulation in both neuroblastoma cell cultures and mice brain regions. Moreover, the *de novo* DNA methylation activity was modulated in both neuroblastoma cells and mice, and resulted decreased in B vitamin deficiency conditions and increased by SAM supplementation, whilst the activity of a putative DNA demethylase (MBD2) showed an opposite tendency [167]. Nutritional B vitamin restriction was also used to study the variation of protein expression profile in mice brain regions. A group of proteins mainly involved in neuronal plasticity and mitochondrial functions was identified as modulated by one-carbon metabolism [168]. The same group also analysed 588 genes of the central nervous system in SK-N-BE neuroblastoma cells, observing that only 7 genes were modulated by SAM treatment (and therefore by DNA methylation); 3 were up-regulated and 4 down-regulated [169]. Moreover, SAM administration reduced amyloid production, increased spatial memory in mice and inhibited the upregulation of B vitamin deficiency-induced PSEN1 and BACE1 expression and Tau phosphorylation. Furthermore, SAM treatment reduced plaque spreading independently on B vitamin deficiency [170].

***Environmental Stressors and Aβ Treatment***

Oxidative stress and inflammation have been largely involved as the pathological components of AD, and mitogen-activated protein kinase (MAPK) family members, JNK and p38 MAPK subfamilies are relevant to the response of environmental stress, inflammatory stimuli, or other insults. Human neuroblastoma SH-SY5Y cells were treated with anisomycin, an activator of stress-related MAPKs (JNK and p38 MAPK). A significant increase of intracellular Aβ level in anisomycin-treated SH-SY5Y cells was observed. The expression of APP, BACE1, and PSEN1 was upregulated by demethylation of their gene promoters associated with the reduction of DNMTs and histone H3 hyperacetylation, thus leading to Aβ overproduction [171]. Others observed that the Aβ peptide induces global DNA hypomethylation in murine cerebral endothelial cells, but specifically leads to hypermethylation of the gene encoding neprilysin (*NEP*), one of the enzymes responsible for Aβ degradation, thus suppressing *NEP* expression in mRNA and protein levels [172]. The Swedish double mutant of APP (K670M/N671L or APP-sw) is one of the mutations leading to excess Aβ production and an early onset of AD [173]. A genome-wide analysis of promoter CpG DNA methylation and gene expression in an APP-sw-expressing AD model cell line revealed three target genes epigenetically modified: hypomethylated *CTIF* (CBP80/CBP20-dependent translation initiation factor) and *NXT2* (nuclear exporting factor 2), and hypermethylated *DDR2* (discoidin domain receptor 2). Treatment with the demethylating agent 5-aza-2'-deoxycytidine restored mRNA expression of these three genes, suggesting that their epigenetic regulatory mechanism may contribute to the pathogenesis of AD [174]. A recent study in mouse BV-2 microglial cells revealed that methylation of two CpG sites at positions +298 and +351 in the 5'-untranslated region (5'-UTR) of the *BACE1* gene resulted in a complete silencing of the promoter activity, providing additional evidence that *BACE1* gene expression is regulated by DNA methylation [175]. It was also shown that Aβ treatment of human neuroblastoma cells modifies the methylation status of several CpG island of *BACE1* promoter [8].

### *Lead (Pb) Exposure*

Also metal exposure has been suggested to be able to epigenetically modulate the expression of AD related genes. Particularly, the analysis of rodents and monkeys exposed to lead (Pb) during embryogenesis and/or early postnatal period revealed that early life exposure to this metal could result in epigenetic modifications of AD related genes, such as *APP*, and to their subsequent deregulated expression later in life [176, 177]. Particularly, it was observed that developmental exposure of rats to lead resulted in a delayed over-expression (20 months later) of the amyloid precursor protein and its amyloidogenic Aβ product. Similarly, aged monkeys exposed to lead as infants also responded in the same way [176, 177]. These data indicate that environmental influences occurring during brain development predetermine the expression and regulation of APP later in life, potentially influencing the course of amyloidogenesis, and suggesting that early life neurogenesis is likely to be a critical step for environmentally induced epigenetic modifications [178]. The same group measured the latent expression of a wide array of brain-specific genes and explored whether epigenetic pathways mediated such latent molecular and pathological changes in 23-year-old primate brains exposed to Pb as infants. Developmental Pb exposure altered the gene expression of the arrayed genes, and was accompanied by a significant decrease in the protein levels of Dnmts, methyl CpG binding protein 2 (MeCP2), and proteins involved in histone modifications [179]. Moreover, differentiated SH-SY5Y cells were exposed to a series of Pb concentrations (5-100 μM) for 48 h and were analyzed for the protein expression of APP, BACE1, specificity protein 1 and 3 (Sp1, Sp3) and epigenetic intermediates like Dnmt1, Dnmt3a, and MeCP2 involved in DNA methylation, six days after the exposure had ceased. Western blot analysis indicated a significant latent elevation in AD biomarkers as well as the transcription factors Sp1 and Sp3, accompanied by a significant reduction in the protein levels of DNA methylating enzymes. RT-PCR analysis of Dnmt1, Dnmt3a and MeCP2 indicated a significant down-regulation of the mRNA levels [180]. Others analyzed the DNA methylation patterns of *APP* gene in rat pheochromocytoma (PC12) cells after Pb acetate exposure, observing that the exposure resulted in *APP* promoter hypomethylation, and impaired DNMT1 expression [181].

**Studies in Humans**

Despite evince of possible epigenetic modifications of AD-related genes obtained in neuronal cell cultures as well as in rodents and primates, epigenetic studies in AD patients are still scarce (Table **7**), likely depending on the difficulty to obtain human post-mortem brain specimens. Mastroeni and co-workers examined global DNA methylation in post-mortem brain regions of monozygotic twins discordant for AD, observing significantly reduced levels of DNA methylation in the temporal neocortex of the AD twin [182]. The same group analyzed brain tissues from 20 AD patients and 20 age-matched controls, and used immunocytochemistry for two markers of DNA methylation (5-methylcytosine and 5-methylcytidine) and eight methylation maintenance factors in the entorhinal cortex layer II, again observing significant decrements in AD cases [183]. However, the analysis of DNA methylation in selected regions of *MAPT* (the gene encoding for microtubule associated tau protein), *APP*, and *PSEN1* in post-mortem frontal cortex and hippocampus of healthy controls and various stages of AD progression, observing no differences in the percentage of CpG methylation in any region analyzed [184]. Others analysed the methylation status of a CpG-island in the *APP* gene in DNA extracted from the cortex regions and the cerebellum using material from six familial AD cases. Bisulfite sequencing showed no methylation in any sample [185]. Siegmund and co-workers examined the DNA methylation status at 50 loci, encompassing primarily 5' CpG islands of genes related to CNS growth and development, in temporal neocortex of 125 subjects ranging in age from 17 weeks of gestation to 104 years old, and including a cohort of AD patients [186]. Only two loci showed AD-related changes (*SORBS3* and *S100A2*, encoding a cell adhesion molecule and a calcium binding protein, respectively), which appeared to reflect an acceleration of age-related changes [186]. Wang and collaborators [187] analyzed 12 potential AD susceptibility loci in peripheral lymphocytes and post-mortem brain samples of LOAD patients and matched controls. From the 12 analyzed CpG-rich regions, only one was hypermethylated, and was the only region analyzed outside of a promoter region (DNA methylation can also occur in CpG sites outside the promoter region "gene-body methylation", and is usually associated with transcriptional activation). Overall, the authors failed to find significant differences in DNA methylation between LOAD cases and controls [187]. However, four of the studied genes (*PSEN1*, *APOE*, *MTHFR,* and *DNMT1*) showed a significant inter-individual epigenetic variability, and LOAD brain samples showed an "age-specific epigenetic drift" associated with unusual methylation patterns [187]. More recently, Bakulsky and co-workers investigated human postmortem frontal cortex genome-wide DNA methylation profiles between 12 LOAD and 12 cognitively normal age- and gender-matched subjects. Their analysis revealed 948 CpG sites representing 918 unique genes. Across these 948 sites the subtle mean methylation difference between LOAD cases and controls was 2.9%. The promoter of the gene encoding for transmembrane protein 59 (*TMEM59*) was 7.3% hypomethylated in cases. The *TMEM59* gene is implicated in amyloid-β protein precursor post-translational processing, supporting a role for epigenetic change in LOAD pathology [188]. Others assessed whether CpG methylation of nucleolar rRNA gene (rDNA) promoter was dysregulated across the AD spectrum, and analyzed brain samples from 10 mild cognitive impairment (MCI), 23 AD, and, 24 age-matched control individuals using bisulfite mapping. The cytosine methylation of total genomic

DNA was similar in AD, MCI, and control samples, but the rDNA promoter became hypermethylated in cerebro-cortical samples from MCI and AD groups. In parietal cortex, the rDNA promoter was hypermethylated more in MCI than in advanced AD. Consistent with a notion that hypermethylation-mediated silencing of the nucleolar chromatin stabilizes rDNA loci, preventing their senescence-associated loss, genomic rDNA content was elevated in cerebrocortical samples from MCI and AD groups [189]. Bollati and co-workers analyzed the methylation pattern of repetitive elements (*i.e.,* Alu, LINE-1 and α-satellite DNA) in the DNA obtained from blood samples of 43 AD patients and 38 matched controls, observing higher levels of LINE-1 methylation in AD subjects with respect to controls [190].

## CONCLUSIONS

Meta-analyses of published studies (Tables **1-4**) indicate that plasma Hcy levels are significantly higher in AD patients with respect to controls. On the contrary, plasma folate values are significantly reduced in AD patients and the levels of vitamin B12 tend to be lower in AD individuals with respect to matched controls, whilst little is still known for other B vitamins [37, 191]. There is also indication from prospective cohort studies that HHcy and the resultant low serum folate values might represent risk factors for the development of AD [37]. Interaction among those biochemical markers, polymorphisms of metabolic genes, and AD risk has been also proposed (Table **5**). Moreover, there is indication of impaired DNA methylation potential in AD brains and CSF (Table **6**). However, the results of randomized trials in humans following dietary B vitamins failed to give the expected improvements of cognitive functions [46, 47]. As a result, it is still largely debated whether or not the impairment in one-carbon metabolism observed in AD individuals is really a disease risk factor or merely a consequence of the impairment of other pathways within the disease course [37].

Among the several hypotheses linking HHcy and the resultant impairment in one-carbon metabolism to the pathological mechanisms of AD one of the most recent and promising is the epigenetic hypothesis, suggesting that a reduced DNA methylation potential, resulting from HHcy, might induce changes in promoter DNA methylation patterns and overexpression of AD related genes [144, 145].

Studies in rodents and in neuroblastoma cell cultures have clearly shown that diets lacking B vitamins lead to demethylation and increased expression of AD-relevant genes, such as *PSEN1* [145], with subsequent increased production of PSEN1 and BACE1 proteins, impairments of DNMTs activities, and increased formation and deposition of the Aβ peptide [145, 162, 163, 167]. By contrast, SAM administration was able to restore promoter methylation and expression of those genes and to protect against Aβ peptide deposition [144, 167, 170]. Moreover, other environmental factors, such as for example lead exposure [176-181] or the exposure to the Aβ peptide itself [1, 172, 174], were able to modify the methylation levels and the expression of genes involved in Aβ peptide production and turnover in both cell cultures and disease animal models. However, the analysis of post-mortem AD brains, although not excluding that some genes might be differentially methylated in AD *versus* control subjects [182-188], have been quite controversial and failed to unequivocally replicate the initial findings of a marked reduction of DNA methylation in disease samples [182-183].

One of the most important things that we still need to clarify is whether or not environmentally induced epigenetic modifications of AD-related genes are reversible and could be modulated through dietary SAM or B vitamin supplementation in humans. The studies performed on mice with differential amounts of dietary SAM or B vitamins point to a possible modulation [162, 163, 167]. Moreover, studies in rodents and primates exposed to Pb in early life suggest that there is a window of time during differentiation when the brain is particularly vulnerable to epigenetic modifications [176, 177]. Particularly, those data indicate that environmental influences occurring during brain development predetermined the expression and regulation of AD-related genes later in life [176, 177]. However, no epigenetic modification of AD-related genes was observed when animals were exposed to Pb later in life [176]. All these observations should lead to the following reflections: a) What are the vulnerable pre-and post-natal periods in humans when the developing brain is particularly susceptible to epigenetic modifications? b) How many environmental and/or dietary factors are able to induce epigenetic changes in the differentiating brain? c) What are the epigenetic consequences of environmental exposure in adulthood?

d) Are all these phenomena reversible? e) Can a dietary intervention occurring during adulthood restore the methylation pattern of a gene which has been epigenetically modified during brain differentiation? f) Shall we use the same dietary intervention for everybody irrespective of the individual metabolic profile (*i.e.,* gene mutations, polymorphisms, *etc.*)? g) What might be the risk associated with folate or SAM supplementation for human health (*i.e.,* induction of cancer or other diseases)? h) And last but not least, what is the role of 5-hmC in the brain, is it only an intermediate during demethylation processes or something else [158]? Unfortunately, most of these questions are still unsolved, and what happens in the human brain remains largely a mystery. Given the inaccessibility of brain samples until death, there is no available data in humans that correlate plasma Hcy or serum folate values, or any other environmental exposure, to the methylation profile of any specific gene in the AD brain.

Moreover, the Aβ peptide itself exerts epigenetic properties and therefore another arguing question is the following: is the impairment in one-carbon metabolism that fosters the epigenetic changes observed in AD brains, or rather they are a consequence of an increased Aβ production in AD? Again we do not have an answer, and probably both factors concur to epigenetically modify the expression of genes in the brain and there is no a unique answer that can apply to everybody. However, if the Aβ peptide is able to epigenetically modify the expression of genes relevant for its production and turnover, the analysis of post-mortem AD brains is unlikely to be able to detect some of the earliest epigenetic changes, if any, that could have triggered the disease, likely because the Aβ peptide itself might have erased or potentiated some of the initial epigenetic marks.

Despite several questions that are still open, results obtained in cell cultures and AD animal models (Table **7**) are encouraging and suggest that, at least in the experimental conditions, B vitamins and SAM supplements can be used to turn on/off the expression of genes relevant to AD pathogenesis. Similarly, several investigators have manipulated histone tail acetylation with histone deacetylase inhibitors in AD animal models, often observing prevention of cognitive deficits and memory recovery [192-196]. Taken overall, the above studies indicate that "epigenetic molecules" are an emerging and promising field of investigation in AD treatment. There is however still some concern for their application to prevent dementia in humans, mainly because of the possible induction of cancer. For example, folic acid supplementation has been associated with an increased risk of prostate cancer [197], and a dual effect was proposed for colorectal cancer with a protective role in healthy subjects and a promoting effect in individuals harbouring preneoplastic lesions [198]. Moreover, complex interactions have been observed among polymorphisms of genes involved in one-carbon metabolism, B vitamin supplementation, and altered methylation of cancer related genes (reviewed in [198]), and something similar might also happen in the brain for genes involved in neurodegeneration.

Another serious concern emerges from studies on the effect of folic acid fluor fortification on other diseases than cancer. If it is true that too little folic acid results in nervous tissue damage, as is accepted by the scientific community in regard to neural tube defects, then it seems plausible that too much folic acid may result in nervous tissue damage associated with autism. Indeed, there is some evidence that children who develop autism may be receiving a massive dose of folic acid *in utero*, as well as, after birth [199]. There is also indication that high folate and vitamin B12 levels during pregnancy are associated with increased prevalence of atopic dermatitis in the offspring [200], suggesting that potential risks of high folate and vitamin B12 concentrations on allergic outcomes should be evaluated when discussing mandatory fortification programs. In a few words folic acid fortification might be good for some people, but not for everybody. For example, the clinical consequences of vitamin B12 deficiency may be worsened by high folic acid intakes, perhaps as a direct result of unmetabolized folic acid, which does not occur naturally in body tissues. Indeed, Morris *et al.* [201] attempted to correlate anemia and cognitive test performance to circulating unmetabolized folic acid or 5-MTHF, observing that circulating unmetabolized folic acid was related to an increased odd of anemia in alcohol users. In seniors with a serum vitamin B12 concentration <148 pmol/L the presence of detectable circulating unmetabolized folic acid was related to lower cognitive test scores and lower mean cell volume. In the same subgroup, higher serum 5-MTHF was related to an increased odd of anemia and a marginally significantly decreased odds of macrocytosis. In seniors with a normal vitamin B12 status, a higher serum 5-MTHF concentration was related to higher cognitive test scores [201]. This example clearly shows the complex interactions among different metabolites, and again points to the fact that not everybody responds in the same way to high folic acid intake. A recent *in vitro* study in lymphoblastoid cells demonstrated that folic acid supplementation dysregulates by ≥ four-fold up or down the expression of more than 1000 genes, and gene dysregulation was observed in a concentration-dependent manner, even at a level that is lower than what is achievable in the blood through folic acid fortification guidelines [202], again suggesting that some caution should be taken prior to recommend a widespread supplementation to humans.

In conclusion, studies in AD mouse models have clearly demonstrated that SAM supplementation delays Aβ and tau pathology [170, 203], and there is also indication that it might delay the onset of motor neuron pathology in a murine model of amyotrophic lateral sclerosis [204]. SAM can also exert a positive direct effect on the activity of antioxidand glutathione S-transferase (GST) enzymes, again pointing to a possible role of SAM as a neuroprotective dietary supplement for AD patients, and leading some authors to suggest future intervention trials using measures of dietary supplementation (dietary polyunsaturated fatty acids and SAM plus B vitamin supplementation) to determine if such supplements will reduce the risk for cognitive decline in very mild AD and mild cognitive impairment [204]. Overall, this is an emerging and very promising field of research for dementia treatment and prevention. However, given the enormous epigenetic potential of SAM, and taking into account the above described possible adverse effects of folate fortification to some individuals, additional studies are required prior to recommend a widespread use of SAM supplements to prevent age-related neurodegenerative diseases. We still need to clarify who could benefit from such a treatment, but also who could experience some serious adverse consequences.

## ACKNOWLEDGEMENTS

FC acknowledges the Italian Ministry of Health for funding his project on “Folate metabolism, epigenetics, and Alzheimer’s disease”.

## CONFLICT OF INTEREST

The author(s) confirm that this chapter content has no conflict of interest.

## DISCLOSURE

The present is an invited chapter following the publication of the manuscript "One-carbon metabolism and Alzheimer's disease: focus on epigenetics. Curr Genomics 2010; 1: 246-60" [1]. The author was invited to make an updated version of that article to be published in this volume series. As such, the manuscript maintains its original structure, but the text is updated including the most recent findings and hypotheses in the field.

## ABBREVIATIONS

| | | |
|---|---|---|
| Aβ | = | amyloid beta |
| AD | = | Alzheimer's disease |
| APOE | = | apolipoprotein E |
| APP | = | amyloid precursor protein |
| BACE1 | = | β-secretase |
| CBS | = | cystathionine β-synthase |
| CSF | = | cerebrospinal fluid |
| DNMTs | = | DNA methyltransferases |
| GSH | = | glutathione |
| Hcy | = | homocysteine |
| HHcy | = | hyperhomocysteinemia |
| 5-hmC | = | 5-hydroxymethylcytosine |
| MAT | = | methionine adenosyltransferase |
| MeCP2 | = | methyl CpG binding protein 2 |
| 5-MTHF | = | 5-methyltetrahydrofolate |
| MTHFD1 | = | methylenetetrahydrofolate dehydrogenase |
| MTHFR | = | methylenetetrahydrofolate reductase |
| MTR | = | methionine synthase |

| | | |
|---|---|---|
| MTRR | = | methionine synthase reductase |
| NFT | = | neurofibrillary tangles |
| PSEN1 | = | presenilin 1 |
| PSEN2 | = | presenilin 2 |
| RFC1 | = | reduced folate carrier |
| SAH | = | S-adenosylhomocysteine |
| SAM | = | S-adenosylmethionine |
| SP | = | senile plaques |
| TC | = | transcobalamin |
| THF | = | tetrahydrofolate |
| TYMS | = | Thymidylate synthase |

## REFERENCES

[1] Coppedè, F. One-carbon metabolism and Alzheimer's disease: focus on epigenetics. *Curr Genomics* **2010**; *11*: 246-60.

[2] Brookmeyer, R., Johnson, E., Ziegler-Graham, K., Arrighi, H.M. Forecasting the global burden of Alzheimer's disease. *Alzheimers Dement* **2007**; *3*: 186-91.

[3] Wenk, G.L. Neuropathologic changes in Alzheimer's disease. *J Clin Psychiatry* **2003**; *64*: 7- 10.

[4] Ertekin-Taner, N. Genetics of Alzheimer's disease: a centennial review. *Neurol Clin*, **2007**; *25*: 611-67.

[5] Migliore, L., Coppedè, F. Genetics, environmental factors and the emerging role of epigenetics in neurodegenerative diseases. *Mutat Res* **2009**; *667*: 82-97.

[6] Bertram, L., McQueen, M.B., Mullin, K., Blacker, D, Tanzi, R.E. Systematic meta-analyses of Alzheimer disease genetic association studies: the AlzGene database. *Nat Genet* **2007**; *39*: 17-23.

[7] Bailey, L.B, Gregory, J.F3rd. Folate metabolism and requirements. *J Nutr* **1999**; *129*: 779-82.

[8] Mastroeni, D., Grover, A., Delvaux, E, Whiteside, C., Coleman, PD., Rogers, J. Epigenetic mechanisms in Alzheimer's disease. *Neurobiol Aging* **2011**; *32* : 1161-80.

[9] Coppedè, F. The complex relationship between folate/homocysteine metabolism and risk of Down syndrome. *Mutat Res* **2009**; *682*: 54-70.

[10] Regland, B., Abrahamsson, L., Gottfries, C.G, Magnus E. Vitamin B12 analogues, homocysteine, methylmalonic acid, and transcobalamins in the study of vitamin B12 deficiency in primary degenerative dementia. *Dement Geriatr Cogn Disord* **1990**; *1*: 272-77.

[11] Joosten, E., Lesaffre, E., Riezler, R., *et al.* Is metabolic evidence for vitamin B-12 and folate deficiency more frequent in elderly patients with Alzheimer's disease? *J Gerontol A Biol Sci Med Sci* **1997**; *52*: 76-79.

[12] Clarke, R., Smith, A.D., Jobst, K.A., Refsum, H, Sutton, L., Ueland, P.M. Folate, vitamin B12, and serum total homocysteine levels in confirmed Alzheimer disease. *Arch Neurol* **1998**; *55*: 1449-55.

[13] Postiglione, A., Milan, G., Ruocco, A., Gallotta, G., Guiotto, G., Di Minno, G. Plasma folate, vitamin B(12), and total homocysteine and homozygosity for the C677T mutation of the 5,10-methylene tetrahydrofolate reductase gene in patients with Alzheimer's dementia. A case-control study. *Gerontology* **2001**; *47*: 324-9.

[14] McIlroy, S.P., Dynan. K,B., Lawson, J.T, Patterson, C.C., Passmore, A.P. Moderately elevated plasma homocysteine, methylenetetrahydrofolate reductase genotype, and risk for stroke, vascular dementia, and Alzheimer disease in Northern Ireland. *Stroke* **2002**; *33*: 2351-6.

[15] Selley, M.L, Close, D.R., Stern, S.E. The effect of increased concentrations of homocysteine on the concentration of (E)-4-hydroxy-2-nonenal in the plasma and cerebrospinal fluid of patients with Alzheimer's disease. *Neurobiol Aging* **2002**; *23*: 383-8.

[16] McCaddon, A., Hudson, P., Hill, D, *et al.* Alzheimer's disease and total plasma aminothiols. *Biol Psychiatry* **2003**; *53*: 254-60.

[17] Selley, M.L. Increased concentrations of homocysteine and asymmetric dimethylarginine and decreased concentrations of nitric oxide in the plasma of patients with Alzheimer's disease. *Neurobiol Aging* **2003**; *24*: 903-7.

[18] Gallucci, M,, Zanardo, A, De Valentin, L, Vianello, A. Homocysteine in Alzheimer disease and vascular dementia. *Arch Gerontol Geriatr Suppl* 2004; *9*: 195-00.

[19] Malaguarnera, M., Ferri, R., Bella, R., Alagona, G., Carnemolla, A., Pennisi, G. Homocysteine, vitamin B12 and folate in vascular dementia and in Alzheimer disease. *Clin Chem Lab Med* **2004**; *42*: 1032-5.

[20] Mizrahi, E.H., Bowirrat, A., Jacobsen, D.W, *et al.* Plasma homocysteine, vitamin B12 and folate in Alzheimer's patients and healthy Arabs in Israel. *J Neurol Sci* **2004**; *227*: 109-13.

[21] Agnati, L.F., Genedani, S., Rasio, G., *et al.* Studies on homocysteine plasma levels in Alzheimer's patients. Relevance for neurodegeneration. *J Neural Transm* **2005**, *112*: 163-9.

[22] de Silva, H.A., Gunatilake, S.B., Johnston, C., Warden, D., Smith, A.D. Medial temporal lobe atrophy, apolipoprotein genotype, and plasma homocysteine in Sri Lankan patients with Alzheimer's disease. *Exp Aging Res* **2005**; *31*: 345-54.

[23] Zhang, Y.D., Ke, X.Y., Shen, W., Liu, Y. Relationship of homocysteine and gene polymorphisms of its related metabolic enzymes with Alzheimer's disease. *Chin Med Sci J* **2005**; *20*: 247-51.

[24] da Silva, V.C., Ramos, F.J., Freitas, E.M, *et al.* Alzheimer's disease in Brazilian elderly has a relation with homocysteine but not with MTHFR polymorphisms. *Arq Neuropsiquiatr* **2006**; *64*: 941-5.

[25] Guidi, I., Galimberti, D., Lonati, S, *et al.* Oxidative imbalance in patients with mild cognitive impairment and Alzheimer's disease. *Neurobiol Aging* **2006**; *27*: 262-9.

[26] Köseoglu, E., Karaman, Y. Relations between homocysteine, folate and vitamin B12 in vascular dementia and in Alzheimer disease. *Clin Biochem* **2007**; *40*: 859-63.

[27] Bi ,X.H., Zhao, H.L., Zhang, Z.X., Zhang, J.W. Association of RFC1 A80G and MTHFR C677T polymorphisms with Alzheimer's disease. *Neurobiol Aging* **2009**; *30*: 1601-7.

[28] Linnebank, M., Popp, J., Smulders, Y., *et al.* S-adenosylmethionine is decreased in the cerebrospinal fluid of patients with Alzheimer's disease. *Neurodegener Dis* **2010**; *7*: 373-8.
[29] Coppedè, F., Tannorella, P., Pezzini, I, *et al.* Folate, Homocysteine, Vitamin B12, and Polymorphisms of Genes Participating in One-Carbon Metabolism in Late-Onset Alzheimer's Disease Patients and Healthy Controls. *Antioxid Redox Signal.* **2011**; in press.
[30] Fekkes, D, van der Cammen, T.J, van Loon, C.P., *et al.* Abnormal amino acid metabolism in patients with early stage Alzheimer dementia. *J Neural Transm* **1998**; *105*: 287-94.
[31] Hogervorst, E., Ribeiro, H.M, Molyneux, A., Budge, M., Smith, A.D. Plasma homocysteine levels, cerebrovascular risk factors, and cerebral white matter changes (leukoaraiosis) in patients with Alzheimer disease. *Arch Neurol* **2002**; *59*: 787-93.
[32] Mizrahi, E.H., Jacobsen, D.W, Debanne, S.M., *et al.* Plasma total homocysteine levels, dietary vitamin B6 and folate intake in AD and healthy aging. *J Nutr Health Aging* **2003**; *7*: 160- 65.
[33] Quadri, P., Fragiacomo, C., Pezzati, R., *et al.* Homocysteine, folate, and vitamin B-12 in mild cognitive impairment, Alzheimer disease, and vascular dementia. *Am J Clin Nutr* **2004**; *80*: 114-22.
[34] Ario ul, S., Cankurtaran, M., Da li,. N., Khalil, M., Yavuz, B. Vitamin B12, folate, homocysteine and dementia: are they really related? *Arch Gerontol Geriatr* **2005**; *40*: 139-46.
[35] Tu, M.C., Huang, C.W., Chen, N.C, *et al.* Hyperhomocysteinemia in Alzheimer dementia patients and cognitive decline after 6 months follow-up period. *Acta Neurol Taiwan* **2010**; *19*: 168-77.
[36] Galimberti, G., Conti, E., Zini, M., *et al.* Post-methionine load test: A more sensitive tool to reveal hyperhomocysteinemia in Alzheimer patients? *Clin Biochem* **2008**; *41*: 914-6.
[37] Zhuo, J.M., Wang, H., Praticò, D. Is hyperhomocysteinemia an Alzheimer's disease (AD) risk factor, an AD marker, or neither? *Trends Pharmacol Sci* **201**; *32*: 562-71.
[38] Wald, D.S., Kasturiratne, A., Simmonds, M. Serum homocysteine and dementia: meta-analysis of eight cohort studies including 8669 participants. *lzheimers Dement* **2011**; *7*: 412-7.
[39] Ho, R.C., Cheung, M.W., Fu, E., *et al.* Is high homocysteine level a risk factor for cognitive decline in elderly? A systematic review, meta-analysis, and meta-regression. *Am J Geriatr Psychiatry* **2011**, *19*: 607-17.
[40] Rajagopalan, P., Hua, X., Toga, A.W., Jack, C.R. Jr, Weiner, M.W., Thompson, P.M. Homocysteine effects on brain volumes mapped in 732 elderly individuals. *Neuroreport* **2011**; *22*: 391-5.
[41] Selley, M.L., Close, D.R,, Stern, S.E. The effect of increased concentrations of homocysteine on the concentration of (E)-4-hydroxy-2-nonenal in the plasma and cerebrospinal fluid of patients with Alzheimer's disease. *Neurobiol Aging* **2002**; *23*: 383-88.
[42] Hasegawa, T., Ukai, W., Jo, D.G, *et al.* Homocysteic acid induces intraneuronal accumulation of neurotoxic Abeta42: implications for the pathogenesis of Alzheimer's disease. *J Neurosci Res* **2005**; *80*: 869-76.
[43] Isobe, C., Murata, T., Sato, C., Terayama, Y. Increase of total homocysteine concentration in cerebrospinal fluid in patients with Alzheimer's disease and Parkinson's disease. *Life Sci* **2005**; *77*: 1836-43.

[44] Smach, M.A., Jacob, N., Golmard, J.L, *et al.* Folate and homocysteine in the cerebrospinal fluid of patients with Alzheimer's disease or dementia: a case control study. *Eur Neurol* **2011**; *65*: 270-78.

[45] Moore, E., Mander, A., Ames, D., Carne, R., Sanders, K., Watters, D. Cognitive impairment and vitamin B12: a review. *Int Psychogeriatr* **2012**, in press.

[46] Balk, E.M., Raman, G., Tatsioni, A., Chung, M., Lau, J., Rosenberg, I.H. Vitamin B6, B12, and folic acid supplementation and cognitive function: a systematic review of randomized trials. *Arch Intern Med* **2007** *167*: 21-30.

[47] Ford ,A.H, Almeida, O.P. Effect of homocysteine lowering treatment on cognitive function: a systematic review and meta-analysis of randomized controlled trials. *Alzheimers Dis* **2012**; *29*: 133-49.

[48] Wang, H., Odegaard, A., Thyagarajan, B, *et al.* Blood folate is associated with asymptomatic or partially symptomatic Alzheimer's disease in the Nun study. *J Alzheimers Dis* **2012**; *28*: 637-45.

[49] Weisberg, I,, Tran, P., Christensen, B., Sibani, S., Rozen, R. A second genetic polymorphism in methylenetetrahydrofolate reductase (MTHFR) associated with decreased enzymeactivity. *Mol Genet Metab* **1998**; *64*: 169-72.

[50] Ueland, P.M., Hustad, S., Schneede, J., Refsum, H., Vollset, S.E. Biological and clinical implications of the MTHFR C677T polymorphism. *Trends Pharmacol Sci* **2001**; *22*: 195-01.

[51] Dekou, V., Whincup, P., Papacosta, O., *et al.* The effect of the C677T and A1298C polymorphisms in the methylenetetrahydrofolate reductase gene on homocysteine levels in elderly men and women from the British regional heart study. *Atherosclerosis* **2001**; *154*: 659-66.

[52] Strandhagen, E., Zetterberg, H., Aires, N., *et al.* The methylenetetrahydrofolate reductase C677T polymorphism is amajor determinant of coffee-induced increase of plasma homocysteine: a randomized placebo controlled study. *Int J Mol Med* **2004**; *13*: 811-15.

[53] Pollak, R.D, Pollak, A., Idelson, M., Bejarano-Achache, I., Doron, D., Blumenfeld, A. The C677T mutation in the methylenetetrahydrofolate reductase (MTHFR) gene and vascular dementia. *J Am Geriatr Soc* **2000**; *48*: 664-68.

[54] Brunelli, T., Bagnoli, S., Giusti, B., *et al.* The C677T methylenetetrahydrofolate reductase mutation is not associated with Alzheimer's disease. *Neurosci Lett* **2001**; *315*: 103-05.

[55] Wakutani, Y., Kowa, H., Kusumi, M., *et al.* Genetic analysis of vascular factors in Alzheimer's disease. *Ann NY Acad Sci* **2002**; *977*: 232-38.

[56] Religa, D., Styczynska, M., Peplonska, B., *et al.* Homocysteine, apolipoproteine E and methylenetetrahydrofolate reductase in Alzheimer's disease and mild cognitive impairment. Dement Geriatr Cogn Disord **2003**; *16*: 64-70.

[57] Seripa, D., Forno, G.D., Matera, M.G, *et al.* Methylenetetrahydrofolate reductase and angiotensin converting enzyme gene polymorphisms in two genetically and diagnostically distinct cohort of Alzheimer patients. *Neurobiol Aging* **2003**; *24*: 933-39.

[58] Anello, G., Guéant-Rodríguez, R.M., Bosco, P., *et al.* Homocysteine and methylenetetrahydrofolate reductase polymorphism in Alzheimer's disease. *Neuroreport* **2004**; *15*: 859-61.

[59] Wang, B., Jin, F., Kan, R., *et al.* Association of MTHFR gene polymorphism C677T with susceptibility to late-onset Alzheimer's disease. *J Mol Neurosci* **2005**; *27*: 23-7.

[60] Keikhaee, M.R., Hashemi, S.B., Najmabadi, H., Noroozian, M. C677T methylentetrahydrofulate reductase and angiotensin converting enzyme gene polymorphisms in patients with Alzheimer's disease in Iranian population. *Neurochem Res* **2006**; *31*: 1079-83.

[61] Kageyama, M., Hiraoka, M., Kagawa, Y. Relationship between genetic polymorphism, serum folate and homocysteine in Alzheimer's disease. *Asia Pac J Public Health* **2008**; *20*: 111-17.

[62] Kim, J.M., Stewart, R., Kim, S.W., *et al.* Methylenetetrahydrofolate reductase gene and risk of Alzheimer's disease in Koreans. *Int J Geriatr Psychiatry* **2008**; *23*: 454-59.

[63] Zhang, M.Y., Miao, L., Li, Y.S., Hu, G.Y. Meta-analysis of the methylenetetrahydrofolate reductase C677T polymorphism and susceptibility to Alzheimer's disease. *Neurosci Res* **2010**; *68*: 142-50.

[64] Hua, Y., Zhao, H., Kong, Y., Ye, M. Association between the MTHFR gene and Alzheimer's disease: a meta-analysis. *Int J Neurosci* **2011**; *12*: 462-71.

[65] Cascalheira, J.F., João, S.S., Pinhanços, S.S., *et al.* Serum homocysteine: interplay with other circulating and genetic factors in association to Alzheimer's type dementia. *Clin Biochem* **2009**; *42*: 783-90.

[66] Laumet, G., Chouraki, V., Grenier-Boley, B, *et al.* Systematic analysis of candidate genes for Alzheimer's disease in a French, genome-wide association study. *J Alzheimers Dis* **2010**; *20*: 1181-88.

[67] Mansoori, N., Tripathi, M., Luthra, K., *et al.* MTHFR (677 and 1298) and IL-6-174 G/C genes in pathogenesis of Alzheimer's and vascular dementia and their epistatic interaction. *Neurobiol Aging* **2012**; *33*: 1003.e1-1003.e8.

[66] Li, K., Liu, S., Yao, S., Wang, B., Dai, D., Yao, L. Interaction between interleukin-8 and methylenetetrahydrofolate reductase genes modulates Alzheimer's disease risk. *Dement Geriatr Cogn Disord* **2009**; *27*: 286-91.

[69] Rosenberg, N., Murata, M., Ikeda, Y., *et al.* The frequent 5,10-methylenetetrahydrofolate reductase C677T polymorphism is associated with a common haplotype in whites, Japanese, and Africans. *Am J Hum Genet* **2002**; *70*: 758-62.

[70] Ogino, S., Wilson, R.B. Genotype and haplotype distributions of MTHFR677C > T and 1298A > C single nucleotide polymorphisms: a meta-analysis. *J Hum Genet* **2003**; *48*: 1-7.

[71] Yamada, K., Chen, Z., Rozen, R., Matthews, R.G. Effects of common polymorphisms on the properties of recombinant human methylenetetrahydrofolate reductase. *Proc Natl Acad Sci USA* **2001**; *98*:14853-58.

[72] Wakutani, Y., Kowa, H., Kusumi, M., *et al.* A haplotype of the methylenetetrahydrofolate reductase gene is protective against late-onset Alzheimer's disease. *Neurobiol Aging* **2004**; *25*: 291-94.

[73] Wakutani, Y., Kowa, H., Kusumi, M., *et al.* The regulatory region polymorphisms of the MTHFR gene are not associated with Alzheimer's disease. *Dement Geriatr Cogn Disord* **2004**; *17*: 147-50.

[74] Naj, A.C., Beecham, G.W., Martin, E.R, *et al.* Dementia revealed: novel chromosome 6 locus for late-onset Alzheimer disease provides genetic evidence for folate-pathway abnormalities. *PLoS Genet* **2010**; *6*: e1001130.

[75] Ramírez-Lorca, R., Boada, M., Antúnez, C., *et al.* The MTHFD1L gene rs11754661 marker is not associated with Alzheimer's disease in a sample of the Spanish population. *J Alzheimers Dis* **2011**; *25*: 47-50.

[76] Ren, R.J., Wang, L.L., Fang, R., *et al.* The MTHFD1L gene rs11754661 marker is associated with susceptibility to Alzheimer's disease in the Chinese Han population. *J Neurol Sci* **2011**; *308*: 32-4.

[77] Ma, X.Y., Yu, J.T., Wu, Z.C, *et al.* Replication of the MTHFD1L Gene Association with Late- Onset Alzheimer's Disease in a Northern Han Chinese Population. *J Alzheimers Dis* **2012**; in press.

[78] Figgins, J.A., Minster, R.L., Demirci, F.Y., Dekosky, S.T., Kamboh, M.I. Association studies of 22 candidate SNPs with late onset Alzheimer's disease. *Am J Med Genet B Neuropsychiatr* **2009**; *150B*: 520-526.

[79] Li, H., Wetten, S., Li, L., *et al.* Candidate single-nucleotide polymorphisms from a genomewide association study of Alzheimer disease. *Arch Neurol* **2008**; *65*: 45-53.

[80] Reiman, E.M., Webster, J.A., Myers, A.J, *et al.* GAB2 alleles modify Alzheimer's risk in APOE epsilon4 carriers. *Neuron* **2007**; *54*: 713-720.

[81] Christensen, K.E., Rohlicek, C.V., Andelfinger, G.U, *et al.* The MTHFD1 p.Arg653Gln variant alters enzyme function and increases risk for congenital heart defects. *Hum Mutat* **2009**; *30*: 212-220.

[82] Dorszewska, J., Florczak, J., Rozycka. A., *et al.* Oxidative DNA damage and level of thiols as related to polymorphisms of MTHFR, MTR, MTHFD1 in Alzheimer's and Parkinson's diseases. *Acta Neurobiol Exp* **2007**; *67*: 113-129.

[83] Bi, X.H., Zhao, H.L., Zhang, Z.X., Liu, Q., Zhang, J.W. Association analysis of CbetaS 844ins68 and MTHFD1 G1958A polymorphisms with Alzheimer's disease in Chinese. *J Neural Transm* **2010**; *117*: 499-503.

[84] Chango, A., Emery-Fillon, N., de Courcy, G.P., *et al.* A polymorphism (80G>A) in the reduced folate carrier gene and its associations with folate status and homocysteinemia. *Mol Genet Metab* **2000**; *70*: 310-15.

[85] Sunder-Plassmann, G., Fodinger, M. Genetic determinants of the homocysteinenlevel. *Kidney Int* **2003**; *284*: S141-S144.

[86] Devlin, A.M., Clarke, R., Birks, J., Evans, J.G., Halsted, C.H. Interactions among polymorphisms in folate-metabolizing genes and serum total homocysteine concentrations in a healthy elderly population. *Am J Clin Nutr* **2006**; *83*: 708-13.

[87] Lopreato, F.R., Stabler, S.P., Carvalho, F.R., *et al.* Relationships between gene polymorphisms of folate-related proteins and vitamins and metabolites in pregnant women and neonates. *Clin Chim Acta* **2008**; *398*: 134-39.

[88] DeVos, L., Chanson, A., Liu, Z., *et al.* Associations between single nucleotide polymorphisms in folate uptake and metabolizing genes with blood folate, homocysteine, and DNA uracil concentrations. *Am J Clin Nutr* **2008**; *88*: 1149-58.

[89] Fredriksen, A., Meyer, K., Ueland, P.M., Vollset, S.E., Grotmol, T., Schneede, J. Largescale population-based metabolic phenotyping of thirteen genetic polymorphisms related to onecarbon metabolism. *Hum Mutat* **2007**; *28*: 856-65.

[90] Harmon, D.L., Shields, D.C., Woodside, J.V., *et al.* Methionine synthase D919G polymorphism is a significant but modest determinant of circulating homocysteine concentrations. *Genet Epidemiol* **1999**; *17*: 298-09.

[91] Fillon-Emery, N., Chango, A., Mircher, C, *et al.* Homocysteine concentrations in adults with trisomy 21: effect of B vitamins and genetic polymorphisms. *Am J Clin Nutr* **2004**; *80*, 1551-57.

[92] Feix, A., Fritsche-Polanz, R., Kletzmayr, J., *et al.* Increased prevalence of combined MTR and MTHFR genotypes among individuals with severely elevated total homocysteine plasma levels. *Am J Kidney Dis* **2001**; *38*: 956-64.

[93] Laraqui, A., Allami, A., Carrie,A. *et al.* Influence of methionine synthase (A2756G) and methionine synthase reductase (A66G) polymorphisms on plasma homocysteine levels and relation to risk of coronary artery disease. *Acta Cardiol* **2006**; *61*: 51-61.

[94] Biselli, J.M., Goloni-Bertollo, E.M., Haddad, R., Eberlin, M.N., Pavarino-Bertelli, E.C. The MTR A2756G polymorphism is associated with an increase of plasma homocysteine concentration in Brazilian individuals with Down syndrome. *Braz J Med Biol Res* **2008**; *41*: 34-40.

[95] Alessio, A.C., Annichino-Bizzacchi, J.M., Bydlowski, S.P., Eberlin, M.N., Vellasco, A.P., Hoehr, N.F. Polymorphisms in the methylenetetrahydrofolate reductase and methionine synthase reductase genes and homocysteine levels in Brazilian children. *Am J Med Genet A* **2004**; *128*: 256-60.

[96] Beyer, K., Lao, J.I., Latorre, P., *et al.* Methionine synthase polymorphism is a risk factor for Alzheimer disease. *Neuroreport* **2003**; *14*: 1391-94.

[97] Bosco, P., Guéant-Rodríguez, R.M., Anello, G., *et al.* Association of IL-1 RN*2 allele and methionine synthase 2756 AA genotype with dementia severity of sporadic Alzheimer's disease. *J Neurol Neurosurg Psychiatry* **2004**; *75*: 1036-38.

[98] Zhao, H.L., Li, X.Q, Zhang, Z.X, Bi, X.H, Wang ,B., Zhang, J.W. Association analysis of methionine synthase gene 2756 A>G polymorphism and Alzheimer disease in a Chinese population. *Brain Res* **2008**; *1204*: 118-22.

[99] Wilson, A., Platt ,R., Wu, Q., *et al.* A common variant in methionine synthase reductase combined with low cobalamin (vitamin B12) increases risk for spina bifida *Mol Genet Metab* **1999**; *67*: 317-23.

[100] Laraqui, A., Allami, A., Carrié, A. *et al.* Influence of methionine synthase (A2756G) and methionine synthase reductase (A66G) polymorphisms on plasma homocysteine levels and relation to risk of coronary artery disease. *Acta Cardiol* **2006**; *61*: 51-61.

[101] Scazzone, C., Acuto, S., Guglielmini, E., Campisi, G., Bono, A. Methionine synthase reductase (MTRR) A66G polymorphism is not related to plasma homocysteine concentration and the risk for vascular disease. *Exp Mol Pathol* **2009**; *86*: 131-33.

[102] Barbosa, P.R., Stabler, S.P., Machado, A.L, *et al.* Association between decreased vitamin levels and MTHFR, MTR and MTRR gene polymorphisms as determinants for elevated total homocysteine concentrations in pregnant women. *Eur J Clin Nutr* **2008**; *62*:1010-21.

[103] Wolthers, K.R., Scrutton, N.S. Cobalamin uptake and reactivation occurs through specific protein interactions in the methionine synthase-methionine synthase reductase complex. *FEBS J* **2009**; *276*: 1942-51.

[104] Hall, C.A. The carriers of native vitamin B12 in normal human serum. *Clin Sci Mol Med* **1977**; *53*: 453-57.

[105] Carmel, R. The distribution of endogenous cobalamin among cobalamin-binding proteins in the blood in normal and abnormal states. *Am J Clin Nutr* **1985**; *41*: 713-19.

[106] Johnston, C.S, Thomas, J.A. Holotranscobalamin II levels in plasma are related to dementia in older people. *J Am Geriatr Soc* **1997**; *45*: 779-80.

[107] Refsum, H., Smith, A.D. Low vitamin B-12 status in confirmed Alzheimer's disease as revealed by serum holotranscobalamin. *J Neurol Neurosurg Psychiatry* **2003**; *74*: 959-61.

[108] Kivipelto, M., Annerbo, S., Hultdin, J., *et al.* Homocysteine and holo-transcobalamin and the risk of dementia and Alzheimers disease: a prospective study. *Eur J Neurol* **2009**; *16*: 808- 13.

[109] von Castel-Dunwoody, K.M., Kauwell, G.P., Shelnutt, K.P., *et al.* Transcobalamin 776C>G polymorphism negatively affects vitamin B-12 metabolism. *Am J Clin Nutr* **2005**; *81*: 1436-41.

[110] Zetterberg, H., Nexö, E., Regland, B., *et al.* The transcobalamin (TC) codon 259 genetic polymorphism influences holo-TC concentration in cerebrospinal fluid from patients with Alzheimer disease. *Clin Chem* **2003**; *49*: 1195-98.

[111] Zetterberg, H., Nexö, E., Minthon, L., *et al.* The transcobalamin 776C > G polymorphism may be a modifiable genetic risk factor for Alzheimer's disease. *Int Psychogeriatr* **2005**; *17*: 329- 31.

[112] McCaddon, A., Blennow, K., Hudson, P., *et al.* Transcobalamin polymorphism and serum holo-transcobalamin in relation to Alzheimer's disease. *Dement Geriatr Cogn Disord* **2004**; *17*: 215-21.

[113] Kraus, J.P Biochemistry and molecular genetics of cystathionine beta-synthase deficiency. *Eur J Pediatr* **1998**; *157*: S50-S53.

[114] Yamanishi, M., Kabil, O., Sen, S., Banerjee, R. Structural insights into pathogenic mutations in heme-dependent cystathionine-beta-synthase. *J Inorg Biochem* **2006**; *100*: 1988-95.

[115] Sebastio, G., Sperandeo, M.P., Panico, M., de Franchis, R., Kraus, J.P, Andria, G. The molecular basis of homocystinuria due to cystathionine beta-synthase deficiency in Italian families, and report of four novel mutations. *Am J Hum Genet* **1995**; *56*: 1324-1333.

[116] Tsai, M.Y., Bignell, M., Schwichtenberg, K., Hanson, N.Q. High prevalence of a mutation in the cystathionine beta-synthase gene. *Am J Hum Genet* **1996**; *59*: 1262-67.

[117] Franco, R.F., Elion, J., Lavinha, J., Krishnamoorthy, R., Tavella, M.H,, Zago, M.A. Heterogeneous ethnic distribution of the 844ins68 in the cystathionine betasynthase gene. *Hum Hered* **1998**; *48*: 338-42.

[118] Pepe, G., Vanegas, O.C., Rickards, O., *et al.* World distribution of the T833C/844INS68 CBS in *cis* double mutation: a reliable anthropological marker. *Hum Genet* **1999**; *104*: 126-29.

[119] Kluijtmans, L.A., Young, I.S., Boreham, C.A, *et al.* Genetic and nutritional factors contributing to hyperhomocysteinemia in young adults. *Blood* **2003**; *101*: 2483-88.

[120] Belkahla, R., Omezzine, A., Kchok, K., *et al.* Effect of polymorphisms on key enzymes in homocysteine metabolism, on plasma homocysteine level and on coronary artery-disease risk in a Tunisian population. *Ann Cardiol Angeiol* (Paris) **2008**; *57*: 219-24.

[121] Beyer, K., Lao, J.I., Carrato, C., *et al.* Cystathionine beta synthase as a risk factor for Alzheimer disease. *Curr Alzheimer Res* **2004**; *1*: 127-33.

[122] Horie, N., Aiba, H., Oguro, K., Hojo, H., Takeishi, K. Functional analysis and DNA polymorphism of the tandemly repeated sequences in the 5'-terminal regulatory region of the human gene for thymidylate synthase. *Cell Struct Funct* **1995**; *20*: 191-97.

[123] Ulrich, C.M., Bigler, J., Bostick, R., Fosdick, L., Potter, J.D. Thymidylate synthase promoter polymorphism, interaction with folate intake, and risk of colorectal adenomas. *Cancer Res* **2002**; *62*: 3361-64.

[124] Hu, J., Fan, H., Liu, D., Zhang, S., Zhang, F., Xu, H. DNMT3B promoter polymorphism and risk of gastric cancer. *Dig Dis Sci* **2010**; *55*: 1011-16.

[125] Lee, S.J, Jeon, H.S., Jang, J.S., *et al.* DNMT3B polymorphisms and risk of primary lung cancer. *Carcinogenesis* **2005**; *26*: 403-409.

[126] Coppedè, F., Zitarosa, M.T., Migheli, F., *et al. DNMT3B* promoter polymorphisms and risk of late onset Alzheimer's disease. *Curr Alzheimer Res* **2012**; in press.

[127] Bottiglieri, T., Godfrey, P., Flynn, T., Carney, M.W., Toone, B.K., Reynolds, E.H. Cerebrospinal fluid S-adenosylmethionine in depression and dementia: effects of treatment with parenteral and oral S-adenosylmethionine. *J Neurol Neurosurg Psychiatry* **1990**; *53*: 1096- 98.

[128] Morrison, L.D., Smith, D.D., Kish, S.J. Brain, S-adenosylmethionine levels are severely decreased in Alzheimer's disease. *J Neurochem* **1996**; *67*: 1328-31.

[129] Gomes Trolin, C., Regland, B., Oreland, L. Decreased methionine adenosyltransferase activity in erythrocytes of patients with dementia disorders. *Eur Neuropsychopharmacol* **1995**; *5*: 107-14.

[130] Mulder, C., Schoonenboom, N.S., Jansen, E.E., *et al.* The transmethylation cycle in the brain of Alzheimer patients. *Neurosci Lett* **2005**; *386*: 69-71.

[131] Selley, M.L. A metabolic link between S-adenosylhomocysteine and polyunsaturated fatty acid metabolism in Alzheimer's disease. *Neurobiol Aging* **2007**; *28*: 1834-39.

[132] Selley, M.L. Increased homocysteine and decreased adenosine formation in Alzheimer's disease. *Neurol Res* **2004**; *26*: 554-57.

[133] Kennedy, B.P., Bottiglieri, T., Arning, E., Ziegler, M.G., Hansen, L.A., Masliah, E. Elevated Sadenosyhomocysteine in Alzheimer brain: influence on methyltransferases and cognitive function. *J Neural Transm* **2004**; *111*: 547-67.

[134] Hooijmans, C.R., Blom, H.J., Oppenraaij-Emmerzaal, D., Ritskes-Hoitinga, M., Kiliaan, A.J. The analysis of S-adenosylmethionine and S-adenosylhomocysteine levels in the aging brain of APP/PS1 Alzheimer mice. *Neurol Sci* **2009**; *30*: 439-45.

[135] Seshadri, S., Beiser, A., Selhub, J., *et al.* Plasma homocysteine as a risk factor for dementia and Alzheimer's disease. *N Engl J Med* **2002**; *346*: 476-83.

[136] Ho, P.I., Ashline, D., Dhitavat, S., *et al.* Folate deprivation induces neurodegeneration: roles of oxidative stress and increased homocysteine. *Neurobiol Dis* **2003**; *14*: 32-42.

[137] Fowler, B. Homocysteine: overview of biochemistry, molecular biology, and role in disease processes. *SeminVasc Med* **2005**; *5*: 77-86.

[138] Ho, P.I., Ortiz, D., Rogers, E., Shea, T.B. Multiple aspects of homocysteine neurotoxicity: glutamate excitotoxicity, kinase hyperactivation and DNA damage. *J Neurosci Res* **2002**; *70*: 694-02.

[139] Borroni, B., Agosti, C., Panzali, A.F., Di Luca, M., Padovani, A. Homocysteine, vitamin B6, and vascular disease in patients with AD. *Neurology* **2002**; *59*: 1475.

[140] Luchsinger, J.A., Reitz, C., Patel, B., Tang, M.X., Manly, J.J., Mayeux ,R. Relation of diabetes to mild cognitive impairment. *Arch Neurol* **2007**; *64*: 570-75.

[141] Irizarry, M.C, Gurol, M.E., Raju, S, *et al.*. Association of homocysteine with plasma amyloid beta protein in aging and neurodegenerative disease. *Neurology* **2005**; *65*: 1402-08.

[142] Vafai, S.B., Stock, J.B. Protein phosphatase 2A methylation: a link between elevated plasma homocysteine and Alzheimer's Disease. *FEBS Lett* **2002**; *518*: 1-4.

[143] Sontag, E., Nunbhakdi-Craig, V., Sontag, J.M, *et al.* Protein phosphatase 2A methyltransferase links homocysteine metabolism with tau and amyloid precursor protein regulation. *J Neurosci* **2007**; *27*: 2751-59.

[144] Scarpa, S., Fuso, A., D'Anselmi, F., Cavallaro, R.A. Presenilin 1 gene silencing by Sadenosylmethionine: a treatment for Alzheimer disease? *FEBS Lett* **2003**; *541*: 145-48.
[145] Fuso, A., Seminara, L., Cavallaro, R.A., D'Anselmi, F., Scarpa, S. Sadenosylmethionine/ homocysteine cycle alterations modify DNA methylation status with consequent deregulation of PS1 and BACE and beta-amyloid production. *Mol Cell Neurosci* **2005**; *28*: 195-204.
[146] Cedar, H., Bergman, Y. Programming of DNA Methylation Patterns. *Annu Rev Biochem* **2012**; in press.
[147] Jones, P.A, Liang, G. Rethinking how DNA methylation patterns are maintained. *Nat Rev Genet* **2009**; *10*: 805-11.
[148] Goll, M.G., Bestor, T.H. Eukaryotic cytosine methyltransferases. *Annu Rev Biochem* **2005**; *74*: 481-514.
[149] Goll, M.G., Kirpekar, F., Maggert, K.A, *et al.* Methylation of tRNAAsp by the DNA methyltransferase homolog Dnmt2. *Science* **2006**; *311*: 395-98.
[150] Métivier, R., Gallais, R., Tiffoche, C., *et al.* Cyclical DNA methylation of a transcriptionally active promoter. *Nature* **2008**; *452*: 45-50.
[151] Feng, J., Chang, H., Li, E., Fan, G. Dynamic expression of *de novo* DNA methyltransferases Dnmt3a and Dnmt3b in the central nervous system. *J Neurosci Res* **2005**; *79* :734-46.
[152] Watanabe, D., Uchiyama, K., Hanaoka, K. Transition of mouse *de novo* methyltransferases expression from Dnmt3b to Dnmt3a during neural progenitor cell development. *Neuroscience* **2006**; *142*: 727-37.
[153] Feng, J., Zhou, Y., Campbell, S.L, *et al.* Dnmtl and Dnmt3a maintain DNA methylation and regulate synaptic function in adult forebrain neurons. *Nat Neurosci* **2010**; *13*: 423-3.
[154] Wu, H., Coskun, V., Tao, J, *et al.* Dnmt3a-dependent nonpromoter DNA methylation facilitates transcription of neurogenic genes. *Science* **2010**; *329*: 444-48.
[155] Molfese, D.L. Advancing neuroscience through epigenetics: molecular mechanisms of learning and memory. *Dev Neuropsychol* **2011**; *36*: 810-27.
[156] Gonzalo, S. Epigenetic alterations in aging. *J Appl Physiol* **2010**; *109*: 586-97.
[157] Guo, J.U., Su, Y., Zhong, C, Ming, G.L, Song, H. Hydroxylation of 5-methylcytosine by TET1 promotes active DNA demethylation in the adult brain. *Cell* **2011**; *145*: 423-34.
[158] Chen, H., Dzitoyeva, S., Manev, H. Effect of aging on 5-hydroxymethylcytosine in the mouse hippocampus. *Restor Neurol Neurosci* **2012**; in press.
[159] Dzitoyeva, S., Chen, H., Manev, H. Effect of aging on 5-hydroxymethylcytosine in brain mitochondria. *Neurobiol Aging* **2012**; in press.
[160] van den Hove, D.L., Chouliaras, L., Rutten, B.P. The role of 5-hydroxymethylcytosine in aging and Alzheimer's disease: current status and prospects for future studies. *Curr Alzheimer Res* **2012**; in press.
[161] Leeb, M., Wutz, A. Establishment of epigenetic patterns in development. *Chromosoma* **2012**; in press.
[162] Fuso, A., Nicolia, V., Cavallaro, R.A, *et al.* B-vitamin deprivation induces hyperhomocysteinemia and brain S-adenosylhomocysteine, depletes brain Sadenosylmethionine, and enhances PS1 and BACE expression and amyloid-beta deposition in mice. *Mol Cell Neurosci* **2008**; *37*: 731-46.
[163] Fuso, A., Nicolia, V, Pasqualato, A., *et al.* Changes in Presenilin 1 gene methylation pattern in diet-induced B vitamin deficiency. *Neurobiol Aging* **2011**; *32*: 187-99.

[164] Chan, A., Shea, T.B. Folate deprivation increases presenilin expression, gamma-secretase activity, and Abeta levels in murine brain: potentiation by ApoE deficiency and alleviation by dietary S-adenosyl methionine. *J Neurochem* **2007**; *102*: 753-60.

[165] Chan, A., Tchantchou, F., Rogers, E.J, Shea, T.B. Dietary deficiency increases presenilin expression, gamma-secretase activity, and Abeta levels: potentiation by ApoE genotype and alleviation by S-adenosyl methionine. *J Neurochem* **2009**; *110*: 831-36.

[166] Lin, H.C., Hsieh, H.M., Chen, Y.H., Hu, M.L. S-Adenosylhomocysteine increases beta-amyloid formation in BV-2 microglial cells by increased expressions of beta-amyloid precursor protein and presenilin 1 and by hypomethylation of these gene promoters. *Neurotoxicology* **2009**; *30*: 622-27.

[167] Fuso, A., Nicolia, V., Cavallaro, R.A,, Scarpa, S. DNA methylase and demethylase activities are modulated by one-carbon metabolism in Alzheimer's disease models. *J Nutr Biochem* **2011**; *22*: 242-51.

[168] Borro, M., Cavallaro, R.A., Gentile, G., *et al.* One-carbon metabolism alteration affects brain proteome profile in a mouse model of Alzheimer's disease. *J Alzheimers Dis* **2010**; *22*: 1257-68.

[169] Cavallaro, R.A., Fuso, A., D'Anselmi, F., Seminara, L., Scarpa, S. The effect of Sadenosylmethionine on CNS gene expression studied by cDNA microarray analysis. *J Alzheimers Dis* **2006**; *9*: 415-19.

[170] Fuso, A., Nicolia, V., Ricceri, L., *et al.* S-adenosylmethionine reduces the progress of the Alzheimer-like features induced by B-vitamin deficiency in mice. *Neurobiol Aging* **2012**; in press.

[171] Guo, X., Wu, X., Ren, L., Liu, G., Li, L. Epigenetic mechanisms of amyloid-   production in anisomycin-treated SH-SY5Y cells. *Neuroscience* **2011**; *194*: 272-81.

[172] Chen, K.L., Wang, S.S., Yang, Y.Y., Yuan, R.Y., Chen, R.M., Hu, C.J. The epigenetic effects of amyloid-beta(1-40) on global DNA and neprilysin genes in murine cerebral endothelial cells. *Biochem Biophys Res Commun* **2008**; *378*: 57-61.

[173] Shin, J., Yu, S.B., Yu, U.Y., Jo, S.A, Ahn, J.H. Swedish mutation within amyloid precursor protein modulates global gene expression towards the pathogenesis of Alzheimer's disease. *BMB Reports* **2010**; *43*: 704-09.

[174] Sung, H.Y., Choi ,E.N,, Ahn Jo, S., Oh, S., Ahn, J.H. Amyloid protein-mediated differential DNA methylation status regulates gene expression in Alzheimer's disease model cell line. *Biochem Biophys Res Commun* **2011**; *414*: 700-05.

[175] Byun, C.J., Seo, J,, Jo, S.A., *et al.*. DNA methylation of the 5'-untranslated region at +298 and +351 represses BACE1 expression in mouse BV-2 microglial cells. *Biochem Biophys Res Commun* **2012**; *417*: 387-92.

[176] Basha, M.R., Wei, W., Bakheet, S.A., *et al.* The fetal basis of amyloidogenesis: exposure to lead and latent overexpression of amyloid precursor protein and beta-amyloid in the aging brain. *J Neurosci* **2005**; *25*: 823-29.

[177] Wu, J., Basha, M.R., Brock, B., *et al.* Alzheimer's disease (AD)-like pathology in aged monkeys after infantile exposure to environmental metal lead (Pb): evidence for a developmental origin and environmental link for AD. *J Neurosci* **2008**; *28*: 3-9.

[178] Zawia, N.H., Lahiri, D.K., Cardozo-Pelaez, F. Epigenetics, oxidative stress, and Alzheimer disease. *Free Radic Biol Med* **2009**; *46*: 1241-49.

[179] Bihaqi, S.W., Huang, H., Wu, J., Zawia, N.H. Infant exposure to lead (Pb) and epigenetic modifications in the aging primate brain: implications for Alzheimer's disease. *J Alzheimers Dis* **2011**; *27*: 819-33.

[180] Bihaqi, S.W., Zawia, N.H. Alzheimer's disease biomarkers and epigenetic intermediates following exposure to Pb *in vitro*. *Curr Alzheimer Res* **2012**; in press.

[181] Li, Y.Y., Chen, T., Wan, Y., Xu, S.Q. Lead exposure in pheochromocytoma cells induces persistent changes in amyloid precursor protein gene methylation patterns. *Environ Toxicol* **2010**; in press.

[182] Mastroeni, D., McKee, A., Grover, A., Rogers, J., Coleman, P.D. Epigenetic differences in cortical neurons from a pair of monozygotic twins discordant for Alzheimer's disease. *PLoS One* **2009**; *4*: e6617.

[183] Mastroeni, D., Grover, A, Delvaux, E., Whiteside, C., Coleman, P.D., Rogers, J. Epigenetic changes in Alzheimer's disease: Decrements in DNA methylation. *Neurobiol Aging* **2010**; *31*: 2025-37.

[184] Barrachina, M., Ferrer, I. DNA methylation of Alzheimer disease and tauopathy-related genes in postmortem brain. *J Neuropathol Exp Neurol* **2009**; *68*: 880-91.

[185] Brohede, J., Rinde, M., Winblad, B., Graff, C. A DNA methylation study of the amyloid precursor protein gene in several brain regions from patients with familial Alzheimer disease. *J Neurogenet* **2010**; *24*: 179-81.

[186] Siegmund, K.D,, Connor, C.M., Campan, M., *et al.* DNA methylation in the human cerebral cortex is dynamically regulated throughout the life span and involves differentiated neurons. *PLoS One*. **2007**; *2*: e895.

[187] Wang, S.C., Oelze, B., Schumacher, A. Age-specific epigenetic drift in late-onset Alzheimer's disease. *PLoS One* **2008**: *3*: e2698.

[188] Bakulski, K.M., Dolinoy, D.C., Sartor, M.A., *et al.* Genome-Wide DNA Methylation Differences Between Late-Onset Alzheimer's Disease and Cognitively Normal Controls in Human Frontal Cortex. *J Alzheimers Dis* **2012**; in press.

[189] Pietrzak, M., Rempala, G., Nelson, P.T., Zheng, J.J, Hetman, M. Epigenetic silencing of nucleolar rRNA genes in Alzheimer's disease. *PLoS One* **2011**; *6*: e22585.

[190] Bollati, V., Galimberti, D., Pergoli, L., *et al.* DNA methylation in repetitive elements and Alzheimer disease. *Brain Behav Immun*. **2011**; *25*: 1078-83.

[191] Van Dam, F., Van Gool, W.A. Hyperhomocysteinemia and Alzheimer's disease: A systematic review. *Arch Gerontol Geriatr* **2009**; *48*: 425-30.

[192] Francis, Y.I., Fà, M., Ashraf, H., *et al.* Dysregulation of histone acetylation in the APP/PS1 mouse model of Alzheimer's disease. *J Alzheimers Dis* **2009**; *18*: 131-39.

[193] Kilgore, M., Miller, C.A., Fass, D.M, *et al.* Inhibitors of class 1 histone deacetylases reverse contextual memory deficits in a mouse model of Alzheimer's disease. *Neuropsychopharmacology* **2009**; *35*: 870-80.

[194] Nuutinen, T., Suuronen, T., Kauppinen, A., Salminen, A. Valproic acid stimulates clusterin expression in human astrocytes: Implications for Alzheimer's disease. *Neurosci Lett* **2010**; *475*: 64-68.

[195] Govindarajan, N., Agis-Balboa, R.C., Walter, J., Sananbenesi, F., Fischer, A. Sodium Butyrate Improves Memory Function in an Alzheimer's Disease Mouse Model When Administered at an Advanced Stage of Disease Progression. *J Alzheimers Dis* **2011**; *26*: 187-97.

[196] Ricobaraza, A., Cuadrado-Tejedor, M., Marco, S., Pérez-Otaño, I., García-Osta, A. Phenylbutyrate rescues dendritic spine loss associated with memory deficits in a mouse model of Alzheimer disease. *Hippocampus* **2010**. In Press.

[197] Wien, T.N., Pike, E., Wisløff, T., Staff, A., Smeland, S., Klemp, M. Cancer risk with folic acid supplements: a systematic review and meta-analysis. *BMJ Open* **2012**; *2*: e000653.

[198] Coppedè, F. Epigenetic biomarkers of colorectal cancer: Focus on DNA methylation. *Cancer Lett* **2011**; in press.

[199] Beard, C.M., Panser, L.A., Katusic, S.K. Is excess folic acid supplementation a risk factor for autism? *Med Hypotheses* **2011**; *77*: 15-7.

[200] Kiefte-de Jong, J.C., Timmermans, S., Jaddoe, V.W, *et al.* High circulating folate and vitamin B-12 concentrations in women during pregnancy are associated with increased prevalence of atopic dermatitis in their offspring. *J Nutr* **2012**; *142*: 731-38.

[201] Morris, M.S., Jacques, P.F., Rosenberg, I.H., Selhub, J. Circulating unmetabolized folic acid and 5-methyltetrahydrofolate in relation to anemia, macrocytosis, and cognitive test performance in American seniors. *Am J Clin Nutr* **2010**; *91*:1733-44.

[202] Junaid, M.A., Kuizon, S., Cardona, J., *et al.* Folic acid supplementation dysregulates gene expression in lymphoblastoid cells--implications in nutrition. *Biochem Biophys Res Commun* **2011**; *412*: 688-92.

[203] Lee, S., Lemere, C.A, Frost, J.L., Shea, T.B. Dietary supplementation with S-adenosyl methionine delayed amyloid- and tau pathology in 3xTg-AD mice. *J Alzheimers Dis* **2012**; *28*: 423-31.

[204] Suchy, J., Lee, S., Ahmed, A., Shea, T.B. Dietary supplementation with S-adenosyl methionine delays the onset of motor neuron pathology in a murine model of amyotrophic lateral sclerosis. *Neuromolecular Med* **2010**; *12*: 86-97.

[205] Coppedè, F., Grossi, E., Migheli, F., Migliore, L. Polymorphisms in folate-metabolizing genes, chromosome damage, and risk of Down syndrome in Italian women: identification of key factors using artificial neural networks. *BMC Med Genomics* **2010**; *3*: 42.

CHAPTER 3

# Mechanisms and Signals for the Nuclear Import of Proteins

**Natália Freitas and Celso Cunha***

*Unidade de Microbiologia Médica, Centro de Malária e outras Doenças Tropicais, Instituto de Higiene e Medicina Tropical, Universidade Nova de Lisboa, Rua da Junqueira, 100 1349-008 Lisboa, Portugal*

**Abstract:** In eukaryotes, the nuclear membrane provides a physical barrier to the passive diffusion of macromolecules from and into the cytoplasm. Nucleocytoplasmic traffic occurs through highly specialized structures known as nuclear pores, and involves the participation of a special class of transport proteins. Active transport across the nuclear pores is an energy-dependent process that relies on the activity of Ran-GTPases both in the nuclear and cytoplasmic compartments.

Nuclear import of proteins is an essential step in regulating gene expression and the replication cycle of several viruses. In this review, the key mechanisms, pathways, and models underlying the transport of proteins across nuclear pores are analysed.

**Keywords:** Nuclear pore complex, nucleoporin, nuclear localization signal, protein import, importin, karyopherin.

## NUCLEAR PORE COMPLEX

In eukaryotic cells the nucleus is physically separated from the cytoplasm by a double membrane structure, the nuclear envelope (NE). The NE is crossed by multiple supramolecular structures specialized in mediating the bidirectional traffic of molecules between the nucleus and the cytoplasm. These structures are designated the nuclear pore complex (NPC). The number of NPCs per cell is variable between species and is dependent on cell size and transcriptional activity. It is generally estimated that the rate of translocation through NPCs may achieve 1000 molecules per second. The nucleus of yeast cells usually contains about 200

***Address correspondence to Celso Cunha:** Unidade de Microbiologia Médica, Centro de Malária e outras Doenças Tropicais, Instituto de Higiene e Medicina Tropical, Universidade Nova de Lisboa, Rua da Junqueira, 100 1349-008 Lisboa, Portugal; Tel: +351 213652600 ; Fax: +351213632105;
E-mail: ccunha@ihmt.unl.pt

**Christian Neri (Ed)**

NPCs while human cells and mature *Xenopus* oocytes may contain $5x10^3$ – $5x10^7$ NPCs per nucleus [1]. Electron microscopy observations have shown that, in most cases, the three dimensional structure of the NPC is conserved even in evolutionary distant species [2]. NPCs display a cylindrical tripartite structure which is about 90 nm in length and 50 nm wide (Fig. **1**). The central structure of

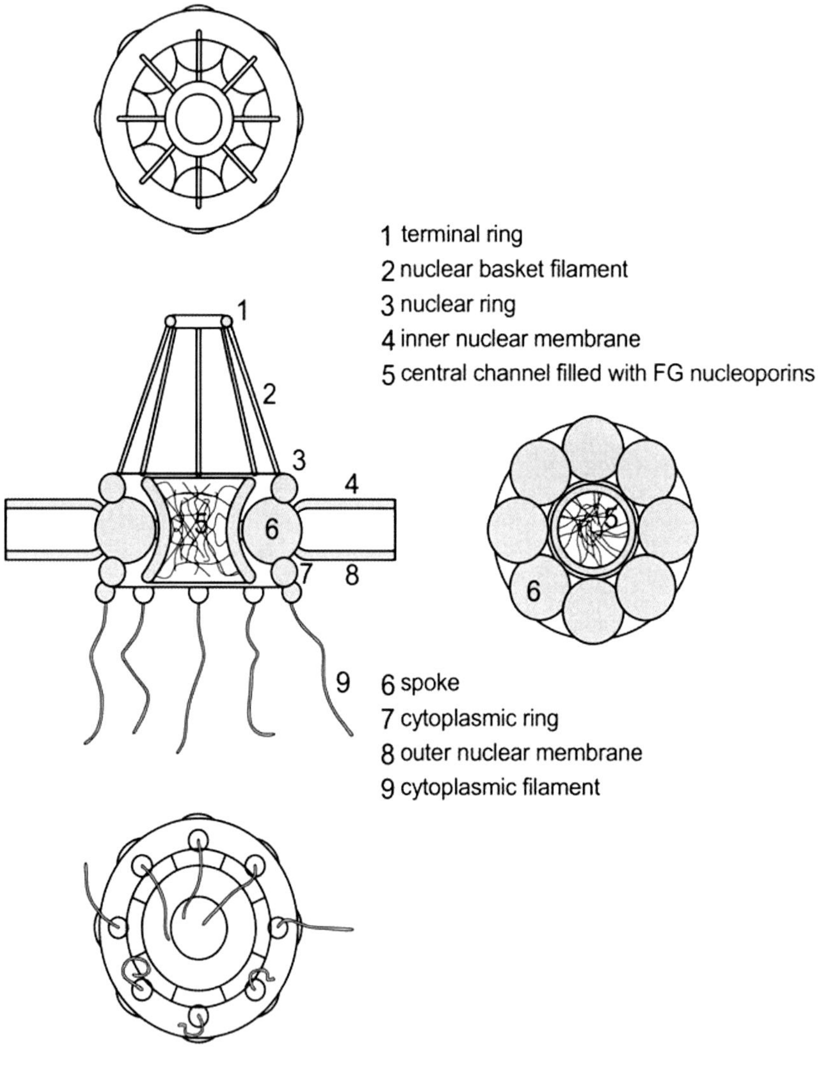


**Figure 1:** Schematic view of the nuclear pore complex. The cross-section of the nuclear ring, central structure, and cytoplasmic ring are displayed on the top, middle, and bottom of the figure, respectively.

the NPC is anchored between the inner and outer layers of the NE. This central structure consists of eight subunits that constitute the nuclear and cytoplasmic rings [3]. Eight protein filaments of 50 nm in length originate from the cytoplasmic ring towards the cytoplasm. From the nuclear ring also originate eight filaments. The filaments in the nuclear side, however, are 75 nm in length and converge to a ring-like structure named the nuclear basket [3]. The central structure of the NPC contains a central channel [3]. This channel is about 30 nm in diameter and allows the transport, by passive diffusion, of ions and small molecules, including proteins with a molecular mass up to 40 KDa. The traffic of larger molecules through the NPCs requires, however, the involvement of energy dependent active transport mechanisms [4].

Early estimates of the number of proteins in NPCs seemed to indicate that these complex structures may contain up to 100 different proteins. More recently, proteomic approaches have been used to analyse the protein composition of NPCs, both in yeast and vertebrates. Surprisingly, these experiments revealed that NPCs are constituted by only 30 different proteins which are generally designated as nucleoporins [5, 6]. About one third of the identified nucleoporins contain a characteristic structural motif constituted by multiple repeats of phenylalanine and glycine residues (FG motifs) [5]. The majority of nucleoporins are symmetrically distributed relative to the middle plan of the NE [6]. Based on the relative abundance and molecular mass of each nucleoporin it was possible to estimate that the molecular mass of the NPC may range between 44 and 60 MDa in yeast and vertebrates, respectively [5, 6]. These values are distinct from those initially obtained by transmission-scanning microscopy which ascribed molecular masses of 60 MDa for the yeast NPC and 125 MDa for the vertebrate NPC [7, 8].

## PROTEIN TRANSPORT RECEPTORS

For the majority of macromolecules, the nucleocytoplasmic transport through the NPCs is an energy-dependent process mediated by soluble transport receptors that generally belong to a family proteins designated β-karyopherins [9]. β-karyopherins mediate both the import and export of all proteins displaying dimensions over the size exclusion limit for simple diffusion through the NPCs [9]. Besides this, β-karyopherins also mediate the transport of non-coding cellular

RNAs [10]. The formation of import or export complexes is dependent on the interaction of β-karyopherins with small peptide motifs present in protein cargos. These motifs are generally called nuclear localization signals (NLS) or nuclear export signals (NES). The molecular mechanisms underlying RNA transport are usually more complex since they include the additional participation of adaptor proteins that interact with the transport receptor and the RNA [10].

β-karyopherins are acidic proteins with molecular masses ranging from 90 to 145 KDa. In eukaryotes, over 20 β-karyopherins were identified of which 10 were shown to directly participate in the nuclear import of proteins [9]. β-karyopherins are characterized by the ability to directly interact with both the Ran GTPase and the FG domains of nucleoporins [11]. Ran GTPase is a monomeric protein of 24 KDa that belongs to the Ras superfamily. Similar to other GTPases, Ran can be found in two distinct forms: GTP-bound and GDP-bound. The two forms of Ran are asymmetrically distributed between the nucleus and the cytoplasm. Ran-GTP localizes predominantly in the nucleus while Ran-GDP is observed mainly in the cytoplasm [12]. The asymmetric distribution of the two forms of Ran is maintained by specific regulatory proteins localized in the nuclear and cytoplasmic compartments. Usually, as in the case of other members of the Ras superfamily, the intrinsic GTPase activity of Ran, promoting the hydrolysis of GTP into GDP, is slow. However, this activity can be accelerated by two cytoplasmic proteins: Ran-GAP and RanBP1 (Fig. **2-II**) [12]. The reverse reaction, conversion of Ran-GDP into Ran-GTP is stimulated by the nuclear protein RCC1. RCC1 binds to Ran-GDP promoting GDP dissociation and subsequent binding of GTP in the active centre of Ran (Fig. **2-V**). The combined action of Ran regulatory proteins creates and maintains a Ran-GTP gradient across the NE. This gradient is a key element in establishing the direction of nucleocytoplasmic transport. In fact, the interaction of β-karyopherins with the respective cargo substrates is regulated by the Ran-GTP gradient [13, 14].

Along with the action of Ran regulatory proteins the Ran-GTP gradient is also maintained by the nuclear import of Ran-GDP. The nuclear transport of Ran-GDP is regulated by the NFT2 import factor [15, 16]. NTF2 is a conserved 15 KDa protein that specifically binds to the GDP-bound form of Ran [15]. The nuclear import of the complex Ran-GDP-NFT2 is not energy dependent. Once in the nucleus, this complex dissociates after conversion of Ran-GDP into Ran-GTP by the action of the nuclear protein RCC1 (Fig. **2-V**) [15].

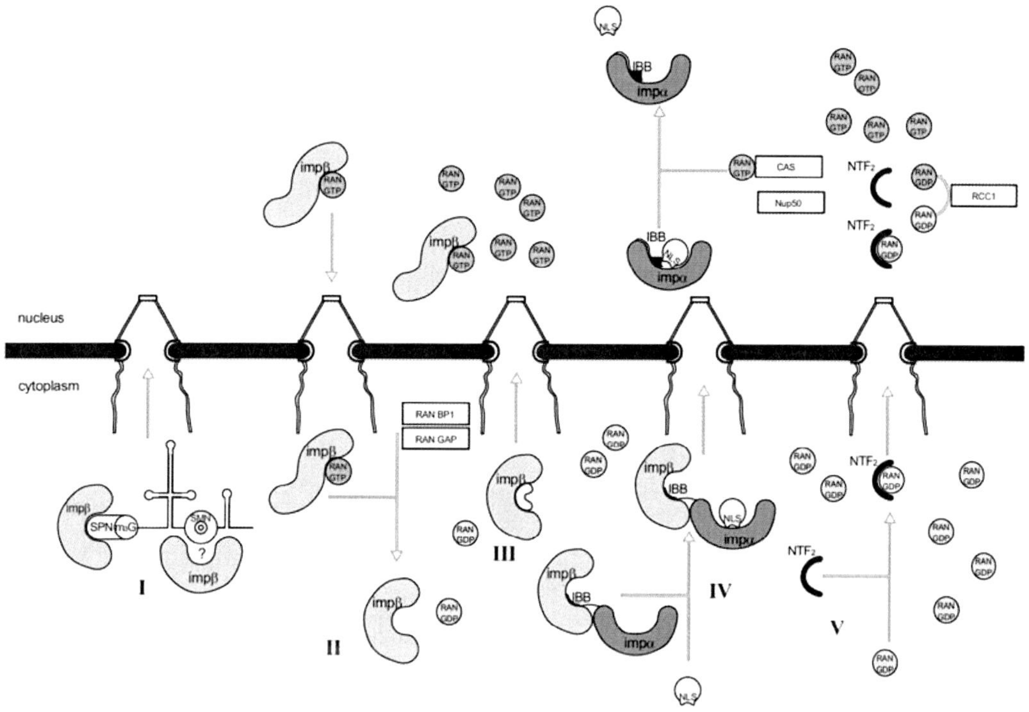


**Figure 2:** Schematic representation of protein nuclear import pathways. **I**- Import of UsnRNPs is mediated by importin-β which associates with the SMN complex coupled to UsnRNAs; **II**- The RanGTP gradient regulates the shuttling of importin-β between the nucleus and the cytoplasm; **III**- The direct interaction of importin-β with cargos promotes translocation through nuclear pores of several proteins; **IV**- Importin-α recognizes the NLS in the cargos and interacts with importin-β through the IBB domain to promote nuclear import; **V**- The nuclear import of Ran-GDP is mediated by the nuclear transport factor NTF2.

## NUCLEAR TRANSPORT MODELS

In spite of the progress made towards the understanding of the architecture and protein composition of the NPC, the mechanisms by which this specialized protein complex act as a selective barrier still remains elusive.

Nevertheless, it has been shown that the interaction of transport receptors with the FG motifs of nucleoporins is an essential step for translocation through NPCs [17, 18]. According to this view, it was also demonstrated that removal of the FG binding motifs of importin-β results in the total abolishment of the import pathways mediated by this transport receptor [19, 20]. While it is firmly accepted that FG domains constitute the physical basis underlying the permeability barrier, it is still highly controversial how FG nucleoporins are spatially arranged in order to exert their function in the nuclear transport process.

Based, at least partially, on experimental evidences, a number of models were proposed in an attempt to explain the mechanisms of transport through NPCs. In 2000, Rout *et al.* proposed a model where the presence of numerous nucleoporin filaments on both sides of the central channel of the NPC is thought to confer resistance to passive diffusion for large macromolecules [6]. In addition, the authors postulated that the complex of macromolecules and the respective transport receptors may become prone to reside for increased periods of time in the vicinity of the central channel of the NPC, and thus augmenting the probability to cross it. The vectorial transport through NPCs is then established by the asymmetric distribution of Ran-GTP and some nucleoporins. In summary, this model supports the existence of a physical barrier constituted by the nucleoporin filaments on both sides of the central channel of the NPC. This barrier hampers the transport of proteins and RNAs that are not linked to the respective transport receptors [21].

One year later, Ribbeck and Gorlich proposed an alternative selective phase model. According to this model, the central channel of the NPC contains FG- rich nucleoporins that interact to form a hydrophobic network [22]. This network acts as a mesh to exclusively allow the transport of small molecules. For molecules of larger size the transport is allowed only when are linked to transport receptors. These cargo-receptors complexes may then recognize the FG motifs in nucleoporins and subsequently cause the local disruption of the network. In support of this model, Frey *et al.* showed that the FG motifs of the yeast nucleoporin Nsp1p may form *in vitro*, hydrogel-like structures [23]. These structures are elastic and mechanically stable. The formation of this type of hydrogels is dependent on the presence of FG motifs. Moreover, in yeasts the substitution of phenilalanine by serine residues in the FG motifs of Nsp1p does not allow the formation of hydrogels and was shown to be lethal [23].

The observations that the binding affinity of importin-β to the FG motifs of nucleoporins progressively increases along the inner side of the NPC, and that importin-β contains more than one interaction domain with FG nucleoporins, led Ben-Efraim and Gerace to propose a third model for nuclear transport [24]. According to this model, the movement of the cargo-receptor complex is favoured in the cytoplasm-nucleus direction due to the increasing affinity of importin-β to the nucleoporins localized in the nuclear side of the NPC. However, the mechanism by which the import complexes are released from the first nucleoporin binding site in order to further bind to the following nucleoporins, still needs to be clarified. It is possible that the simultaneous interaction of import complexes with two nucleoporins may help promoting the dissociation of the complex with the nucleoporin displaying less affinity. This dissociation would simultaneously promote the subsequent tight association with the next nucleoporin. This affinity gradient model anticipates that the export complexes move across the NPC in a way similar to that above described for the importin-β mediated import pathways. This model is based on the asymmetric distribution of some of the NPC components, and suggests that the transport direction is imposed by the NPC itself. However, a number of observations do not seem to be in accordance with this model. First, the majority of FG nucleoporins is symmetrically distributed in the NPCs. Second, the asymmetrically distributed nucleoporins were found to be usually dispensable for the majority of nuclear transport pathways [25]. Finally, the observation that the direction of nuclear transport may be inverted in the presence of high concentrations of Ran-GTP in the cytoplasm indicates that the NPC does not directly rule the direction of the nucleocytoplasmic traffic [26].

In 2006, Lim *et al.*, using atomic force microscopy, showed that Nup153, a FG containing nucleoporin, when attached to planar gold layers assumed a configuration compatible with a brush-like structure but not gels [27]. The authors observed that in the presence of increasing concentrations of importin-β the FG brushes collapse adopting a more compact configuration that can be reverted by addition of RanGTP. Similar configurations where observed for Nup153 *in vivo*, leading to the proposal of the reversible collapse model in which the selective gating of the NPC is regulated by the configuration of the FG domains that undergo reversible collapse upon binding of transport factors drawing the transport complexes into the transport channel [27].

Later in 2010, Yamada *et al.* analysed all FG domains of yeast Nups and found that the FG domains are structurally and chemically heterogeneous and adopt distinct categories of disordered structures [28]. Some FG domains show low charge content and adopt collapsed coil configurations. These FG domains display attraction towards one another forming low affinity binding interactions. Others are highly charged and adopt more dynamic, extended coil configurations, avoiding interactions with each other.

Furthermore, the authors identified some FG Nups that exhibited a bimodal distribution of charged amino acids and consequently adopted a bipartite topology in the FG domains, where one portion adopts collapsed-coil configurations while the other adopts relaxed or extended-coil configurations. Based on these configurations, Yamada *et al.* proposed a model of FG domain distribution within the NPC that resembles a forest with shrubs and trees [28]. The shrubs represent the FG nups with collapsed coil configuration where the FG domains are positioned close to NPC scaffold whereas the trees feature the FG containing nups with extended configuration. Most importantly, both the shrubs and the tip of the trees are able to interact laterally and therefore form two distinct regions with hydrogel-like properties, one close to the inner wall of the NPC and the other converging at the NPC center forming a tunnel structure with an open channel. The proposed model features two zones of traffic with different physiochemical properties. The central channel is expected to be hydrophobic and used by karyopherin-β family carrying large cargos. The alternative transport zone is formed by the extended and relaxed coil regions of the FG nups and displays a hydrophilic character. This transport zone appears to be used by smaller molecules such as transport receptors free of cargos [28].

Finally, the reduction of dimensionality model, proposed by Peters in 2009 [29], implies the existence of a narrow tube located at the middle of the central pore channel. According to this model the FG-domains form a coherent hydrophobic layer coating the wall of pore channel and the surface of the filaments. Moreover, this model assumes that at physiological conditions the FG domains are saturated with transport receptors through hydrophobic interactions and those cross the channel by a two-dimensional random walk on the FG-layer [29].

## NUCLEAR LOCALIZATION SIGNALS

The first evidence that the nuclear transport of proteins is mediated by specific peptide signals came from the observation that the proteolytic hydrolysis of the C-terminal amino acids of nucleoplasmin hampers its migration to the nucleus [30]. Moreover, it was also observed that last 50 C-terminal amino acids of nucleoplasmin were able to cross the nuclear pores, accumulating in the nucleus of *Xenopus laevis* oocytes [30]. Taken together, these results suggested that the nuclear import of proteins is a selective process which, in the particular case of nucleoplasmin, is dependent on the presence of a peptide signal present in the carboxylic terminal region of the protein. In spite of these findings, the first nuclear localization signal (NLS) to be identified, at the molecular level, was the NLS of the SV40 large T antigen [31]. The sequence of this NLS was found to consist of seven amino acids PKKKRKV (Table **1**). This sequence is rich in basic amino acids, and was demonstrated to be necessary and sufficient to promote the nuclear import of heterologous cytoplasmic proteins, namely β-galactosidase and pyruvate kinase [31].

A few years later, Dingwall *et al.* analysed the intracellular distribution of pyruvate kinase fused with C-terminal aminoacid sequences of nucleoplasmin [32]. The authors found that the minimal aminoacid sequence necessary to promote the nuclear import of nucleoplasmin consists of eighteen residues KRPAATKKAGQAKKKKLD. A detailed analysis of this NLS showed that the two domains rich in basic amino acids are essential to maintain the import activity [33].

The NLSs of the SV40 large T antigen and of nucleoplasmin are now considered as prototypes and are designated as classical or conventional. They consist of a single or bipartite sequence rich in basic amino acids. Experimental approaches based on the use of fusion protein constructs led subsequently to the identification of a significant number of sequence distinct NLSs (see Table **1**). According to a

**Table 1. Examples of different types of nuclear localization signals (NLSs). The two stretches of basic amino acids in bipartite NLSs as well as the putative critical motifs in Proline/Tyrosine NLSs are shown underlined. The numbers indicate the position in the protein sequence of the last aminoacid of the NLS.**

| NLS Type | Protein | NLS Amino Acid Sequence | Ref. |
|---|---|---|---|
| **Conventional NLSs** | SV40 large T-Ag | PKKKRKV $^{132}$ | [31] |
| | Polyoma large T-Ag | VSRKRPRP $^{196}$ | [86] |
| | Hepatitis D virus antigen | EGAPPAKRAR $^{75}$ | [85] |
| | murine p53 | PPQPKKKPLDGE $^{322}$ | [87] |
| | NF-κB p50 | QRKRQK $^{372}$ | [88] |
| | NF-κB p65 | EEKRKR $^{286}$ | [89] |
| | Human c-myc | PAAKRVKLD $^{328}$ / RQRRNELKRSF $^{374}$ | [35] |
| **Bipartite NLSs** | *Xenopus* nucleoplasmin | KRPAATKKAGQAKKKKLD $^{171}$ | [32] |
| | Rat glucocorticoid receptor | YRKCLQAGMNLEARKTKKKIKGIQQATA $^{524}$ | [90] |
| | RCC1 | MSPKRIAKRRSPPADAIPKSKKVKVSHR $^{28}$ | [91] |
| **Arginine rich NLSs** | HTLV-1 Rex protein | MPKTRRRPRRSQRKRPPT $^{18}$ | [83] |
| | HIV-1 Rev protein | RQARRNRRRRWR $^{46}$ | [93] |
| **Proline-Tyrosine NLSs** | Human hnRNP A1 | NQSSNFGPMKGGNFGGRSSGPYGGGGQ-YFAKPRNQGGY $^{305}$ | [96] |
| | hnRNP D | YGDYSNQQSGYGKVSRRGGHQNSYKPY $^{355}$ | [38] |
| | TAP | SSRLEEDDGDVAMSDAQDGPRVRYNPY $^{75}$ | [38] |
| | hnRNP M | GEGERPAQNEKRKEKNIKRGGNRFEPY $^{64}$ | [38] |
| | PQBP-1 | RDRGYDKADREEGKERRHHRREELAPY $^{187}$ | [38] |
| **Atypical NLSs** | SREBP2 | RSSINDKIIELKDLVMGTDAKMHKSGVLRK-AIDYIKYLQQVNHKLRQENMVLKLANQKNKL$^{403}$ | [95] |
| | Matα2 | MNKIPIKDLLNPQ $^{13}$/ VRILESWFAKNI $^{159}$ | [94] |
| | Hepatitis B virus core antigen | SKCLGWLWG $^{29}$ | [92] |
| | Human rpL23a | VHSHKKKKIRTSPTFTTPKTLRLRRQPKYPR-KSAPRRNKLDHY $^{74}$ | [36] |

bioinformatics analysis of the proteome of *S. cerevisiae*, 45% of the proteins annotated in GenBank™ and 57% of the proteins known to be localized in the nucleus, contain a conventional NLS [34]. There are, however, some exceptions. The human c-myc protein contains a NLS constituted by a single stretch of nine amino acids, PAAKRVKLD, of which only three are basic residues [35]. Another example of non-conventional NLSs is represented by the aminoacid sequence responsible for the nuclear import of the ribosomal protein rpL23a. This NLS, also called BIB, consists of a complex 42 aminoacid sequence rich in basic amino acids [36]. Besides the NLS of the c-myc protein and the ribosomal protein rpL23a, there are also a number of already identified NLSs that do not match the classical sequences.

One the best characterised is the NLS of the hnRNP A1 protein. This NLS, designated M9, is constituted by 38 amino acids. Surprisingly, this sequence was found to be essential for both the nuclear import and export of the protein [37]. More recently, Lee *et al.* while studying the rules for nuclear localization sequence recognition by karyopherin β2, found that hnRNP A1 NLS belongs to a new group of NLSs the so called Proline-Tyrosine NLSs (PY-NLSs) [38]. PY-NLSs, just like conventional NLSs, have an overall basic character and lack evident secondary structure features when unbound to the receptor. PY-NLSs are usually longer than conventional NLSs, typically consist of a stretch of at least 30 amino acids, and despite little or no significant sequence similarity among them PY-NLSs display a C-terminal consensus sequence that includes an Arginine (R)/Lysine (K)/Histidine (H) residue followed by 2 to 5 amino acids that can be represented by any residue, and then the Proline (P)/Tyrosine (Y) motif (see Table **1**). Diverse PY-NLS sequences can be further subdivided into two subclasses according to the composition of the central region. The first subclass includes NLSs from hnRNP A1, hnRNP D and TAP proteins, where the central region is composed of four hydrophobic residues located 11-13 residues upstream of the PY motif. In the second subclass the central region of PY-NLSs consists of a stretch of amino acids enriched in basic residues. This type of PY-NLSs was found in proteins like hnRNP M and PQBP-1. A bioinformatics analysis led to the identification of 81 potential candidates in the human proteome that contain PY-NLSs, and apparently use the karyopherin β2 mediated pathway to access the nucleus [38]. A third subclass of linear NLSs may be typified by the arginine-serine dipeptide repeats that form the RS domain of a number of splicing factors. Nuclear import of splicing factors such as ASF/SF2 or SC35 relies on direct binding of a specific transport receptor, transportin-SR, to the RS domain present in the C-terminal region of these proteins [39-41]. Similar to conventional NLS sequences, the RS domains serve as nuclear targeting signals when attached to heterologous proteins [39-41]. Despite the evidence in support that RS domains may represent an independent class of NLSs, it is still unknown whether all RS domains can function as NLSs.

Not all of the existing NLSs are simple linear sequences. There is evidence that conformational epitopes might be recognized as nuclear targeting signals and thus act as NLSs. Structural domains such as the leucine-zipper domain of c-Jun protein or the compact 5-helix domain of the Gag polyprotein of Rous Sarcoma Virus were shown to be involved in direct binding to importins [42].

A structural element bearing properties similar to NLSs was identified in the transcription factor STAT1/STAT1 homodimers and STAT1/STAT2 heterodimers. Nuclear import of STAT1/STAT1 and STAT1/STAT2 dimers is dependent on two elements rich in arginine and lysine residues. These elements are localized in the DNA binding domain of each subunit [43, 44]. The formation of dimers between the STAT transcription factors is catalysed by phosphorylation of a single tyrosine residue in the carboxylic end of the protein by Janus kinases [45]. Janus kinases can be activated by interferon and other extracellular signals. In non-stimulated cells, unphosphorylated STAT transcription factors are predominantly localized in the cytoplasm [46]. However, it was observed that some unphosphorylated STAT proteins can cross the nuclear pore by a mechanism that does require the interaction with transport receptors [46]. In this case, the import is presumably consequence of a direct interaction with the FG motifs of nucleoporins.

There are a number of evidences supporting the hypothesis that some proteins that do not contain an NLS may also be imported to the nucleus [47-50]. These proteins would thus not be able to interact with transport receptors. They seem to be imported through interaction with other proteins that contain a functional NLS.

This piggyback mechanism seems to contribute to the nuclear localization of an import defective mutant of the hepatitis D virus antigen [50].

## NUCLEAR IMPORT PATHWAYS

The import pathways characterized in more detail involve the participation of the importin-β receptor. Importin-β is thought to be responsible for the nuclear import of all proteins that contain a classical NLS. However, the interaction of importin-β with the aminoacid sequence of the NLS occurs indirectly and involves the participation of other proteins which are members of the importin-α family [51]. The central region of importin-α contains 10 arginine-rich motif (ARM) repetitions in tandem that form a NLS-binding domain [52]. The three-dimensional resolution of importin-α, bound to both monopartite and bipartite NLSs, showed that the central domain contains two specific NLS binding sites [52, 53]. The first binding site, located between the ARM motifs 1-4, directly contacts with the amino acids of monopartite NLSs and the longer aminoacid sequence of bipartite NLSs. The second binding site of importin-α is located in the ARM motifs 7-8. It binds to the shorter aminoacid sequence of bipartite NLSs [11]. According to data obtained from structural analysis of the protein, mutations in amino acids of the first NLS binding site severely hamper the interaction of importin-α with both types of classical NLSs [54]. Furthermore, aminoacid substitutions in the second NLS binding site reduce the affinity between importin-α and bipartite NLSs. However, these substitutions do not seem to affect the interaction with monopartite NLSs [54].

Although importin-α directly recognizes the aminoacid sequence of the NLS, the formation of functional import complexes requires the additional interaction between importin-α and importin-β. The interaction between the two importins occurs through the importin beta binding (IBB) domain localized in the N-terminal region of importin-α (Fig. **2-IV**) [55, 56]. Moroianu *et al.* could demonstrate, for the first time that the IBB domain contains a sequence of basic amino acids similar to a NLS [57]. This sequence has the ability to bind to the central domain of importin-α. Several experimental evidences, based on competition assays, support a model where proteins containing a classical NLS preferentially bind to importin-α in association with importin-β, but with less affinity to free importin-α [58].

The import complexes cross the nuclear pore by a mechanism involving the interaction between importin-β and the FG domains of nucleoporins [17]. Once in the inner face of the NPC, the import complexes are dissociated due the presence of Ran-GTP in the nucleus [58]. As a consequence, the cargos are released and importins are exported back to the cytoplasm (Fig. **2-II**). The dissociation of import complexes starts with the binding of Ran-GTP to three different regions of importin-β [59]. This interaction results in changes of the conformation of importin-β that ultimately hampers the interaction with the IBB domain of importin-α [60]. *In vitro*, the dissociation rate of importin-α/NLS complexes is slow. However, this rate can be accelerated in the presence of exportin CAS bound to Ran-GTP (Fig. **2-IV**) [60]. Exportin CAS is the transport receptor that mediates export of importin-α [61].

There are also evidences pointing to a possible participation of nucleoporins in the dissociation of import complexes [60, 62, 63]. Mouse and *S. cerevisiae* Nup2 and Nup50 nucleoporins, respectively, were shown to promote the dissociation of import complexes through direct interaction with importin-α [62, 63]. Structural data and site-directed mutagenesis experiments additionally demonstrated that Nup2 and Nup50 bind to two specific sites in importin-α that are essential for the interaction with exportin CAS bound to Ran-GTP [62]. Upon dissociation of import complexes, importin-β is exported back to the cytoplasm in association with Ran-GTP (Fig. **2-II**). The nuclear import cycle thus ends with the export of importins to the cytoplasm. In this compartment, the GTPase activity of Ran is stimulated by RanGAP and RanBP1 proteins, and after the subsequent GTP hydrolysis, Ran is dissociated from importin-β and exportin CAS (Fig. **2-II**) [61].

All organisms analysed to date were shown to possess a single gene encoding for importin-β [64]. In contrast, the human importin-α family consists of six proteins, encoded by six different genes that are included in three sub-families according to the respective aminoacid sequence homology [51]. With the exception of importin-α6, which can only be detected in testis, all the remaining importin-α proteins are expressed in the majority of tissues analysed [65, 66]. However, the relative expression levels of the different importins-α varies depending on the tissue or cell line [65-67]. All importin-α proteins can efficiently bind to importin-β and exportin CAS. Moreover, they are able to promote the nuclear import of nucleoplasmin and BSA conjugated with the NLS of the large SV40 T antigen in digitonin semi-permeabilized cells [66]. A more recent work, described the effects of inactivation of different importin-α genes in the nuclear import [64]. The obtained results showed that nucleoplasmin is imported to the nucleus independent of the importin-α gene inactivated. This indicates that, at least for some import pathways, different importin-α proteins may play identical roles. In contrast, other proteins like RCC1, seem to preferentially use one single importin-α as receptor to travel to the nucleus [68, 69]. *In vitro* experiments showed that RCC1 binds with higher affinity to importin-α3 [66, 69]. Additionally, silencing of the importin-α3 gene results in inhibition of the nuclear import of RCC1 in about 50% of the analysed cells [64]. These results suggest that the preferential use of an importin-α is determined by the specificity of the interaction with the substrate. However, the aminoacid sequences of the NLS are not sufficient *per se* to determine this specificity [70]. The use of fusion constructs of nucleoplasmin and the NLS of RCC1 showed that this chimeric protein has an increased affinity for importin-α3 when compared with the wild type protein [70]. In contrast, wild type RCC1 fused with the NLS of nucleoplasmin is unable to interact with importin-α3 [70]. Altogether, these results suggest that the three-dimensional conformation of the protein containing the NLS, contributes to promote the specific binding of importin-α to the substrate.

Besides being responsible for the nuclear import of proteins bearing a classical NLS, importin-β additionally participates in other import pathways that are independent of importin-α. These include the import of UsnRNPs and replication protein A (RPA) [71, 72]. In the case of UsnRNPs, it is thought that nuclear import may involve the direct binding of importin-β or, alternatively, may be mediated by a different from importin-α, adaptor protein [71]. The observation that the nuclear import of UsnRNPs is inhibited in the presence of saturable amounts of analogs of the 5' RNA cap structure (m7GpppG) and in the presence of UsnRNP central domains, suggests a possible participation of two transport receptors or two transport adaptors [73]. Huber *et al.* found that binding of importin-β to the hypermethylated cap structure of UsnRNAs is mediated by the protein snurportin [74]. Snurportin contains a 65 aminoacid N-terminal domain which is responsible for the direct interaction with importin-β [74]. This domain shares a high degree of homology with the IBB domain of importin-α (Fig. **2-I**) [74]. Similar to importin-α, the basic aminoacid residues of the IBB-like domain of snurportin are essential to promote binding to importin-β [75, 76]. Additionally, the snurportin IBB-like domain contains a second importin-β binding sequence which is homologous to a small region in the nucleoporin Nup153 [76]. Nup153 is localized in the inner face of the NPC, and interacts with several transport receptors, namely with importin-β [77]. Importin-β was found to bind with higher affinity to Nup153 than to snurportin. This affinity difference could account for the dissociation of snurportin and importin-β, in a process similar to the observed for importin-α bound to Nup2/Nup50 [76]. Apart from the IBB domain, snurportin shares no sequence homology with importin-α. However, it was demonstrated that similar to importin-α, snurportin forms an intramolecular interaction between a motif within the IBB domain (amino acids 48-52) and an element located in the c-terminus [78]. Mutations that disrupt this interaction increase the affinity of snurportin for importin-β, suggesting a potential auto-inhibitory function similar to that of importin-α [78].

The factors that recognize the NLS in the central domain of UsnRNPs and mediate the nuclear import still remain to be identified. However, some evidences point to a possible participation of the survival of motor neurons (SMN) protein complex [79]. One of the roles of the SMN complex is the coordination of the association of Sm proteins to UsnRNAs. This complex remains associated to UsnRNPs upon binding of snurportin to the hypermethylated cap structure of UsnRNAs (Fig. **2-I**) [79, 80]. Interestingly, it was found that the SMN complex directly interacts with importin-β, *in vitro* [79]. The SMN complex remains associated with UsnRNPs during biogenesis of these ribonucleoproteins [80]. This observation, together with the fact that SMN proteins can establish a direct interaction with importin-β, led to the hypothesis that the SMN complex may also participate in nuclear import of UsnRNPs (Fig. **2-I**). In fact, it was possible to demonstrate that the proteins of the SMN complex and importin-β can promote the nuclear import of UsnRNPs in the absence of additional cellular proteins [81].

Moreover, depletion of intracellular SMN complexes hampers the nuclear import of U1snRNP, *in vitro* [81]. Further addition of purified, functional cellular SMN complexes restores the nuclear import of U1snRNP [81]. However, the use of recombinant SMN proteins in the same import assays did not promote the import of U1snRNP thus pointing to a possible participation of additional cellular factors in this process [81]. Supporting this hypothesis, it was possible to demonstrate *in vivo*, that the interaction between importin-β and the SMN complex is indirect and dependent on the presence of RNA [81].

A nuclear localization signal was recently identified in Gemin4, one of the core components of the SMN-Gemin complex. It was demonstrated that Gemin4 harbours a bipartite NLS between amino acids 193 and 243, and both elements of the bipartite import signal are required for nuclear localization of Gemin4 [82]. This is the first functional NLS that has been identified in the SMN-Gemin complex. However, there is no evidence that Gemin4 is the missing adaptor that bridges the interaction between the SMN complex and importin-β. Moreover, the presence of a bipartite NLS suggests that importin-α may mediate the nuclear import of Gemin4. As importin-α is not required for the nuclear import of the UsnRNPs is still uncertain whether Gemin4 is the link between the SMN complex and importin-β.

Similar to spliceosomal UsnRNPs, the nuclear import of RPA also occurs *via* importin-β and independently of the presence of importin-α [72]. In this case, the adaptor protein is XRP1α [72]. XRP1α binds to importin-β through an arginine-rich basic aminoacid sequence localized in the N-terminal region of the protein. Depletion of RPA results in an almost complete removal of XRP1α from cellular extracts, suggesting that RPA may be the only substrate for XRP1α [72].

Importin-β is the only transport receptor of the karyopherin-β family that uses adaptor proteins to interact with the respective substrates. However, as the other members of the karyopherin-β family, importin-β also participates in nuclear import pathways that do not require adaptor proteins to interact with the substrates. One well studied example is represented by the Rex protein of HTLV-1. Rex is imported by a mechanism that involves the direct binding of importin-β to an arginine-rich signal peptide [83]. Histones and ribosomal proteins represent additional examples of proteins containing NLSs that are directly recognized by importin-β, without the participation of additional adaptor proteins [36, 84]. These proteins, however, do not depend exclusively on importin-β to travel to the nucleus [36, 84]. Several proteins of the karyopherin-β family were found to participate in the nuclear transport of both histones and ribosomal proteins. One these proteins are transportin. Transportin participates in the nuclear transport of over 20 proteins, like hnRNP A1 and TAP, involved in mRNA processing [38].

In summary, the regulation of nucleocytoplasmic traffic is one of the key steps in controlling gene expression in eukaryotes and several viruses. This traffic occurs through nuclear pore complexes and is regulated by a Ran-GTP gradient. It involves the participation of specialized proteins, named importins or exportins which recognize particular domains in the respective cargos. The recent use of proteomic approaches decisively contributed to draw a detailed picture of the NPC components and structure. Ultimately, this allowed proposing and supporting new models of nuclear import mechanisms. However, these models still need to be tested with further robust experiments before being unequivocally accepted. One of the key features when investigating the active transport through NPCs is the identification of structural motifs in both cargo and receptor proteins. In this context, the advent of genomics provided invaluable tools for the search and identification of transport domains in eukaryotic and virus proteins. In infected cells, several virus proteins are imported to the nucleus where they participate in essential steps of the replication cycle. The detailed identification of mechanisms and signals, both in the cargos and receptors that mediate nuclear import may contribute to identify potential targets for the development of new therapies capable of inhibiting this essential step.

## CONFLICT OF INTEREST

The author(s) confirm that this chapter content has no conflict of interest.

## ACKNOWLEDGEMENTS

The authors are grateful to Dr. Pedro Santos for assistance and support over the years. N.F. was a recipient of a FCT grant (SFRH/BD/17556/2004).

## DISCLOSURE

This is an updated report of the earlier review article By N. Freitas and C. Cunha "Mechanisms and signals for the nuclear import of proteins". Current Genomics, 2009 10: 550-557.

## REFERENCES


[1] Lim R.Y. H.; Aebi U.; Fahrenkrog B. Towards reconciling structure and function in the nuclear pore complex. *Histochem. Cell Biol.* **2008**, *129*: 105-116.

[2] Adam S.A. The nuclear pore complex. *Genome Biol.* **2001**, *2*: 7.1-7.6

[3] Suntharalingam M.; Wente S.R. Peering the pore: nuclear pore complex structure, assembly, and function. *Develop. Cell.* **2003**, *4*: 775-789.

[4] Fahrenkrog B.; Aebi U. The nuclear pore complex: nucleocytoplasmic transport and beyond. *Mol. Cell Biol.* **2003**, *4*: 757-766.

[5] Cronshaw J.M.; Krutchinsky A.N.; Zhang W.; Chait B.T.; Matutis M.J. Proteomic analysis of the mammalian nuclear pore complex. *J. Cell Biol.* **2002**, *158*: 915-927.

[6] Rout M.P.; Aitchison J.D.; Suprapto A.; Hjertaas K.; Zhao Y.; Chait B.T. The yeast nuclear pore complex: composition, architecture, and transport mechanism. *J. Cell Biol.* **2000**, *148*: 635-651.

[7] Reichelt R.; Holzenburg A.; Buhle Jr. E.L.; Jarnik M.; Engel A.; Aebi U. Correlation between structure and mass distribution of the nuclear pore complex and of distinct pore complex components. *J. Cell Biol.* **1990**, *110*: 883-894.

[8] Yang Q.; Rout M.P.; Akey C.W. Three-dimensional architecture of the isolated yeast nuclear pore complex: functional and evolutionary implications. *Mol. Cell.* **1998**, *1*: 223-234.

[9] Mosammaparast N.; Pemberton L.F. Karyopherins: from the nuclear-transport mediators to nuclear-function regulators. *Trends Cell Biol.* **2004**, *14*: 547-556.

[10] Kohler A.; Hurt E. Exporting RNA from the nucleus to the cytoplasm. *Mol. Cell Biol.* **2007**, *8*: 761-773.

[11] Chook Y.M.; Blobel G. Karyopherins and nuclear import. *Cur. Opi. Stru. Biol.* **2001**, *11*: 703-715.

[12] Kalab P.; Heald R. The RanGTP gradient-a GPS for the mitotic spindle. *J. Cell. Sci.* **2008**, *121*: 1577-1586.

[13] Gorlich D.; Kutay U. Transport between the cell nucleus and the cytoplasm. *Ann. Rev. Cell Dev. Biol.* **1999**, *15*: 607-660.

[14] Madrid A.S.; Weis K. Nuclear transport is becoming crystal clear. *Chromosoma* **2006**, *115*: 98-10.

[15] Ribbeck K.; Lipowsky G.; Kent H.M.; Stewart M.; Gorlich D. NTF2 mediates nuclear import of Ran. *EMBO J.* **1998**, *17*: 6587-6598.

[16] Quimby B.B.; Lamitina T.; L'Hernault S.W.; Corbett A.H. The mechanism of Ran import into the nucleus by nuclear transport factor 2. *J. Biol. Chem.* **2000**, *275*: 28575-28582.

[17] Weis K. Regulating access to the genome: nucleocytoplasmic transport throughout the cell cycle. *Cell* **2003**, *112*: 441-451.

[18] Bednenko J.; Cingolani G.; Gerace L. Nucleocytoplasmic transport: navigating the channel. *Traffic* **2003**, *4*: 127-135.

[19] Bednenko J.; Cingolani G.; Gerace L. Importin β contains a COOH-terminal nucleoporin binding region important for nuclear import. *J. Cell Biol.* **2003**, *162*: 391-401.

[20] Bayliss R.; Littlewood T.; Stewart M. Structural basis for the interaction between FxFG nucleoporin repeats and importin-beta in nuclear trafficking. *Cell* **2000**, *102*: 99-108.

[21] Rout M.P.; Aitchison J.D.; Magnasco M.O.; Chait B.T. Virtual gating and nuclear transport: the hole picture. *Trends Cell Biol.* **2003**, *13*: 622-628.

[22] Ribbeck K.; Gorlich D. Kinetic analysis of translocation through nuclear pore complexes. *Embo J.* **2001**, *20*: 1320-1330.

[23] Frey S.; Richter R.P.; Gorlich D. FG-rich repeats of nuclear pore proteins form a three-dimensional meshwork with hydrogel-like properties. *Science* **2006**, *314*: 815-817.

[24] Ben-Efraim I.; Gerace L. Gradient of increasing affinity of importin β for nucleoporins along the pathway of nuclear import. *J. Cell Biol.* **2001**, *152*: 411-417.

[25] Zeitler B.; Weis K. The FG-repeat asymmetry of the nuclear pore complex is dispensable for bulk nucleocytoplasmic transport *in vivo*. *J. Cell Biol.* **2004**, *167*: 583-590.

[26] Nachury M.V.; Weis K. The direction of transport through the nuclear pore can be inverted. *Proc. Natl. Acad. Sci. USA.* **1999**, *96*: 9622-9627.

[27] Lim R. Y. H.; Fahrenkrog B.; Koser J.; Schwarz-Herion K.; Deng J.; Aebi U. Nanomechanical basis of selective gating by the nuclear pore complex. *Science.* **2007**, *318*: 640-643.

[28] Yamada J.; Phillips J. L.; Goldfien G.; Calestagne-Morelli A.; Huang H.; Reza R.; Acheson J.; Krishnan V. V.; Newsam S.; Gopinatham A.; Lau E. Y.; Colvin M. E.; Uversky V. N.M; Rexach M. F. A bimodal distribution of two distinct categories of intrinsically-disordered structures with separate functions in FG nucleoporins. *Mol. Cell. Proteomics.* **2010**, *9*: 2205-2224.

[29] Peters R. Translocation through the nuclear pore: kaps pave the way. *BioEssays.* **2009**, *31*: 466-477.

[30] Dingwall C.; Sharnick S.V.; Laskey R.A. A polypeptide domain that specifies migration of nucleoplasmin into the nucleus. *Cell* **1982**, *30*: 449-458.

[31] Kalderon D.; Roberts B.L.; Richardson W.D., Smith A. E. A short amino acid sequence able to specify nuclear location. *Cell* **1984**, *39*: 499-509.

[32] Dingwall C.; Robbins J.; Dilworth S.M.; Roberts B.; Richardson W.D. The nucleoplasmin nuclear location sequence is larger and more complex than that of SV-40 large T antigen. *J. Cell Biol.* **1988**, *107*: 841-849.

[33] Robbins J.; Dilworth S.M.; Laskey R.A.; Dingwall C. Two interdependent basic domains in nucleoplasmin nuclear targeting sequence: identification of a class of bipartite nuclear targeting sequence. *Cell* **1991**, *64*: 615-623.

[34] Lange A.; Mills R.E.; Lange C.J.; Stewart M.; Devine S.E.; Corbett A.H. Classical nuclear localization signals: definition, function, and interaction with importin- α. *J. Biol. Chem.* **2007**, *282*: 5101-5105.

[35] Dang C.V.; Lee W.M. Identification of the human c-myc protein nuclear translocation signal. *Mol. Cell. Biol.* **1988**, *8*: 4048-4054.

[36] Jakel S.; Gorlich D. Importin β, transportin, RanBP5 and RanBP7 mediate nuclear import of ribosomal proteins in mammalian cells. *EMBO J.* **1988**, *17*: 4491-4502.

[37] Siomi H.; Dreyfuss G. A nuclear localization domain in the hnRNP A1 protein. *J. Cell Biol.* **1995**, *129*: 551-560.

[38] Lee B. J.; Cansizoglu A. E.; Suel K. E., Louis T. H; Zhang Z.; Chook Y. M. Rules for nuclear localization sequence recognition by karyopherinβ2. *Cell.* **2006**, *126*: 543-558.

[39] Kataoka N.; Bachorik J. L.; Dreyfuss G. Transportin-SR, a nuclear import receptos for the RS proteins. *J. Cell Biol.* **1999**, *145*: 1145-1152.

[40] Lai M.-C.; Lin R.-I.; Huang S.-Y.; Tsai C.-W.; Tarn W.-Y. A human importin-family protein, transportin-SR2 interacts with the phosphorylated RS domain of SR proteins. *J. Biol. Chem.* **2000**, *275*: 7950-7957.

[41] Gama-Carvalho M.; Carvalho M. P.; Kehlenbach A.; Valcárcel J.; Carmo-Fonseca M. Nucleocytoplasmic Shuttling of Heterodimeric Splicing Factor U2AF. *J. Biol. Chem.* **2001**, *276*: 13104 –13112.

[42] Chook Y. M.; Suel K. E. Nuclear import by karyopherin-βs: recognition and inhibition. *Biochimica et Biophysica Acta.* **2011**, *1813*: 1593-1606.

[43] Melén K.; Kinnunen L.; Julkunen I. Arginine/lysine-rich structural element is involved in interferon-induced nuclear import of STATs. *J. Biol. Chem.* **2001**, *276*: 16447-1455.

[44] Fagerlund R.; Melén K.; Kinnunen L.; Julkunen I. Arginine/lysine-rich nuclear localization signals mediate interactions between dimeric STATs and importin α5. *J. Biol. Chem.* **2002**, *277*: 30072-30078.

[45] Meyer T.; Vinkemeier U. Nucleocytoplasmic shuttling of STAT transcription factors. *Eur. J. Biochem.* **2004**, *271*: 4606-4612.

[46] Vinkemeier U. Getting the message across, STAT! Design principles of a molecular signaling circuit. *J. Cell Biol.* **2004**, *167*: 197-201.

[47] Leslie D.M.; Zhang W.; Timney B.L.; Chait B.T.; Rout M.P.; Wozniak R.W.; Aitchison, J.D. Characterization of karyopherin cargoes reveals unique mechanisms of kap121-p mediated nuclear import. *Mol. Cell. Biol.* **2004**, *24*: 8487-8503.

[48] Jans D.A.; Briggs L.J.; Gustin S.E.; Jans P.; Ford S.; Young I.G. The cytokine interleukin-5 (IL-5) effects cotransport of its receptor subunits to the nucleus *in vitro*. *FEBS letters* **1997**, *410*: 368-372.

[49] Shiota C.; Coffey J.; Grimsby J.; Grippo J.F.; Magnuson M.A. Nuclear import of hepatic glucokinase depends upon glucokinase regulatory protein, whereas export is due to a nuclear export signal sequence in glucokinase. *J. Biol. Chem.* **1999**, *274*: 37125-37130.

[50] Xia Y.-P.; Yeh C.-T.; Ou J.-H.; Lai M.M.C. Characterization of nuclear targeting signal of hepatitis delta antigen: nuclear transport as a protein complex. *J. Virol.* **1992**, *66*: 914-921.

[51] Goldfarb D.S.; Corbett A. H.; Mason D.A.; Harreman M.T.; Adam S.A. Importin α: a multipurpose nuclear-transport receptor. *Trends Cell Biol.* **2004**, *14*: 505-514.

[52] Conti E.; Uy M.; Leighton L.; Blobel G.; Kuriyan J. Crystallographic analysis of the recognition of a nuclear localization signal by the nuclear import factor karyopherin alpha. *Cell* **1998**, *94*: 193-204.

[53] Fontes M.R.; Teh T.; Kobe B. Structural basis of recognition of monopartite and bipartite nuclear localization sequences by mammalian importin-alpha. *J. Mol. Biol.* **2000**, *297*: 1183-1194.

[54] Leung S.W.; Harreman M.T.; Hodel M.R.; Hodel A.E.; Corbett A.H. Dissection of the karyopherin α nuclear localization signal (NLS)-binding groove. Functional requirements for NLS binding. *J. Biol. Chem.* **2003**, *278*: 41947-41953.

[55] Gorlich D.; Henklein P.; Laskey R.A.; Hartmann E. A 41 amino acid motif in importin-alpha confers binding to importin-beta and hence transit into the nucleus. *EMBO J.* **1996**, *15*: 1810-1817.

[56] Weis K.; Ryder U.; Lamond A.I. The conserved amino-terminal domain of hSRP1 is essential for nuclear protein import. *EMBO J.* **1996**, *15*: 1818-1825.

[57] Moroianu J.; Blobel G.; Radu A. The binding site of karyopherin α for karyopherin β overlaps with a nuclear localization sequence. *Proc. Natl. Acad. Sci. USA.* **1996**, *93*: 6572-6576.

[58] Stewart M. Molecular mechanism of the nuclear protein import cycle. *Mol. Cell. Biol.* **2007**, *8*: 195-208.

[59] Lee S.J.; Matsuura Y.; Liu S.M.; Stewart M. Structural basis for nuclear import complex dissociation by RanGTP. *Nature* **2005**, *435*: 693-696.

[60] Gilchrist D.; Mykytka B.; Rexach M. Accelerating the rate of disassembly of karyopherin-cargo complexes. *J. Biol. Chem.* **2002**, *277*: 18161-18172.

[61] Kutay U.; Bischoff F.R.; Kostka S.; Kraft R.; Gorlich D. Export of importin α from the nucleus is mediated by a specific nuclear transport factor. *Cell* **1997**, *90*: 1061-1071.

[62] Matsuura Y.; Stewart M. Nup50/Npap60 function in nuclear protein import complex disassembly and importin recycling. *EMBO J.* **2005**, *24*: 3681-3689.

[63] Matsuura Y.; Lange A.; Harreman M.T.; Corbett A.H.; Stewart M. Structural basis for Nup2p function in cargo release and karyopherin recycling in nuclear import. *EMBO J.* **2003**, *22*: 5358-5369.

[64] Quensel C.; Friedrich B.; Sommer T.; Hartmann E.; Kohler M. *In vivo* analysis of importin α proteins reveals cellular proliferation inhibition and substrate specificity. *Mol. Cell. Biol.* **2004**, *24*: 10246-10255.

[65] Kohler M.; Ansieau S.; Prehn S.; Leutz A.; Haller H.; Hartmann E. Cloning of two novel human importin-α subunits and analysis of the expression pattern of the importin-α protein family. *FEBS letters* **1997**, *417*: 104-108.

[66] Kohler M.; Speck C.; Christiansen M.; Bischoff F.R.; Prehn S.; Haller H.; Gorlich D.; Hartmann E. Evidence for distinct substrate specificities of importin α family members in nuclear protein import. *Mol. Cell. Biol.* **1999**, *19*: 7782-7791.

[67] Kamei Y.; Yuba S.; Nakayama T.; Yoneda Y. Three distinct classes of the α-subunit of the nuclear pore-targeting complex (importin-α) are differentially expressed in adult mouse tissues. *J. Histochem. Cytochem.* **1999**, *47*: 363-372.

[68] Miyamoto Y.; Imamoto N.; Sekimoto T.; Tachibana T.; Seki T.; Tada S.; Enomoto T.; Yoneda Y. Differential modes of nuclear localization signal (NLS) recognition by three distinct classes of NLS receptors. *J. Biol. Chem.* **1997**, *272*: 26375-26381.

[69] Talcott B.; Moore M.S. The nuclear import of RCC1 requires a specific nuclear localization sequence receptor, karyopherin α3/Qip. *J. Biol. Chem.* **2000**, *275*: 10099-10104.

[70] Friedrich B.; Quensel C.; Sommer T.; Hartmann E.; Kohler M. Nuclear localization signal and protein context both mediate importin α specificity of nuclear import substrates. *Mol. Cell. Biol.* **2006**, *26*: 8697-8709.

[71] Palacios I.; Hetzer M.; Adam S.A.; Mattaj I.W. Nuclear import of U snRNPs requires importin β. *EMBO J.* **1997**, *16*: 6783-6792.

[72] Jullien D.; Gorlich D.; Laemmli U.K.; Adachi Y. Nuclear import of RPA in *Xenopus* egg extracts requires a novel protein XRIPα but not importin α. *EMBO J.* **1999**, *18*: 4348-4358.

[73] Fischer U.; Sumpter V.; Sekine M.; Satoh T.; Luhrmann R. Nucleo-cytoplasmic transport of U snRNPs: definition of a nuclear location signal in the Sm core domain that binds a receptor independently of the $m_3G$ cap. *EMBO J.* **1993**, *12*: 573-583.

[74] Huber J.; Croshagen U.; Kadokura M.; Marshallsay C.; Wada T.; Sekine M.; Luhrmann R. Snurportin1, an $m_3G$-cap-specific nuclear import receptor with a novel domain structure. *EMBO J.* **1998**, *17*: 4114-4126.

[75] Huber J.; Dickmanns A.; Luhrmann R. The importin-β binding domain of snurportin1 is responsible for the Ran- and energy-independent nuclear import of spliceosomal U snRNP *in vitro*. *J. Cell Biol.* **2002**, *156*: 467-479.

[76] Mitrousis G.; Olia A.S.; Walker-Kopp N.; Cingolani G. Molecular basis for the recognition of snurportin 1 by importin-β. *J. Biol. Chem.* **2008**, *283*: 7877-7884.

[77] Walther T.C.; Fornerod M.; Pickersgill H.; Goldberg M.; Allen T.D.; Mattaj I.W. The nucleoporin Nup153 is required for nuclear pore basket formation, nuclear pore complex anchoring and import of a subset of nuclear proteins. *EMBO J.* **2001**, *20*: 5703-5714.

[78] Ospina J. K; Gonsalvez G. B.; Bednenko J.; Darzynkiewicz E.; Gerace L.; Matera A. G. Cross-talk between snurportin1 subdomains. *Mol. Biol. Cell.* **2005**, *16*: 4660-4671.

[79] Narayanan U.; Ospina J.K.; Frey M.R.; Hebert M.D.; Matera A.G. SMN, the spinal muscular atrophy protein, forms a pre-import snRNP complex with snurportin1 and importin β. *Human Molecular Genetics* **2002**, *11*: 1785-1795.

[80] Massenet S.; Pellizzoni L.; Paushkin S.; Mattaj I.W.; Dreyfuss G. The SMN complex is associated with snRNPs throughout their cytoplasmic assembly pathway. *Mol. Cell. Biol.* **2002**, *22*: 6533-6541.

[81] Narayanan U.; Achsel T.; Luhrmann R.; Matera A.G. Coupled *in vitro* import of U snRNPs and SMN, the spinal muscular atrophy protein. *Mol. Cell.* **2004**, *16*: 223-234.

[82] Lorson M. A.; Dickson A. M.; Shaw D. J.; Todd A. G.; Young E. C.; Morse R.; Wolstencroft C.; Lorson C. L; Young P. J. Identification and characterization of a nuclear localization signal in the SMN associated protein, Gemin4. *Biochemical and Biophysical Research Communications*. **2008**, *375*: 33-37.

[83] Palmeri D.; Malim M.H. Importin β can mediate the nuclear import of an arginine-rich nuclear localization signal in the absence of importin α. *Mol. Cell. Biol.* **1999**, *19*: 1218-1225.

[84] Muhlhausser P.; Muller E.-C.; Otto A.; Kutay U. Multiple pathways contribute to nuclear import of core histones. *EMBO reports* **2001**, *2*: 690-696.

[85] Alves, C.; Freitas, N.; Cunha, C. Characterization of the nuclear localization signal of the hepatitis delta virus antigen. *Virology* **2008**, *370*: 12-21.

[86] Richardson W.D.; Roberts B.L.; Smith A.E. Nuclear location signals in polyoma virus large-T. *Cell* **1986**, *44*: 77-85.

[87] Shaulsky G.; Goldfinger N.; Ben-Ze'ev A.; Rotter V. Nuclear accumulation of p53 protein is mediated by several nuclear localization signals and plays a role in tumorigenesis. *Mol. Cell Biol.* **1990**, *10*: 6565-6577.

[88] Henkel T.; Zabel U.; Zee K.V.; Müller J.M.; Fanning E.; Baeuerle P.A. Intramolecular masking of the nuclear location signal and dimerization domain in the precursor for the p50 NF-κB subunit. *Cell* **1992**, *68*: 1121-1133.

[89] Zabel U.; Henkel T; Silva M.S.; Baeuerle P.A. Nuclear uptake control of NF-kappa B by MAD-3, and I kappa B protein present in the nucleus. *EMBO J.* **1993**, *12*: 201-211.

[90] Picard D.; Yamamoto K.R. Two signals mediate hormone-dependent nuclear localization of the glucocorticoid receptor. *EMBO J.* **1987**, *6*: 3333-3340.

[91] Nemergut M.E.; Macara I.G. Nuclear import of the Ran exchange factor, Rcc1, is mediated by at least two distinct mechanisms. *J. Cell Biol.* **2000**, *149*: 835-850.

[92] Ou J.-H.; Yeh C.-T.; Yen T.B. Transport of hepatitis B virus precore protein into the nucleus after cleavage of its signal peptide. *J. Virol.* **1989**, *63*: 5238-5243.

[93] Truant R.; Cullen B.R. The arginine-rich domains present in human immunodeficiency vírus type 1 Tat and Rev function as direct importin β-dependent nuclear localization signals. *Mol. Cell. Biol.* **1999**, *19*: 1210-1217.

[94] Hall M. L.; Hereford L.; Herskowitz I. Targeting of E. coli β-galactosidase to the nucleus in yeast. *Cell* **1984**, *36*: 1057-1065.

[95] Nagoshi E.; Yoneda Y. Dimerization of sterol regulatory element-binding protein 2 *via* the helix-loop-helix-leucine zipper domain is a prerequisite for its nuclear localization mediated by importin β. *Mol. Cell. Biol.* **2001**, *21*: 2779-2789.

[96] Siomi, H.; Dreyfuss, G. A nuclear localization domain in the hnRNP A1 protein. *Journal Cell Biol.* **1995**, *129*: 551-560.

Send Orders for Reprints to reprints@benthamscience.net



# CHAPTER 4

# Cancer Proneness in Nijmegen Breakage Syndrome Carriers

**Alessandra di Masi[*,1,2], Francesco Berardinelli[2], Domenica Cilli[1,2] and Antonio Antoccia[2]**

*[1]Interdepartmental Laboratory of Electron Microscopy, University Roma Tre, Via della Vasca Navale 79, I-00146 Roma, Italy; [2]Department of Biology, University Roma Tre, Viale Guglielmo Marconi 446, I-00146 Roma, Italy*

**Abstract:** Biallelic mutations in the *NBN* gene are responsible for the Nijmegen breakage syndrome (NBS), a rare autosomal recessive disorder characterized by chromosome instability and hypersensitivity to ionising radiation (IR). Epidemiological data evidence that the *NBN* gene can be considered a susceptibility factor for cancer development, as demonstrated by the fact that almost 40% of NBS patients have developed a malignancy before the age of 21. Interestingly, also NBN heterozygotes, which are clinically asymptomatic, display an elevated risk to develop some types of malignant tumours, especially breast, prostate and colorectal cancers, lymphoblastic leukaemia, and non-Hodgkin's lymphoma (NHL). So far, nine mutations in the *NBN* gene have been found, at the heterozygous state, in cancer patients. Among them, the 657del5, the I171V and the R215W mutations are the most frequently described. The pathogenicity of these mutations is presumably connected with their occurrence in the highly conserved BRCT tandem domains of the NBN protein, which are present in a large superfamily of proteins and are recognized as major mediators of processes related to cell-cycle checkpoints and DNA repair. This review will focus on the current state-of-knowledge regarding the correlation between carriers of *NBN* gene mutations and the proneness to the development of malignant tumours.

**Keywords:** NBN, nijmegen breakage syndrome, NBN carriers, 657del5 founder mutation, R215W, I171V, BRCT domain, DNA repair.

## INTRODUCTION

Mutations at the homozygous status in the *NBN* gene (formerly known as *NBS1*) are responsible for a rare disease known as Nijmegen breakage syndrome (NBS; OMIM 251260), an autosomic recessive disorder whose signs are a distinct

***Address correspondence to Alessandra di Masi:** Department of Biology, University Roma Tre, Viale Guglielmo Marconi 446, I-00146 Roma, Italy; Tel: +39-06-57333621; Fax: +39-06-57336321; E-mail: dimasi@uniroma3.it

**Christian Neri (Ed)**

facial appearance, microcephaly, immunodeficiency, chromosome rearrangements and sensitivity to ionising radiation (IR) [1]. In NBS, a defective response to DNA damage is associated with chromosomal instability, and in turn with a strong predisposition to develop malignancy, in particular lymphomas [2, 3]. The majority of them are non Hodgkin lymphoma (NHL) (with a high incidence of diffuse large B-cell lymphomas), lymphoblastic anaemia, and Hodgkin lymphoma (HL) [3, 4]. Among solid tumours, medulloblastoma has been observed in four patients [5], and rhabdomiosarcoma of the perianal region in three others [6]. Since the latter is extremely uncommon among children, a strong association with NBS has been suggested [6]. All the 11 disease-causing mutations so far identified in *NBN* gene have been found within exons 6-10 (Fig. **1**), and eight of them result in premature truncation of the NBN protein, with the possible synthesis of NBN variants of lower molecular weight [7, 8].

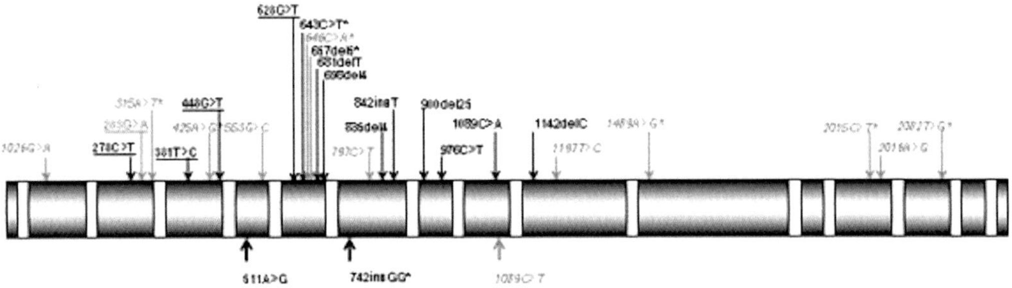


**Figure 1:** Coding sequence variants of NBN gene identified to date (modified from http://www.nijmegenbreakagesyndrome.net). Grey: high frequency polymorphisms; grey*: low frequency polymorphisms; black: mutations found in NBS patients; black*: mutations also found in cancer patients at the heterozygous state; black: mutations found in cancer patients only; grey: mutation found both in cancer patients and in healthy controls.

As a measure of cancer incidence in NBS, the Polish registry report that 40% of patients suffering for NBS developed lymphoma within the first two decades of life [3]. Though NBS is a recessive disease and one would not expect any cellular feature or clinical symptom, a growing number of papers report higher spontaneous and induced chromosome instability and an increased incidence of tumours among NBS carriers. The FISH chromosome painting analysis revealed that NBS carriers display a 3-fold higher rate of chromosome translocations compared with non-carriers [9]. Furthermore, the same authors reported that after irradiation the response of cells from heterozygous carriers was intermediate between that of NBS homozygous and normal individuals, and could be clearly differentiated from those of the other groups in double-coded studies. Moreover, NBS heterozygosity can be distinguished from other genotypes by the number of the long-lived stable aberrations in NBS cells [10]. Radiation hypersensitivity in NBS carriers seems anyhow restricted to cells irradiated in the $G_1$-phase, whereas the number of chromatid-aberrations scored in $G_2$-phase-treated NBS heterozygous cells is in the range of normal cells or slightly higher [8].

The NBN protein consists of 754 amino acids, has a molecular weight of ~95 kDa, and is composed of three regions (Fig. **2**). The *N*-terminus contains the fork-head associated (FHA) domain (amino acids 24-109), and two breast cancer 1 (BRCA1) carboxy-terminal (BRCT) tandem domains (*i.e.*, BRCT1, amino acids 114-183; and BRCT2, amino acids 221-291) [11]. The central region of NBN contains a consensus sequence encompassing the Ser343 residue that, together with the Ser278 residue located within the BRCT2 domain, is phosphorylated by the ataxia-telangiectasia mutated (ATM) kinase in response to ionizing radiation (IR) [12-15]. Finally, the *C*-terminus of NBN contains the MRE11-binding domain and the ATM-binding motif [16-18]. Notably, the BRCT domains are present as a tandem repeat in a large super-family of proteins that are the major mediators of phosphorylation-dependent protein-protein interactions in processes related to cell cycle checkpoints and DNA repair functions [4, 19-24]. Mutations within tandem BRCT domains of NBN are responsible for cancer susceptibility, both at the homozygous and heterozygous status [4, 22, 25, 26].

Nine mutations localized in the coding sequence of the *NBN* gene have been found, at the heterozygous state, in cancer patients (Fig. **1**). The 657del5 (or founder mutation), the 511A>G (I171V), the 643C>T (R215W), and the 742insGG mutations were found both in NBS patients and in cancer patients. Four mutations have been found in cancer patients only (278C>T, 381C>T, 448G>T, and 628G>T). The 283G>A (D95N) mutation has been identified both in cancer patients and in healthy controls [27], but is not listed as known *NBN* polymorphism by the NCBI website [28].

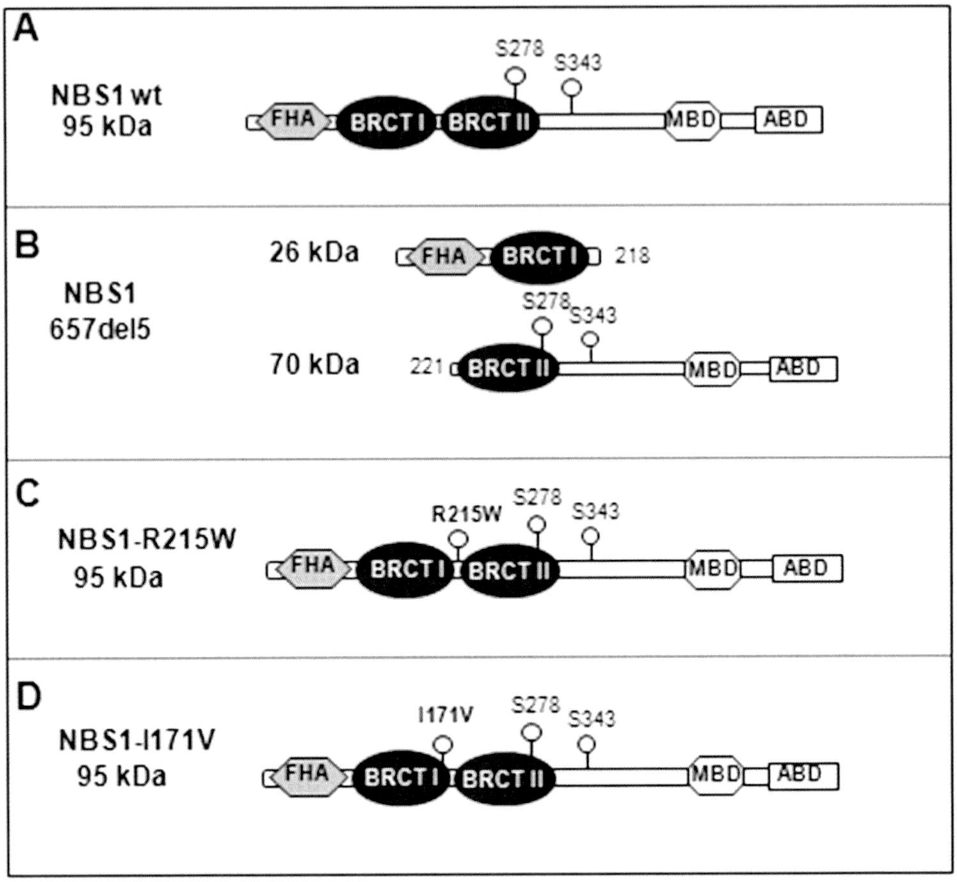


**Figure 2:** Wild type and mutated NBN proteins. (**A**) Structure of the NBN wild-type protein. (**B**) The 657del5 mutation, which splits up the tandem BRCT domains, determines the expression of two truncated proteins of 26 and 70kDa. (**C**) The R215W mutation occurs in the linker region that connects the two BRCT domains, and determines the substitution at position 215 of an arginine (R) with a tryptophan (W). (**D**) The I171V missense mutation occurs in the first BRCT domain, and determines the substitution at position 171 of an isoleucine (I) with a valine (V). (MBD: MRE11 binding domain; ABD: ATM binding domain).

## NBN MUTATIONS AND CANCER RISK

The first evidence of a possible correlation between NBN mutation carriers and cancer risk came from family data studies, indicating that blood relatives of NBS patients with the 657del5 founder mutation had a high probability to develop malignancy [29]. From 1998, several studies evaluated the frequency of the *NBN* mutations in cancer patients. In Table **1**, we have collected all the existing data relative to the frequency of cancer onset in carriers of three of the most frequent

NBN mutations, *i.e.* 657del5, R215W, and I171V. Remarkably, all these mutations have been defined susceptibility factor in cancer development. In Table **2** we have performed a statistical elaboration of the whole epidemiological data summarized in Table **1**: all the carriers for each NBN mutation among tumor patients were compared with a control population derived from the sum of all the healthy carriers of each mutation. Adults and children were considered separately.

## The 657del5 Mutation

The 657del5 hypomorphic germ line mutation in exon 6 of *NBN* gene accounts for more than 90% of all mutant alleles in NBS. The highest frequency of heterozygous carriers of the 657del5 mutation has been found in the Slavic population of Central Europe, with an average frequency of 1/177 [30]. A total number of 130 carriers of 11725 cancer patients (1.1%), have been so far identified (Table **1**). Among heterozygous carriers of the 657del5 mutation, the following type of cancers were observed: gastrointestinal lymphoma (10.8%), familial prostate cancer (8.9%), medulloblastoma (2.9%), NHL (2.4%), non familial prostate cancer (2.3%), ovarian (1.6%), ALL (1.6%), melanoma (1%), astrocytic cancer (0.8%), breast cancer (0.9%), colorectal cancer (0.6%) (Table **2**) [31-47].

The evidence of a strong correlation between carriers of the 657del5 mutation and cancer risk has been strengthened by a large study on 344 blood relatives (first- through fourth-degree) of NBS patients, in 24 different NBS families of Czech Republic and Slovakia, from 1998 to 2003 [48]. Thirteen blood relatives developed malignancies of any type, among them 11 were carriers of the 657del5 *NBN* mutation, compared with six expected. In this study, the most frequently observed type of cancer was stomach cancer and colorectal cancer. Breast cancers were also reported, though at a lower frequency [48].

The cancer risk of the 657del5 mutation carriers has been also assessed in cancer patients with no NBS cases in the family. It was found that carriers of the 657del5 mutation were about twice more frequent among cancer patients than among matched controls [35]. In this study, Steffen and co-workers observed that most of the 657del5 carriers were found among patients with melanoma (3.8%; OR: 6.376, p=0.0081), NHL (4.8%; OR: 8.05, p=0.0351), breast cancer (1.8%; OR: 2.927,

**Table 1. Frequency of the 657del5, R215W, and I171V NBN mutations among NBN carriers, affected or not by cancer (NHL: non-Hodgkin lymphoma; ALL: acute lymphoblastic leukaemia; HL: Hodgkin lymphoma; OR: odd ratio) (nad: not available data). Data were considered significant when p<0.05.**

| *NBN* Mutation | Patient Age | Cancer Type | n° of *NBN* Mutation Carriers Among Tumour Patients | n° of *NBN* Mutation Carriers Among Control Subjects | Statistical Analysis | Ref. |
|---|---|---|---|---|---|---|
| **657del5** | **adult** | colorectal | 3/234 (1.3%) | 10/1620 (0.6%) | OR: 2.091, p=0.2197 | [35] |
| | | | 3/750 (0.4%) | 5/1411 (0.35%) | OR: 0.95, p=0.946 | [43] |
| | | breast | 4/224 (1.8%) | 10/1620 (0.6%) | OR: 2.927, p=0.0795 | [35] |
| | | | 11/562 (2%) | 1/162 (0.6%) | OR: 3.21, p = 0.0107 | [66] |
| | | | 1/477 (0.2%) | 1/866 (0.1%) | OR: 1.8, p=0.76 | [31] |
| | | | 7/873 (0.8%) | 2/692 (0.3%) | OR: 2.8, p=0.32 | [36] |
| | | | 17/2012 (0.8%) | 18/4000 (0.5%) | OR: 1.9, p=0.09 | [37] |
| | | | 2/150 (1.3%) | 3/530 (0.6%) | $\chi^2$ test, p<0.0001 | [37] |
| | | | 3/80 (3.8%) | 3/530 (0.6%) | $\chi^2$ test, p<0.0001 | [37] |
| | | | 2/270 (0.7%) | 2/295 (0.7%) | OR: 1.11, p=1.00 | [42] |
| | | | 16/2664 (0.6%) | 1/2031 (0.05%) | OR: 12.26, p=0.98 | [41] |
| | | melanoma | 4/105 (3.8%) | 10/1620 (0.6%) | OR: 6.376, p=0.0081 | [35] |
| | | | 1/376 (0.3%) | nad | nad | [40] |
| | | NHL | 2/42 (4.8%) | 10/1620 (0.6%) | OR: 8.05, p=0.0351 | [35] |
| | | | 8/228 (3.5%) | 10/1620 (0.6%) | OR: 5.85, p=0.0001 | [66] |
| | | prostate | 7/305 (2.2%) (non familial) | 9/1500 (0.6%) | OR: 3.9, p=0.01 | [34] |
| | | | 5/56 (9%) (familial) | 9/1500 (0.6%) | OR: 16, p>0.0001 | [34] |
| | | gastrointestinal lymphoma | 4/37 (10.8%) | 10/1620 (0.6%) | OR: 19.52, p=0.0002 | [66] |
| | | ovarian | 1/64 (1.6%) | nad | nad | [44] |
| | **children** | NHL | 1/68 (1.5%) | 11/2261 (0.5%) | $\chi^2$ test, p<0.0001 | [33] |
| | | | 2/212 (0.9%) | 42/6984 (0.6%) | OR: 1.57, p=0.041 | [38] |
| | | ALL | 3/270 (1.1%) | 42/6984 (0.6%) | OR: 1.85, p=0.035 | [38] |
| | | | 1/68 (1.5%) | 11/2261 (0.5%) | $\chi^2$ test, p<0.0001 | [33] |
| | | | 8/400 (2%) | 53/7706 (0.6%) | OR: 3.01, p=0.004 | [47] |

*(Table 1) contd…..*

| *NBN* Mutation | Patient Age | Cancer Type | n° of *NBN* Mutation Carriers Among Tumour Patients | n° of *NBN* Mutation Carriers Among Control Subjects | Statistical Analysis | Ref. |
|---|---|---|---|---|---|---|
| | | ALL+NHL+HL | 5/545 (0.9%) | 42/6984 (0.6%) | OR: 1.48, p=0.037 | [38] |
| | | ALL+NHL | 5/482 (1.03%) | 42/6984 (0.6%) | OR: 1.73, p=0.029 | [38] |
| | | medulloblastoma | 3/104 (2.9%) | 74/12484 (0.6%) | OR: 4.86, p=0.0028 | [45] |
| | | astrocytic | 1/127 (0.8%) | 74/12484 (0.6%) | OR: 1.33, p=0.777 | [46] |
| **R215W** | **adult** | colorectal | 3/234 (1.3%) | 4/1620 (0.2%) | OR: 5.247, p=0.0472 | [35] |
| | | breast | 18/2664 (0.7%) | 7/2031 (0.3%) | OR: 1.9, p=0.18 | [41] |
| | | melanoma | 1/376 (0.3%) | nad | nad | [40] |
| | | NHL | 2/186 (1.1%) | 10/1620 (0.6%) | nad | [66] |
| | | prostate | 6/338 (1.8%) (non familial) | 3/208 (1.4%) | OR: 1.24, p =0.77 | [55] |
| | | | 2/139 (1.4%) (familial) | 3/208 (1.4%) | OR: 1, p=1 | [55] |
| | **children** | HL | 1/39 (2.6%) | nad | nad | [54] |
| | | ALL | 1/47 (2.1%) | nad | nad | [27] |
| | | astrocytic | 1/127 (0.8%) | 7/2815 (0.3%) | OR: 3.2, p=0.254 | [46] |
| **I171V** | **adult** | larynx cancer | 4/176 (2.3%) | 1/500 (0.2%) | OR: 11.7, p=0.0175 | [57, 60] |
| | | breast | 20/1636 (1.2%) | 18/1014 (1.8%) | OR: 0.68, p=0.3 | [41] |
| | | | 10/1048 (0.9%) | 7/1017 (0.7%) | OR: 1.39, p=0.7 | [41] |
| | | | 5/270 (1.8%) | 1/500 (0.2%) | OR: 9.42, p=0.02 | [58, 60] |
| | | head and neck | 5/81 (6.2%) | 1/600 (0.2%) | OR: 39.41, p=0.0001 | [60] |
| | | colorectal carcinoma | 3/131(2.3%) | 1/600 (0.2%) | OR: 14.39, p=0.0196 | [60] |
| | **children** | medulloblastoma | 4/104 (3.8%) | 54/4227 (1.3%) | OR: 3.0, p=0.0241 | [45] |

p=0.0795), and colorectal cancer (1.3%; OR: 2.091, p=0.2197) [35]. Moreover, malignant tumors among parents and siblings of 657del5 carriers were twice more frequent (14/77) than in control population [35]. Interestingly, in a study on 2400 healthy NBN heterozygous Polish women, emerged a frequency of 96/10603 (8.8%) malignant tumors among parents and siblings. This suggested that first-

**Table 2. Frequency of the 657del5, R215W, and I171V NBN mutations among either adult or young carriers, affected or not by cancer (NHL: non-Hodgkin lymphoma; ALL: acute lymphoblastic leukaemia; HL: Hodgkin lymphoma; OR: odd ratio). The number of the NBN mutation carriers among tumor patients and control subjects were obtained pooling data presented in Table 1. Statistical analysis was performed by $\chi^2$ test, and data were considered significant when $p<0.05$.**

| *NBN* Mutation | Cancer Type | N° of *NBN* Mutation Carriers Among Tumour Patients | | N° of *NBN* Mutation Carriers Among Control Subjects | | | |
|---|---|---|---|---|---|---|---|
| | | Cases (%) | Frequency | Cases (%) | Frequency | OR | p |
| *Adults* | | | | | | | |
| **657del5** | colorectal | 6/984 (0.6%) | 0,006 | 103/21737 (0.5%) | 0,005 | 1,289 | 0,71 |
| | breast | 63/7312 (0.9%) | 0,009 | 103/21737 (0.5%) | 0,005 | 1,825 | 0,0002 |
| | melanoma | 5/481 (1%) | 0,010 | 103/21737 (0.5%) | 0,005 | 2,206 | 0,15 |
| | NHL | 10/270 (3.7%) | 0,037 | 103/21737 (0.5%) | 0,005 | 8,078 | < 0.0001 |
| | prostate (familial) | 5/56 (8.9%) | 0,089 | 103/21737 (0.5%) | 0,005 | 20,592 | < 0.0001 |
| | prostate (non familial) | 7/305 (2.3%) | 0,023 | 103/21737 (0.5%) | 0,005 | 4,934 | < 0.0001 |
| | gastrointestinal lymphoma | 4/37 (10.8%) | 0,108 | 103/21737 (0.5%) | 0,005 | 25,459 | < 0.0001 |
| | ovarian | 1/64 (1.6%) | 0,016 | 103/21737 (0.5%) | 0,005 | 3,334 | 0,72 |
| **R215W** | colorectal | 3/234 (1.3%) | 0,013 | 24/5479 (0.4%) | 0,004 | 2,952 | 0,17 |
| | breast | 18/2664 (0.7%) | 0,007 | 24/5479 (0.4%) | 0,004 | 1,546 | 0,22 |
| | melanoma | 1/376 (0.3%) | 0,003 | 24/5479 (0.4%) | 0,004 | 0,606 | 0,93 |
| | NHL | 2/186 (1.1%) | 0,011 | 24/5479 (0.4%) | 0,004 | 2,471 | 0,48 |
| | prostate (familial) | 2/139 (1.4%) | 0,014 | 24/5479 (0.4%) | 0,004 | 3,318 | 0,28 |
| | prostate (non familial) | 6/338 (1.8%) | 0,018 | 24/5479 (0.4%) | 0,004 | 4,108 | 0,003 |
| **I171V** | larynx cancer | 4/176 (2.3%) | 0,023 | 31/5231 (0.6%) | 0,006 | 3,901 | 0,02 |
| | multiple primary tumors | 5/93 (5.4%) | 0,054 | 31/5231 (0.6%) | 0,006 | 9,531 | < 0.0001 |
| | breast | 35/2954 (1.2%) | 0,012 | 31/5231 (0.6%) | 0,006 | 2,011 | 0,006 |
| | head and neck | 5/81 (6.2%) | 0,062 | 31/5231 (0.6%) | 0,006 | 11,036 | < 0.0001 |
| | colorectal carcinoma | 3/131 (2.3%) | 0,023 | 31/5231 (0.6%) | 0,006 | 3,931 | 0,06 |

*(Table 2) contd.....*

| **NBN Mutation** | **Cancer Type** | **N° of *NBN* Mutation Carriers Among Tumour Patients** | | **N° of *NBN* Mutation Carriers Among Control Subjects** | | | |
|---|---|---|---|---|---|---|---|
| | | **Cases (%)** | **Frequency** | **Cases (%)** | **Frequency** | **OR** | **p** |
| *Children* | | | | | | | |
| **657del5** | | | | | | | |
| | NHL | 3/280 (1.1%) | 0,011 | 223/38680 (0.6%) | 0,006 | 1,868 | 0,49 |
| | ALL | 12/738 (1.6%) | 0,016 | 223/38680 (0.6%) | 0,006 | 2,850 | 0,0006 |
| | medulloblastoma | 3/104 (2.9%) | 0,029 | 223/38680 (0.6%) | 0,006 | 5,122 | 0,015 |
| | astrocytic | 1/127 (0.8%) | 0,008 | 223/38680 (0.6%) | 0,006 | 1,369 | 0,75 |
| **R215W** | | | | | | | |
| | HL | 1/39 (2.6%) | 0,026 | 7/2815 (0.2) | 0,002 | 10,556 | 0,23 |
| | ALL | 1/47 (2.1%) | 0,021 | 7/2815 (0.2) | 0,002 | 8,720 | 0,30 |
| | astrocytic | 1/127 (0.8%) | 0,008 | 7/2815 (0.2) | 0,002 | 3,184 | 0,79 |
| **I171V** | | | | | | | |
| | medulloblastoma | 4/104 (3.8%) | 0,038 | 54/4227 (1.3%) | 0,013 | 3,091 | 0,07 |

degree relatives of the 657del5 mutation carriers may have an elevated risk of cancer [35]. Heterozygotes for the 657del5 mutation are about three times more frequent among non-selected breast cancer patients than expected. Since an elevated risk of breast cancer has been also observed among carriers of mutations in the *BRCA1*, *p53* and *ATM* genes [49-51], and because these gene products interact with each other and with NBN [52], these findings suggest that *NBN* is another gene that might be associated with increased risk of breast cancer in heterozygotes [35].

## The R215W Mutation

The R215W mutation has been considered for a long time a polymorphism of *NBN*, and only recently its severe pathogenicity is emerged with the identification of compound heterozygous 657del5/R215W NBS patients [53].

The R215W missense mutation was first described in a case of acute lymphoblastic leukemia (ALL) and in 9 probands of Slavic origin from a population-based study [27]. Subsequently, a high frequency of heterozygous carriers of the R215W mutation was found not only among children affected by ALL (2.1%), but also by HL (2.6%) (Table **1**) [54].

To date, the R215W mutation has been detected in 35 of 4150 individuals with tumours (0.8%) (Table **1**). Several studies conducted among Poland, Germany, Czech Republic and United Kingdom reported that among carriers of the R215W missense mutation the following type of cancers were observed: ALL (2.1%), HL (2.6%), prostate cancer (1.7%, both familial and non familial forms), colorectal cancer (1.3%), NHL (1.1%), astrocytic cancer (0.8%), and breast cancer (0.7%), melanoma (0.3%) (Table **2**) [35, 39-41, 46, 55].

## The I171V Mutation

The 511A>G (I171V) germ-line mutation was identified for the first time in five of 47 children with ALL [27]. These children were all characterized by late prognoses due to a late relapses [27]. The same mutation, at the homozygous status was detected in a Japanese patient with aplastic anaemia, but with no other clinical signs of NBS [56]. In a large study aimed to assess the frequency of *NBN* mutations in patients with larynx cancer and multiple primary tumours emerged that the frequency of the I171V mutation carriers is significantly higher than in control populations (2.3% in larynx cancer patients; 5.4% in multiple primary tumours) (Tables **1** and **2**). These results imply that the I171V mutation contributes significantly to the overall incidence of larynx carcinoma [57].

To date, the I171V mutation has been detected in 56 of 3539 individuals with tumours (1.6%) (Table **1**). Among heterozygous carriers of the I171V missense mutation the following type of tumors were observed: head and neck cancers (6.2%), multiple primary tumors (5.4%), medulloblastoma (3.8%), colorectal carcinoma (2.3%), larynx cancer (2.3%), breast cancer (1.2%) (Table **2**) [42, 45, 57-60].

An investigation among the Polish population has provided evidence that the I171V mutation could be associated with an increased breast cancer risk (1.8% of I171V carriers). In particular, this association concerns patients with breast cancer, whose first-degree relatives also had diagnosis of those malignancies [58].

Furthermore, it has been observed that the I171V mutation occurred more frequently in the breast cancer group than in the control one, thus indicating that the heterozygosity for this mutation represent a significant risk factor for breast cancer development [42]. However, in an association study in two large hospital-based case-control settings from Germany and Belarus emerged that the I171V missense mutation does not significantly increase the breast cancer risk (0.9%) [59].

Remarkably, the percentage of the I171V mutation carriers is particularly high among patients with neck and head tumours (6.2%) [60], and with medulloblastoma (3.9%) [45]), thus suggesting that the I171V mutation in *NBN* gene may be susceptibility factor in solid tumours.

## BIOCHEMICAL EFFECTS OF THE NBN MUTATION ON THE STRUCTURE AND FUNCTION OF THE NBN PROTEIN

Low expression of abbreviated polypeptides of both *N*-terminal and *C*-terminal NBN has been demonstrated in NBS lymphoblastoid cell lines with different mutations. Particularly, *C*-terminal peptides of lower molecular weight than 95kDa, which maintain the ability to interact with MRE11, has been detected by means of a co-immunoprecipitation assay in lymphoblastoid NBS cell lines [7, 8]. NBS cells characterized by the presence of the classical mutation 657del5 in the *NBN* gene, show two alternative forms of nibrin with a lower molecular weight, of approximately 26 and 70 kDa. In particular, the 5bp deletion in position 657 splits the BRCT tandem domain exactly in the linker region that connects the two BRCT domains. The 26kDa protein includes the region 1-218 of the NBN protein, and comprises the FHA and the first BRCT domains. The 70kDa protein is produced by an alternative initiation of translation upstream the 5bp deletion: after a 18 residue extension at the *N*-terminus, the sequence is identical to that of the wild type NBN, from the amino acid 221 to the end, and contains the second BRCT domain and the *C*-terminal half of NBN [11, 61] (Fig. **2B**).

The R215W missense mutation determines the substitution at position 215 of an arginine (R) with a tryptophan (W). The 215 residue of NBN protein is located at the *C*-terminus of the BRCT1 domain, right before the linker region which connects the two BRCT domains, and seems to be pivotal for the relative orientation of the NBN BRCT domains [62] (Fig. **2C**). Since tryptophan is a hydrophobic and bulky residue, it could lead to a perturbation of the relative geometry of the tandem BRCT domains. It has been demonstrated, in fact, that the R215W mutation in NBN impairs histone γ-H2AX binding after induction of DNA damage, leading to a delay in DNA-DSB rejoining [62].

The pathogenic character of the I171V mutation is presumably connected with its occurrence in the BRCT1 domain of NBN (Fig. **2D**). So, similarly to the R215W mutation, the I171V may perturb the proper geometry of the tandem BRCT domain, thus impairing the binding to γ-H2AX and the delocalization of the MRN complex to the vicinity of the DNA damage site [57, 63].

## CONCLUSIONS

Epidemiological data so far collected point to an increased risk of cancer incidence in heterozygous carriers of the 657del5, R215W, and I171V mutations of the *NBN* gene. The pathogenicity of these *NBN* cancer-related mutations is presumably connected with their occurrence in the BRCT tandem domains of NBN. Interestingly, cytogenetic analysis revealed that NBN heterozygotes exhibit increased spontaneous chromosomal aberrations respect to normal individuals [14]. These data indicate that carriers of *NBN* mutations are less effective in DSBs repair. Notably also a single DSB not repaired or inaccurately repaired can have profound effects, producing aberrant chromosomal rearrangements and chromosomal instability.

The only 657del5 founder mutation frequency among newborns is 1:154 in Czech Republic, 1:182 in Ukraine (Lvov region), 1:190 in Poland, with a mean prevalence of 1:177 for the three populations tested [30], whereas the R215W mutation frequency among Czech newborns is 1:234 [64]. A moderately elevated risk in heterozygous carriers would results in hundreds of new cancer cases in these populations every year. Because it cannot be excluded that cancer patients who carry germ line *NBN* mutations may show a specific sensitivity to treatment with IR or cytostatic drugs, as recently shown [10], systematic studies are now under way to protocol their responses to radio- and chemotherapy.

Experimental support that *NBN* heterozygosity predispose cells to malignancy, come from a study in which the mouse homologue of the human *NBN* gene, *Nbn*, was disrupted in mice [65]. $Nbn^{+/-}$ mice showed a significantly increased occurrence of spontaneous solid tumours (epithelial tumours affecting the liver, prostate and mammary glands, and gonad malignancy) in addition to lymphoma. Moreover, IR dramatically increased cancer formation in $Nbn^{+/-}$ mice, especially thyroid tumours. These data provide a clear relationship between *NBN* heterozygosity, radiation sensitivity and increased cancer risk. Interestingly, examination of the tumours gave no evidence for loss or mutation of the wild-type allele, suggesting that haploinsufficiency is the presumed pathogenic mechanism. In contrast, for human heterozygotes, the possible existence of a truncated protein produced by alternative translation [7], and capable of interaction with MRE11 would be compatible with a dominant negative mechanism.

## CONFLICT OF INTEREST

The author(s) confirm that this chapter content has no conflict of interest.

## ACKNOWLEDGEMENTS

D.C. is a PhD fellow at the Department of Biology, University Roma Tre. This work was partly supported by grants from ISS-NIH 526/R2.

## DISCLOSURE

This is an updated report of the earlier review article by A. di Masi and A. Antoccia NBS1 heterozygosity and cancer risk. Curr. Genomics, 2008, 9: 275-281.

## ABBREVIATIONS

| | | |
|---|---|---|
| ALL | = | Acute lymphoblastic leukaemia |
| ATM | = | Ataxia telangiectasia mutated gene |
| BRCA1 | = | Breast cancer 1 |

| | | |
|---|---|---|
| BRCT | = | BRCA1 C-terminal domain |
| FHA | = | Forkhead associated domain |
| HL | = | Hodgkin lymphoma |
| IR | = | Ionising radiation |
| NBS | = | Nijmegen breakage syndrome |
| NHL | = | Non Hodgkin lymphoma |
| OR | = | Odds ratio |

## REFERENCES

[1] Weemaes, C.M.; Hustinx, T.W.; Scheres, J.M.; van Munster, P.J.; Bakkeren, J.A.; Taalman, R.D. A new chromosomal instability disorder: the Nijmegen breakage syndrome. *Acta Paediatr. Scand.*, **1981**, *70*, 557-564.

[2] The International Nijmegen Breakage Syndrome Study Group, Nijmegen breakage syndrome. The International Nijmegen Breakage Syndrome Study Group. *Arch. Dis. Child*, **2000**, *82*, 400-406.

[3] Digweed, M.; Sperling, K. Nijmegen breakage syndrome: clinical manifestation of defective response to DNA double-strand breaks. *DNA Repair (Amst)*, **2004**, *3*, 1207-1217.

[4] di Masi, A.; Gullotta, F.; Cappadonna, V.; Leboffe, L.; Ascenzi, P. Cancer predisposing mutations in BRCT domains. *IUBMB Life*, **2011**, *63*, 503-512.

[5] Distel, L.; Neubauer, S.; Varon, R.; Holter, W.; Grabenbauer, G. Fatal toxicity following radio- and chemotherapy of medulloblastoma in a child with unrecognized Nijmegen breakage syndrome. *Med. Pediatr. Oncol.*, **2003**, *41*, 44-48.

[6] Meyer, S.; Kingston, H.; Taylor, A.M.; Byrd, P.J.; Last, J.I.; Brennan, B.M.; Trueman, S.; Kelsey, A.; Taylor, G.M.; Eden, O.B. Rhabdomyosarcoma in Nijmegen breakage syndrome: strong association with perianal primary site. *Cancer Genet. Cytogenet.*, **2004**, *154*, 169-174.

[7] Maser, R.S.; Zinkel, R.; Petrini, J.H. An alternative mode of translation permits production of a variant NBS1 protein from the common Nijmegen breakage syndrome allele. *Nat. Genet.*, **2001**, *27*, 417-421.

[8] Tanzarella, C.; Antoccia, A.; Spadoni, E.; di Masi, A.; Pecile, V.; Demori, E.; Varon, R.; Marseglia, G.L.; Tiepolo, L.; Maraschio, P. Chromosome instability and nibrin protein variants in NBS heterozygotes. *Eur. J. Hum. Genet.*, **2003**, *11*, 297-303.

[9] Stumm, M.; Neubauer, S.; Keindorff, S.; Wegner, R.D.; Wieacker, P.; Sauer, R. High frequency of spontaneous translocations revealed by FISH in cells from patients with the cancer-prone syndromes ataxia telangiectasia and Nijmegen breakage syndrome. *Cytogenet. Cell Genet.*, **2001**, *92*, 186-191.

[10] Neubauer, S.; Arutyunyan, R.; Stumm, M.; Dörk, T.; Bendix, R.; Bremer, M.; Varon, R.; Sauer, R.; Gebhart, E. Radiosensitivity of ataxia telangiectasia and Nijmegen breakage syndrome homozygotes and heterozygotes as determined by three-color FISH chromosome painting. *Radiat. Res.*, **2002**, *157*, 312-321.

[11] Becker, E.; Meyer, V.; Madaoui, H.; Guerois, R. Detection of a tandem BRCT in Nbs1 and Xrs2 with functional implications in the DNA damage response. *Bioinformatics*, **2006**, *22*, 1289-1292.

[12] Gatei, M.; Young, D.; Cerosaletti, K.M.; Desai-Mehta, A.; Spring, K.; Kozlov, S.; Lavin, M.F.; Gatti, R.A.; Concannon, P.; Khanna, K. ATM-dependent phosphorylation of nibrin in response to radiation exposure. *Nat. Genet.*, **2000**, *25(1)*, 115-9.

[13] Lim, D.S.; Kim, S.T.; Xu, B.; Maser, R.S.; Lin, J.; Petrini, J.H.; Kastan, M.B. ATM phosphorylates p95/nbs1 in an S-phase checkpoint pathway. *Nature*, **2000**, *404*, 613-617.

[14] Wu, X.; Ranganathan, V.; Weisman, D.S.; Heine, W.F.; Ciccone, D.N.; O'Neill, T.B.; Crick, K.E.; Pierce, K.A.; Lane, W.S.; Rathbun, G.; Livingston, D.M.; Weaver, D.T. ATM phosphorylation of Nijmegen breakage syndrome protein is required in a DNA damage response. *Nature*, **2000**, *405*, 477-482.

[15] Zhao, S.; Weng, Y.C.; Yuan, S.S.; Lin, Y.T.; Hsu, H.C.; Lin, S.C.; Gerbino, E.; Song, M.H.; Zdzienicka, M.Z.; Gatti, R.A.; Shay, J.W.; Ziv, Y.; Shiloh, Y.; Lee, EY. Functional link between ataxia-telangiectasia and Nijmegen breakage syndrome gene products. *Nature*, **2000**, *405*, 473-477.

[16] Wu, L.; Luo, K.; Lou, Z.; Chen, J. MDC1 regulates intra-S-phase checkpoint by targeting NBS1 to DNA double-strand breaks. *Proc. Natl. Acad. Sci. U S A*, **2008**, *105*, 11200-11205.

[17] Kobayashi, J.; Antoccia, A.; Tauchi, H.; Matsuura, S.; Komatsu, K. NBS1 and its functional role in the DNA damage response. *DNA Repair (Amst.)*, **2004**, *3*, 855-861.

[18] di Masi, A.; Beradinelli, F.; Antoccia, A. Nijmegen breakage syndrome: a DNA double strand breaks repair defective disorder. In: Genetic Inheritance Patterns; Hauppauge, Ed.; NY, **2008**.

[19] Koonin, E.V.; Altschul, S.F.; Bork, P. BRCA1 protein products ... Functional motifs... *Nat. Genet.*, **1996**, *13*, 266-268.

[20] Manke, I.A.; Lowery, D.M.; Nguyen, A.; Yaffe, MB. BRCT repeats as phosphopeptide-binding modules involved in protein targeting. *Science*, **2003**, *302*, 636-639.

[21] Yu, X.; Chini, C.C.; He, M.; Mer, G.; Chen, J. The BRCT domain is a phospho-protein binding domain. *Science*, **2003**, *302*, 639-642.

[22] Glover, J.N.; Williams, R.S.; Lee, M.S. Interactions between BRCT repeats and phosphoproteins: tangled up in two. *Trends Biochem. Sci.*, **2004**, *29*, 579-585.

[23] Williams, R.S.; Bernstein, N.; Lee, M.S.; Rakovszky, M.L.; Cui, D.; Green, R.; Weinfeld, M.; Glover, J.N. Structural basis for phosphorylation-dependent signaling in the DNA-damage response. *Biochem. Cell Biol.*, **2005**, *83*, 721-727.

[24] Glover, J.N. Insights into the molecular basis of human hereditary breast cancer from studies of the BRCA1 BRCT domain. *Fam. Cancer*, **2006**, *5*, 89-93.

[25] di Masi, A.; Antoccia, A. NBS1 Heterozygosity and Cancer Risk. *Curr. Genomics*, **2008**, *9*, 275-281.

[26] Gullotta, F.; De Marinis, E.; Ascenzi, P.; di Masi, A. Targeting the DNA double strand breaks repair for cancer therapy. *Curr. Med. Chem.*, **2010**, *17*, 2017-2048.

[27] Varon, R.; Reis, A.; Henze, G.; von Einsiedel, H.G.; Sperling, K.; Seeger, K. Mutations in the Nijmegen Breakage Syndrome gene (NBS1) in childhood acute lymphoblastic leukemia (ALL). *Cancer Res.*, **2001**, *61*, 3570-3572.

[28] http://www.nijmegenbreakagesyndrome.net. (Last access: March 2012).

[29] Seemanova, E. An increased risk of neoplasm in heterozygotes for a syndrome of microcephaly, normal intelligence, growth retardation, remarkable facies, immunodeficiency and chromosomal instability. *Mutat. Res.*, **1990**, *238*, 321-324.

[30] Varon, R.; Seemanova, E.; Chrzanowska, K.; Hnateyko, O.; Piekutowska-Abramczuk, D.; Krajewska-Walasek, M.; Sykut-Cegielska, J.; Sperling, K., Reis, A. Clinical ascertainment of Nijmegen breakage syndrome (NBS) and prevalence of the major mutation, 657del5, in three Slav populations. *Eur. J. Hum. Genet.*, **2000**, *8*, 900-902.

[31] Carlomagno, F.; Chang-Claude, J.; Dunning, A.M.; Ponder, B.A. Determination of the frequency of the common 657del5 Nijmegen breakage syndrome mutation in the German population: no association with risk of breast cancer. *Genes Chromosomes Cancer*, **1999**, *25*, 393-395.

[32] Stanulla, M.; Stümm, M.; Dieckvoss, B.O.; Seidemann, K.; Schemmel, V.; Müller Brechlin, A.; Schrappe, M.; Welte, K.; Reiter, A. No evidence for a major role of heterozygous deletion 657del5 within the NBS1 gene in the pathogenesis of non-Hodgkin's lymphoma of childhood and adolescence. *Br. J. Haematol.*, **2000**, *109*, 117-120.

[33] Resnick, I.B.; Kondratenko, I.; Pashanov, E.; Maschan, A.A.; Karachunsky, A.; Togoev, O.; Timakov, A.; Polyakov, A.; Tverskaya, S.; Evgrafov, O.; Roumiantsev, A.G. 657del5 mutation in the gene for Nijmegen breakage syndrome (NBS1) in a cohort of Russian children with lymphoid tissue malignancies and controls. *Am. J. Med. Genet. A.*, **2003**, *120*, 174-179.

[34] Cybulski, C.; Górski, B.; Debniak, T.; Gliniewicz, B.; Mierzejewski, M.; Masojć, B.; Jakubowska, A.; Matyjasik, J.; Złowocka, E.; Sikorski, A.; Narod, S.A., Lubiński, J. NBS1 is a prostate cancer susceptibility gene. *Cancer Res.*, **2004**, *64*, 1215-1219.

[35] Steffen, J.; Varon, R.; Mosor, M.; Maneva, G.; Maurer, M.; Stumm, M.; Nowakowska, D.; Rubach, M.; Kosakowska, E.; Ruka, W.; Nowecki, Z.; Rutkowski, P.; Demkow, T.; Sadowska, M.; Bidziński, M.; Gawrychowski, K.; Sperling, K. Increased cancer risk of heterozygotes with NBS1 germline mutations in Poland. *Int. J. Cancer*, **2004**, *111*, 67-71.

[36] Buslov, K.G.; Iyevleva, A.G.; Chekmariova, E.V.; Suspitsin, E.N.; Togo, A.V.; Kuligina, E.Sh.; Sokolenko, A.P.; Matsko, D.E.; Turkevich, E.A.; Lazareva, Y.R.; Chagunava, O.L.; Bit-Sava, E.M.; Semiglazov, V.F.; Devilee, P.; Cornelisse, C.; Hanson, K.P.; Imyanitov, EN. NBS1 657del5 mutation may contribute only to a limited fraction of breast cancer cases in Russia. *Int. J. Cancer*, 2005, *114*, 585-589.

[37] Górski, B.; Cybulski, C.; Huzarski, T.; Byrski, T.; Gronwald, J.; Jakubowska, A.; Stawicka, M.; Gozdecka-Grodecka, S.; Szwiec, M.; Urbański, K.; Mituś, J.; Marczyk, E.; Dziuba, J.; Wandzel, P.; Surdyka, D.; Haus, O.; Janiszewska, H.; Debniak, T.; Tołoczko-Grabarek, A.; Medrek, K.; Masojć, B.; Mierzejewski, M.; Kowalska, E.; Narod, S.A.; Lubiński, J. Breast cancer predisposing alleles in Poland. *Breast Cancer Res. Treat.*, **2005**, *92*, 19-24.

[38] Chrzanowska, K.H., Piekutowska-Abramczuk, D., Popowska, E., Gładkowska-Dura, M.; Małdyk, J., Syczewska, M.; Krajewska-Walasek, M.; Goryluk-Kozakiewicz, B.; Bubała, H.; Gadomski, A.; Gaworczyk, A.; Kazanowska, B.; Kołtan, A.; Kuźmicz, M.; Luszawska-Kutrzeba, T.; Maciejka-Kapuścińska, L.; Stolarska, M.; Stefańska, K.; Sznurkowska, K.; Wakulińska, A.; Wieczorek, M.; Szczepański, T.; Kowalczyk, J. Carrier frequency of

mutation 657del5 in the NBS1 gene in a population of Polish pediatric patients with sporadic lymphoid malignancies. *Int. J. Cancer*, **2006**, *118*, 1269-1274.

[39] Steffen, J.; Maneva, G.; Poplawska, L.; Varon, R.; Mioduszewska, O.; Sperling, K. Increased risk of gastrointestinal lymphoma in carriers of the 657del5 NBS1 gene mutation. *Int. J. Cancer*, **2006**, *119*, 2970-2973.

[40] Meyer, P.; Stepelmann, H.; Frank, B.; Varon, R.; Burwinkel, B.; Schmitt, C.; Boettger, M.B.; Klaes, R.; Sperling, K.; Hemminki, K.; Kammerer, S. Molecular genetic analysis of NBS1 in German Melanoma patients. *Melanoma Res.*, **2007**, *17*, 109-116.

[41] Bogdanova, N.; Feshcenko, S.; Schürmann, P.; Waltes, R.; Wieland, B.; Hillemanns, P.; Rogov, Y.I.; Dammann, O.; Bremer, M.; Karstens, J.H.; Sohn, C.; Varon, R.; Dörk, T. Nijmegen Breakage Syndrome mutations and risk of breast cancer. *Int. J. Cancer*, **2008**, *122*, 802-806.

[42] Roznowski, K.; Januszkiewicz-Lewandowska, D.; Mosor, M.; Pernak, M.; Litwiniuk, M.; Nowak, J. I171V germline mutation in the NBS1 gene significantly increases risk of breast cancer. *Breast Cancer Res. Treat.*, **2008**, *110*, 343-348.

[43] Pardini, B.; Naccarati, A.; Polakova, V.; Smerhovsky, Z.; Hlavata, I.; Soucek, P.; Novotny, J.; Vodickova, L.; Tomanova, V.; Landi, S.; Vodicka, P. NBN 657del5 heterozygous mutations and colorectal cancer risk in the Czech Republic. *Mut. Res.*, **2009**, *666*, 64-67.

[44] Suspitsin, E.N.; Sherina, N.Y.; Ponomariova, D.N.; Sokolenko, A.P.; Iyevleva, A.G.; Gorodnova, T.V.; Zaitseva, O.A.; Yatsuk, O.S.; Togo, A.V.; Tkachenko, N.N.; Shiyanov, G.A.; Lobeiko, O.S.; Krylova, N.Y.; Matsko, D.E.; Maximov, S.Y.; Urmancheyeva, A.F.; Porhanova, N.V.; Imyanitov, E.N. High frequency of BRCA1, but not CHEK2 or NBS1 (NBN), founder mutations in Russian ovarian cancer patients. *Hered. Cancer Clin. Pract.* , **2009**, *7*, 5.

[45] Ciara, E.; Piekutowska-Abramczuk, D.; Popowska, E.; Grajkowska, W.; Barszcz, S.; Perek, D.; Dembowska-Bagińska, B.; Perek-Polnik, M.; Kowalewska, E.; Czajńska, A.; Syczewska, M.; Czornak, K.; Krajewska-Walasek, M.; Roszkowski, M.; Chrzanowska, K.H. Heterozygous germ-line mutations in the NBN gene predispose to medulloblastoma in pediatric patients. *Acta Neuropathol.*, **2010**, *119*, 325-334.

[46] Piekutowska-Abramczuk, D.; Ciara, E.; Popowska, E.; Grajkowska, W.; Dembowska-Bagińska, B.; Kowalewska, E.; Czajńska, A.; Perek-Polnik, M.; Roszkowski, M.; Syczewska, M.; Krajewska-Walasek, M.; Perek, D.; Chrzanowska, K.H. The frequency of NBN molecular variants in pediatric astrocytic tumors. *J. Neurooncol.*, **2010**, *96*, 161-168.

[47] Pastorczak, A.; Górniak, P.; Sherborne, A.; Hosking, F.; Trelińska, J.; Lejman, M.; Szczepański, T.; Borowiec, M.; Fendler, W.; Kowalczyk, J.; Houlston, R.S.; Młynarski, W. Role of 657del5 NBN mutation and 7p12.2 (IKZF1), 9p21 (CDKN2A), 10q21.2 (ARID5B) and 14q11.2 (CEBPE) variation and risk of childhood ALL in the Polish population. *Leuk. Res.*, **2011,** *35*, 1534-1536.

[48] Seemanova, E.; Jarolim, P.; Seeman, P.; Varon, R.; Digweed, M.; Swift, M.; Sperling, K. Cancer risk of heterozygotes with the NBN founder mutation. *J. Natl. Cancer Inst.*, **2007**, *99*, 1875-1880.

[49] Kairouz, R.; Clarke, R.A.; Marr, P.J.; Watters, D.; Lavin, M.F.; Kearsley, J.H.; Lee, C.S. ATM protein synthesis patterns in sporadic breast cancer. *Mol. Pathol.*, **1999**, *52*, 252-256.

[50] Broeks, A.; Urbanus, J.H.; Floore, A.N.; Dahler, E.C.; Klijn, J.G.; Rutgers, E.J.; Devilee, P.; Russell, N.S.; van Leeuwen, F.E.; van 't Veer, L.J. ATM-heterozygous germline

mutations contribute to breast cancer-susceptibility. *Am. J. Hum. Genet.*, **2000**, *66*, 494-500.

[51] Antoccia, A.; Kobayashi, J.; Tauchi, H.; Matsuura, S.; Komatsu, K. Nijmegen breakage syndrome and functions of the responsible protein, NBS1. *Genome Dyn.*, **2006**, *1*, 191-205.

[52] Zhong, Q.; Chen, C.F.; Li, S.; Chen, Y.; Wang, C.C.; Xiao, J.; Chen, P.L.; Sharp, Z.D.; Lee, W.H. Association of BRCA1 with the hRad50-hMre11-p95 complex and the DNA damage response. *Science*, **1999**, *285*, 747-750.

[53] Seemanova, E.; Sperling, K.; Neitzel, H.; Varon, R.; Hadac, J.; Butova, O.; Schröck, E.; Seeman, P.; Digweed, M. Nijmegen breakage syndrome (NBS) with neurological abnormalities and without chromosomal instability. *J. Med. Genet.*, **2006**, *43*, 218-224.

[54] Taylor, G.M.; O'Brien, H.P.; Greaves, M.F.; Ravetto, P.F.; Eden, OB. Mutations in the Nijmegen breakage syndrome gene (NBS1) in childhood acute lymphoblastic leukaemia. *Cancer Res.*, **2003**, *63*, 6563-6564.

[55] Hebbring, S.J.; Fredriksson, H.; White, K.A.; Maier, C.; Ewing, C.; McDonnell, S.K.; Jacobsen, S.J.; Cerhan, J.; Schaid, D.J.; Ikonen, T.; Autio, V.; Tammela, T.L.; Herkommer, K.; Paiss, T.; Vogel, W.; Gielzak, M.; Sauvageot, J.; Schleutker, J.; Cooney, K.A.; Isaacs, W.; Thibodeau, SN. Role of the Nijmegen Breakage Syndrome 1 gene in familial and sporadic prostate cancer. *Cancer Epidemiol. Biomarkers Prev.*, **2006**, *15*, 935-938.

[56] Shimada, H.; Shimizu, K.; Mimaki, S.; Sakiyama, T.; Mori, T.; Shimasaki, N.; Yokota, J.; Nakachi, K.; Ohta, T.; Ohki, M. First case of aplastic anemia in a Japanese child with a homozygous missense mutation in the NBS1 gene (I171V) associated with genomic instability. *Hum. Genet.*, **2004**, *115*, 372-376.

[57] Ziolkowska, I.; Mosor, M.; Wierzbika, M.; Rydzanicz, M.; Pernak-Schwarz, M.; Nowak, J. Increased risk of larynx cancer in heterozygous carriers of the I171V mutation of the NBS1 gene. *Cancer Sci.*, **2007**, *98*, 1701-1705.

[58] Kanka, C.; Brozek, I.; Skalska, B.; Siemiatkowska, A.; Limon, J. Germline NBS1 mutations in families with aggregation of Breast and/or ovarian cancer from north-east Poland. *Anticancer Res.*, **2007**, *27*, 3015-3018.

[59] Bogdanova, N.; Schürmann, P.; Waltes, R.; Feshchenko, S.; Zalutsky, I.V.; Bremer, M.; Dörk, T. NBS1 variant I171V and breast cancer risk. *Breast Cancer Res. Treat.*, **2008**, *112*, 75-79.

[60] Nowak, J.; Mosor, M.; Ziołkowska, I.; Wierzbicka, M.; Pernak-Schwarz, M.; Przyborska, M.; Roznowski, K.; Pławski, A.; Słomski, R.; Januszkiewicz, D. Heterozygous carriers of the I171V mutation of the NBS1 gene have a significantly increased risk of solid malignant tumours. *Eur. J. Cancer*, **2008**, *44*, 627-630.

[61] Williams, B.R.; Mirzoeva, O.K.; Morgan, W.F.; Lin, J.; Dunnick, W.; Petrini, J.H. A murine model of Nijmegen breakage syndrome. *Curr.Biol.*, **2002**, *12*, 648-653.

[62] di Masi, A.; Viganotti, M.; Polticelli, F.; Ascenzi, P.; Tanzarella, C.; Antoccia, A. The R215W mutation in NBS1 impairs γ-H2AX binding and affects DNA repair: molecular bases for the severe phenotype of 657del5/R215W Nijmegen breakage syndrome patients. *Biochem Biophys. Res. Commun.*, **2008**, *369*, 835-840.

[63] Kobayashi, J.; Tauchi, H.; Sakamoto, S.; Nakamura, A.; Morishima, K.; Matsuura, S.; Kobayashi, T.; Tamai, K.; Tanimoto, K.; Komatsu, K. NBS1 localizes to gamma-H2AX foci through interaction with the FHA/BRCT domain. *Curr. Biol.*, **2002**, *12*, 1846-1851.

[64] Seemanova, E.; Radvanska, J.; Stary, J.; Seeman, P.; Gebertova, G.; Varon, R.; Sperling, K. Nositele mutaci NBS1 genu mezi pacienty detske onkologie. *Csl. Pediatr.*, **2004**, *59*, 242-245.

[65] Dumon-Jones, V.; Frappart, P.O.; Tong, W.M.; Sajithlal, G.; Hulla, W.; Schmid, G.; Herceg, Z.; Digweed, M.; Wang, Z.Q. Nbn heterozygosity renders mice susceptible to tumor formation and ionizing radiation-induced tumorigenesis. *Cancer Res.*, **2003**, *63*, 7263-7269.

[66] Steffen, J.; Nowakowska, D.; Niwińska, A.; Czapczak, D.; Kluska, A.; Piatkowska, M.; Wiśniewska, A.; Paszko, Z. Germline mutations 657del5 of the NBS1 gene contribute significantly to the incidence of breast cancer in Central Poland. *Int. J. Cancer*, **2006**, *119*, 472-475.

120 *Advances in Genome Science, Vol. 2,* 2013, 120-154

**CHAPTER 5**

# Alternative Splicing and Tumor Progression

**Claudia Ghigna*, Cristina Valacca and Giuseppe Biamonti**

*Institute of Molecular Genetics (IGM), National Research Council (CNR), Via Abbiategrasso 207, 27100 Pavia, Italy*

**Abstract:** Alternative splicing is a key molecular mechanism for increasing the functional diversity of the eukaryotic proteomes. A large body of experimental data implicates aberrant splicing in various human diseases, including cancer. Both mutations in *cis*-acting splicing elements and alterations in the expression and/or activity of splicing regulatory factors drastically affect the splicing profile of many cancer-associated genes. In addition, the splicing profile of several cancer-associated genes is altered in particular types of cancer arguing for a direct role of specific splicing isoforms in tumor progression. Deciphering the mechanisms underlying aberrant splicing in cancer may prove crucial to understand how splicing machinery is controlled and integrated with other cellular processes, in particular transcription and signaling pathways. Moreover, the characterization of splicing deregulation in cancer will lead to a better comprehension of malignant transformation. Cancer-associated alternative splicing variants may be new tools for the diagnosis and classification of cancers and could be the targets for innovative therapeutical interventions based on highly selective splicing correction approaches.

**Keywords:** Alternative splicing, cancer, EMT, splicing correction, splicing factors, biomarkers.

## INTRODUCTION

During evolution the number of genes stopped growing in parallel with the complexity of the proteome. Thus, the human genome contains only 20.000 - 25.000 genes (International Human Genome Sequencing Consortium 2004 [1]), a number not significantly different from that counted in less complex organisms such as sea urchin (23,000) (Sea Urchin Genome Sequencing Consortium 2006) and the nematode worm (19,000) [2]. Moreover, the number of human genes is

***Address correspondence to Claudia Ghigna:** Institute of Molecular Genetics (IGM), National Research Council (CNR), *Via* Abbiategrasso 207, 27100 Pavia, Italy; Tel: +39.0382 546324; Fax: +39.0382.422286; E-mail: arneri@igm.cnr.it

**Christian Neri (Ed)**

not sufficient to account for all the proteins revealed by proteomic analysis. How can these paradoxes be explained?

Recent cDNA sequencing and microarray data have implicated alternative splicing (AS) as the main source of proteomic and functional diversity in metazoan organisms [3]. Together with alternative promoters and polyadenylation sites, RNA editing and post-translational processing, AS gives rise to an estimated number of at least 100,000 different human proteins.

The term "alternative splicing" describes any situation in which a single primary transcript (pre-mRNA) can be spliced in more than one pattern to generate multiple, distinct mature mRNAs leading to expression of protein isoforms with different structural and functional properties. The "record-holder" for alternative splicing is a Drosophila gene called *Dscam*, with 38,000 splice variants, more than the number of Drosophila genes [4]. In humans more than 90% of the genes encode for transcripts that undergo alternative splicing [5], which underscores the importance of this regulatory mechanism in the biology of our species.

Because of its capacity to generate protein diversity, alternative splicing is expected to play a major role in gene expression regulation, a prediction which is substantiated by the observation that appropriate spatio-temporal generation of splicing variants is involved in many cellular and developmental processes (including sex determination, apoptosis, axon guidance, cell excitation and contraction and many others). It is not surprising, therefore, that deregulation of alternative splicing programs is tightly linked to inherited and acquired human genetic disorders [6]. Indeed, works in the last few years have started to recognize inappropriate alternative splicing as a genetic modifier during tumorigenesis [7]. Many cancer-related genes are regulated by alternative splicing. They encode for proteins involved in all major aspects of cancer cell biology, including cell cycle control, proliferation, differentiation, signal transduction pathways, cell death, angiogenesis, invasion, motility and metastasis [8]. A common signature of cancer cells is a general loss of splicing fidelity, with the concomitant reorganization of splicing profiles, and even switching to specific splicing isoforms usually expressed in other cell types. All these events may contribute to carcinogenesis [8]. Notably, there are several cases of alternative splicing that are restricted to specific cancer types, which clearly involve that particular splicing isoform in tumor progression [8, 9]. Cancer-specific splicing events may be also beneficial to therapy, since they generate novel epitopes against which it is possible to raise antibodies for immunotherapy. Hence, there is great interest in discovering the effect of alternative splicing on the transcriptome complexity in cancer cells and in understanding how this regulatory mechanism contributes to tumorigenesis. This review discusses (1) the basic mechanisms of alternative splicing, (2) the present knowledge on the regulatory mechanisms governing alternative splicing and their deregulation in cancer, (3) the biological consequences deriving from the alteration of splicing in some relevant cancer-related genes and (4) cancer-associated alternative splicing variants with a clear diagnostic/prognostic value that can provide potential therapeutic targets.

## ALTERNATIVE SPLICING

Eukaryotic genomes are characterized by the presence of "interrupted" genes consisting of an ordered succession of coding (exon) and non-coding (intron) sequences. In metazoan, particularly in mammals, interrupted genes account for the vast majority of the genes. Thus, RNA splicing occurs as an obligatory, highly regulated, process. Through pre-mRNA splicing, introns are precisely removed and exons are joined together to reconstitute the reading frame and to generate translatable mRNAs that are then exported to the cytoplasm.

The splicing machinery (called spliceosome), is the largest molecular machine so far described in the cells being composed of five small nuclear ribonucleoproteins (snRNPs U1, U2, U4, U5 and U6) and more than 100 different polypeptides. The spliceosome recognizes short, poorly conserved, *cis*-acting sequence elements at exon–intron boundaries (5' and 3' splice sites also known as "donor" and "acceptor" sites) and use them for the cut-and-paste reaction (two sequential *trans*-esterifications) that removes the intron.

By using various combinations of donor and acceptor sites from different exons, through the process of alternative splicing it is possible to produce distinct mRNAs from a single pre-mRNA. Five distinct alternative splicing patterns have been observed: 1) a regulated (cassette) exon, which is sometimes included and sometimes excluded from the mRNA; 2) multiple cassette exons that are mutually exclusive, *i.e.,* the mature mRNA always contains only one of several possible exon choices; 3) in rare cases, a whole intron is retained; 4) alternative donor and 5) acceptor sites can result in exons of different size (see Fig. **1**). Finally, alternative promoters and poly-adenylation sites contribute to the heterogeneity of transcripts encoded by a single gene. In addition to modifying protein features, alternative splicing can affect the stability of transcripts through the nonsense-mediated mRNA decay (NMD) pathway, an mRNA quality-control mechanism which depends on the translation machinery [10]. Recent analyses suggest that approximately 35% of all alternative splicing events in mammalian cells generate mRNA species containing premature termination codons (PTCs), which can be efficiently degraded through NMD [11]. Interestingly, Ni *et al.* [12] have recently shown that exons containing a stop codon are particularly frequent and conserved in genes for splicing factors involved in splicing choices. These exons frequently overlap ultraconserved elements in mammalian genomes.

From a mechanistic point of view the different types of alternative splicing can be simply viewed as a problem of splice-site recognition by the spliceosome where the decision between inclusion and skipping of a particular exon mainly depends on the recognition and utilization of the splice sites that flank the exon. Alternatively spliced exons are often characterized by short and degenerate splice sites. The intrinsic weakness of these sites, which reflects the reduced affinity for spliceosomal proteins, is the main cause of alternative splicing. The recognition of alternative exons is modulated by an additional layer of information provided by an extensive and complex arrays of auxiliary *cis*-acting elements (non-splice site RNA sequence elements), referred to as enhancers and silencers of splicing that respectively promote and inhibit exon recognition (Fig. **2A**). These are short regulatory elements (~10 nucleotides) that can be found isolated or clustered on the pre-mRNA and are present both within exons (ESEs, Exonic Splicing Enhancers and ESSs, Exonic Splicing Silencers) and introns (ISEs, Intronic Splicing Enhancer and ISSs, Intronic Splicing Silencers) [13]. The best-characterized splicing enhancers are typically purine-rich and function by providing binding sites for serine-arginine (SR) factors, a family (about a dozen)

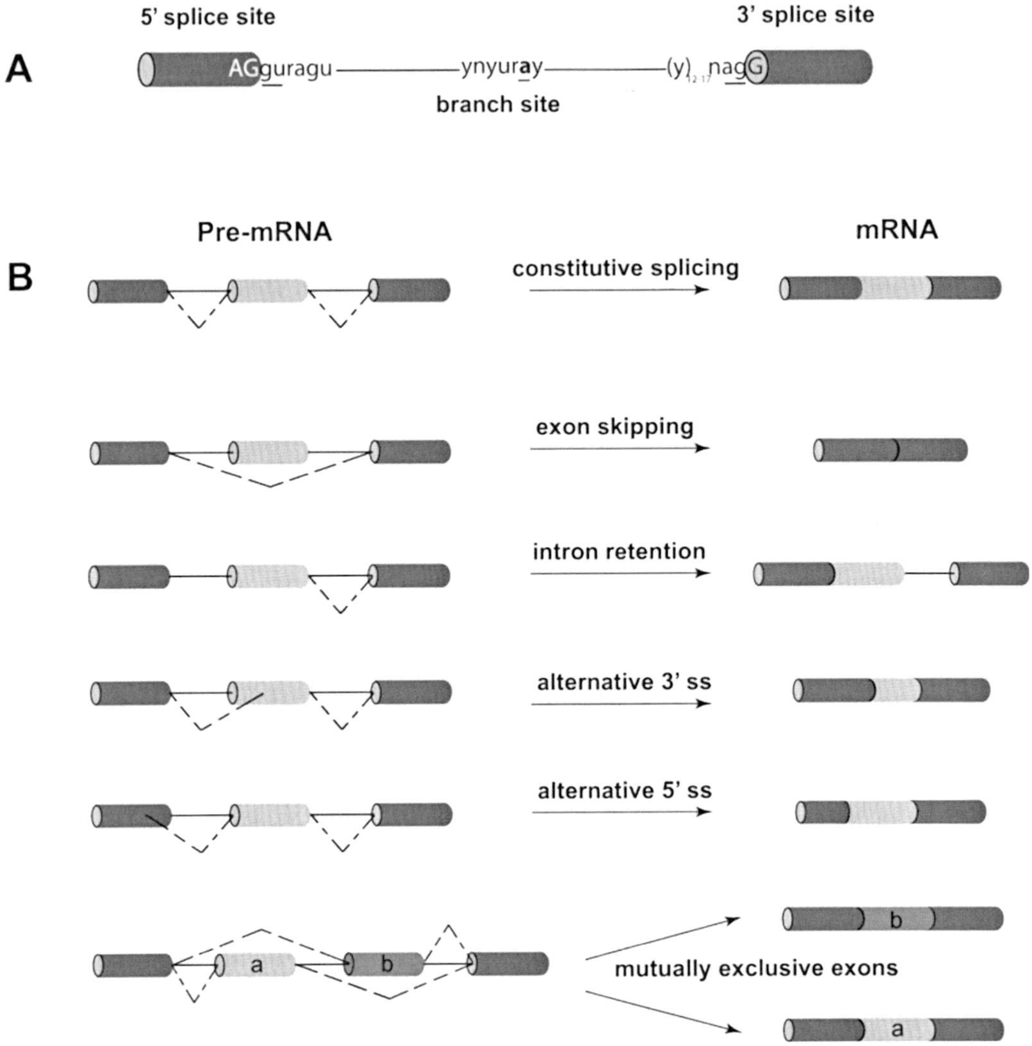


**Figure 1:** *Cis*-acting sequences required for the splicing reaction and different types of alternative splicing events. (**A**) Splicing consensus sequences of a typical eukaryotic gene (exon/intron splice site signals, branch site and polypyrimidine tract). (**B**) Alternatively spliced mRNAs result from exon skipping, intron retention, usage of alternative 3'- (acceptor) or 5'- (donor) sites and from selection of mutually exclusive exons. At the protein level, alternative splicing drastically affects the amino acid sequence by deletion or insertion of domains, frame-shifts or stop codons. Alternative splicing in non-coding regions of the mature mRNA might impact on translation and mRNA stability.

of essential and abundant RNA-binding proteins highly conserved in animal and plant cells [14, 15]. SR factors display multiple roles in constitutive and alternative splicing, as well as in other aspects of gene expression [16]. All members of this family share a modular structure consisting of one or two copies of an N-terminal RNA-recognition motif (RRM) followed by a C-terminal domain of variable length rich in alternating serine-arginine dipeptides (the RS domain). The RRMs determine the RNA-binding specificity, whereas the RS domain mediates specific protein–protein interactions that are essential for the recruitment of the splicing apparatus. However, within the functional spliceosome also the RS domains may directly contact the pre-mRNA. The sequential character of these contacts suggests that RS domain interactions with RNA promote spliceosome assembly [17].

In addition, serine residues of the RS domain are targets of extensive phosphorylation events that influence protein interactions [18], and regulate the activity and sub-cellular distribution of SR proteins [19]. Although several kinases, including SR protein kinases (SRPKs) 1 and 2, CLK/STY, dual-specificity tyrosine-regulated kinase, CRKRS, DNA topoisomerase I, glycogen synthase kinase-3 and AKT, have been shown to phosphorylate SR proteins [19-24], the signal-transduction pathways that regulate alternative splicing are still poorly understood.

Several models have been proposed for the function of ESEs and SR factors (Fig. **2B**). According to one of these models, ESE-bound SR proteins promote exon definition by directly recruiting the splicing machinery through specific protein-protein interactions mediated by the RS domain [13]. Another model predicts that the main function of ESE-bound SR factors is to antagonize the negative effect on splicing of an inhibitory protein that is bound to a juxtaposed silencer element (ESS) (inhibitor model) [13]. Exon inclusion or skipping is determined by balance of these competing activities, which in turn reflect by relative concentrations of the cognate RNA-binding activator and repressor proteins. These models of splicing enhancement are not necessarily mutually exclusive, as they might reflect different requirements in the context of different exons.

Splicing silencers identified to date appear remarkably diverse. They may act as binding sites for factors that block access of the splicing machinery to a splice site. Among the proteins interacting with ESSs and ISSs elements there are heterogeneous nuclear ribonucleoproteins (hnRNPs), a group of RNA-binding proteins initially recognized as factors that interact with RNA polymerase II transcripts to form hnRNP particles [13]. On two dimensional gels approximately 30 spots were described, called with alphabet letters from hnRNP A1 through U. Similarly to SR factors, hnRNP proteins have a modular structure in which one or more RNA binding domains, generally at the N-terminus, are associated to different "auxiliary" domains. Three types of RNA binding domains (RRMs, hnRNP K homology domain and RGG domain, a protein region rich in Arg-Gly-Gly repetitions) have been identified in hnRNP proteins and shown to provide a certain level of RNA binding specificity [13]. The auxiliary domains are very different in sequence and control the sub-cellular localization and the interaction with other proteins. RNA binding specificity and protein-protein interactions contribute to the assembly of the ribonucleoprotein complexes that are the substrates for the ensuing splicing reaction. The mechanism of hnRNPs action depends on the position of their binding sites along the pre-mRNA. In addition to the inhibitor model described above, hnRNPs bound to ISSs flanking an alternative exon may cause the exon to loop out, resulting in skipping of the exon (Fig. **2C**). Alternatively, inhibitory factors bound to ESS may polymerize along the exon and displace the ESE-bound SR proteins [13, 25] (Fig. **2C**).

Only in few cases alternatively spliced exons are controlled by tissue-specific splicing regulators [13]. In the vast majority of cases splicing events appear to be controlled by the relative abundance and/or activity of widely expressed antagonistic SR factors and hnRNP proteins through a combinatorial mechanism, with multiple positive and negative factors and sequence elements influencing the final outcome of the splicing reaction. This is exemplified by antagonistic effects on pre-mRNA splicing of SRSF1 (also known as SF2/ASF or ASF/SF2), an SR factor, and hnRNP A1 proteins: high levels of SRSF1 induce exon inclusion whereas high levels of hnRNP A1 promote exon skipping [26]. Interestingly, the relative expression level of hnRNP A1 and SRSF1 have been shown to change during neoplastic lung growth [27]. Another example is SRSF6 (SRp55) and its antagonistic factor hnRNP I/PTB that control the splicing profile of the *fibroblast growth factor receptor* (*FGFR1*). Skipping of $\alpha$ exon results in the production of the FGFR1-$\beta$ isoform which has a higher affinity for fibroblast growth factors [28]. Increased expression of this isoform correlates with cancer in the pancreas and brain and with poor prognosis in breast tumors [29, 30] and its overexpression promotes tumor formation in nude mice [28]. SRp55 binds to a 69-nucleotide ESE and is required for α-exon inclusion while PTB recognizes a sequence upstream of the α-exon and promotes exon skipping and production of FGFR1-β [31].

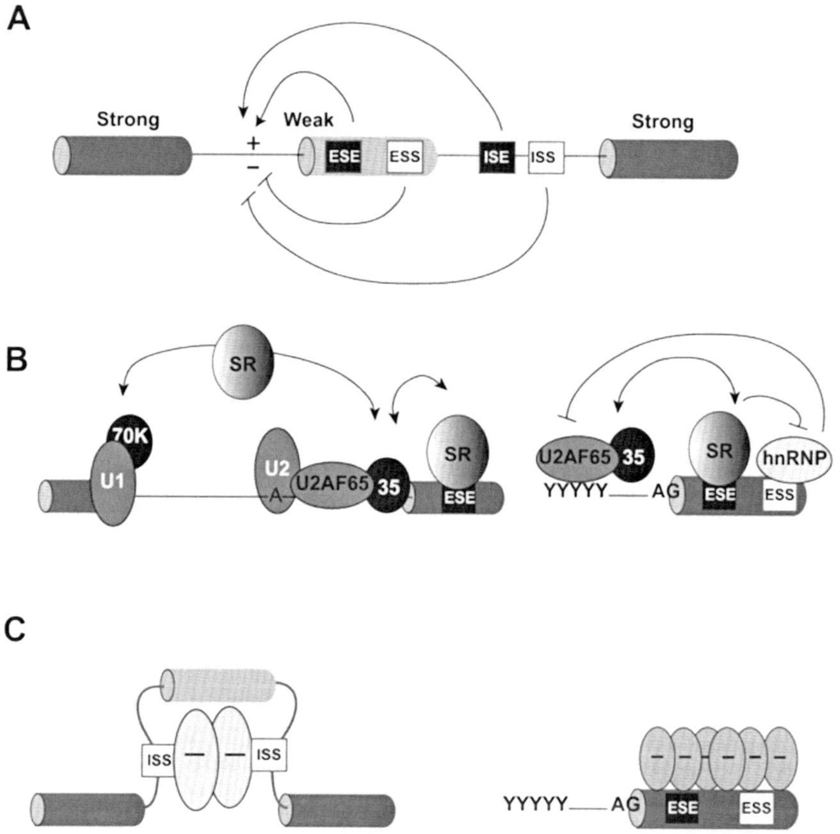


**Figure 2:** *Cis*- and *trans*-acting regulatory elements that control alternative splicing and models for the function of splicing enhancers and silencers. (**A**) Alternatively spliced exons are usually characterized by weak splice sites. Recognition of these sites depends on splicing regulatory elements: exonic splicing enhancers (ESE) and silencers (ESS) and intronic splicing enhancers (ISE) and silencers (ISS). (**B**) ESE elements are bound by splicing factors of the SR family. *Via* interactions with proteins of the splicing apparatus, the RS domain of SR factors promotes the assembly of the spliceosome on the adjacent intron (left). In addition, SR factors can counteract the inhibitory activity of hnRNP proteins bound to ESS elements (right). (**C**) The mechanism of action of the silencer elements depend on their position along the pre-mRNA. In some cases, inhibitory factors (for example hnRNP proteins) bind to ISSs sequences flanking an alternative exon and cause looping out and skipping of this exon (left). Alternatively, inhibitory factors bound to ESSs may polymerize along the exon and displace the ESE-bound SR proteins (right).

Recent studies indicate that signaling pathways may control splicing decisions by affecting the sub-cellular distribution and/or activity of splicing regulators. Many SR factors and hnRNP proteins continuously and rapidly shuttle between the nucleus and the cytoplasm [32], which unveils a cytoplasmic function of these proteins, for example in mRNA translation. In this regard, Michlewski *et al.* [33] showed that the splicing factor SRSF1 stimulates translation initiation by directly recruiting the mammalian target of rapamycin (mTOR) to a subset of mRNAs. Notably, several stress treatments perturb the nucleo-cytoplasmic distribution of some splicing regulators. For instance, exposure of the cells to stress stimuli such as osmotic shock or UVC irradiation, results in a marked cytoplasmic accumulation of hnRNP A1, concomitant with an increase in protein phosphorylation. These effects are mediated by the MKK(3/6)-p38 pathway and correlate with changes in the alternative splicing pattern of an adenovirus E1A pre-mRNA splicing reporter [34]. Moreover, several splicing regulators, including SRSF1, are sequestered in nuclear stress bodies after treatments that activate the heat shock response [35]. This capacity to modulate the activity of splicing regulators opens the exciting possibility that stressing conditions, as those in the tumor microenvironment, may influence the splicing profile of a number of genes and thus affect cell identity.

Our knowledge of the molecular mechanisms underlying alternative splicing is still limited. Indeed, there are many exonic and intronic elements for which trans-acting mediators remain to be identified and, in most of the cases, regulatory circuits that control splicing decisions and the link with signal-transduction pathways are not well understood. Deciphering the complex regulatory network underlying alternative splice site choice remains a major challenge in this area of research.

## ABERRANT ALTERNATIVE SPLICING IN CANCER-RELATED GENES

Genome wide analyses indicate that ~94% of human genes are regulated by alternative splicing [5]. Thus, alternative splicing seems to be the rule for human genes and it is not surprising that splicing defects in specific genes are causatively linked to genetic disorders. Well-characterized examples are spinal muscular atrophy, myotonic dystrophy, retinitis pigmentosa, Frasier syndrome, hemophilia

A, β-thalassemia, atypical cystic fibrosis and some neurodegenerative diseases [6]. It has been calculated that about 15% of point mutations that cause hereditary diseases do alter pre-mRNA processing by affecting canonical 5' and 3' splice sites, branch sites or polyadenylation signals [14, 36]. This figure is probably an underestimation since it does not take into account mutations in splicing regulatory elements (enhancers and silencer) that may prevent the interaction with splicing regulators (hnRNPs and SR factors). Indeed, at least 50% of 50 single-base substitutions that cause exon skipping in human genes disrupt splicing enhancers [14, 36, 37]. Frequently, splicing defects due to gene mutations introduce premature stop codons and target the mRNA to degradation by the nonsense-mediated decay pathway (NMD) [10]. The final effect is that, instead of a truncated polypeptide, no protein is produced.

In the last few years a link has emerged between cancer development and deregulation of alternative splicing [7, 9]. A common feature of cancer cells, in fact, is the general deregulation of splicing, which may lead to the expression of tumor-specific variants and which is most likely promoted by alterations in signaling pathways and by variations in concentration, localization and activity of *trans*-acting splicing regulators [9, 24]. It is still disputed whether these changes in splicing profiles are simply a "noise" occurring in cancer cells or have a direct role in tumorigenesis. As a matter of fact, recent progresses in molecular and cell biology indicate that altered splicing profiles of critical genes may impact on all the major aspects of cancer cell biology [8, 9]. This can be due to inactivation of onco-suppressor [38] or to gain of function of proteins implicated in cancer susceptibility [39] and in tumor progression [8, 9, 40].

## *Cis*-Acting Mutations

These are inherited or somatic mutations that affect the splicing process. According to their position and effect on splicing, these mutations can be divided into two subclasses.

Subclass I (60% of cases) comprises splicing mutations in the invariant splice-sites and completely abolish exon recognition. These mutations are associated with severe diseases. Subclass II is often associated with a relatively mild phenotype and includes mutations in the variant motifs (such as the alternative splicing poly-pyrimidine tract) and intronic mutations that generate cryptic donor or acceptor sites. An example comes from the tumor suppressor *p53* gene, *i.e.,* the most commonly mutated gene in human cancers. Isoforms of *p53* originate through alternative splicing and differ for their tumor suppressor function [41]. Interestingly, "silent" mutations in the *p53* gene are predicted to affect splicing because they create splice sites in the middle of an exon [38]. It is worth considering, however, that many mutation screenings focused on exons and ignored non-coding regions [42]. Below, we discuss other examples of mutations that affect splicing of oncogenes and tumor suppressor genes.

Mutations in the *APC* gene result in familiar adenomatous polyposis (FAP). Recently, it has been reported that an attenuated form of FAP is caused by an insertion of a single T between the second and third position of intron 4 of the *APC* gene that leads to skipping of exon 4 and to the predicted expression of a truncated protein [43]. Two additional mutations in introns 3 and 4 have the identical effect [44].

Also mutations in splicing regulatory elements, such ESE and ESS sequences, can perturb splicing profiles. A good example is the tumor-suppressor *BRCA1* gene that encodes a protein involved in a variety of cellular processes including transcriptional regulation, recombination, DNA repair [45] and cell apoptosis [46]. Mutations in the *BRCA1* gene are well-known markers of susceptibility to ovarian and breast cancer, the last one being the most common malignancy among women. Mazoyer *et al.* [39] demonstrated that an inherited nonsense mutation within exon 18 disrupts an ESE element (the binding site for the SR factor SRSF1) and provokes exon skipping. Computer analysis with the ESEfinder program has identified 23 highly conserved ESEs in the 22 exons of the *BRCA1* gene [25]. About 60% of these motifs contain nucleotide substitutions reported in the Breast Cancer Information Core [47] suggesting the possibility that targeting of splicing profiles may be the mechanisms through which mutations affect the *BRAC1* function [47]. The *BRCA1* gene encodes at least thirty distinct splicing variants that are differentially expressed in tissues [48]. Most of the alternatively spliced transcripts maintain the open reading frame of the original cDNA and have the potential to code for functional proteins. Notably, breast and ovarian cells express a common set of variants suggesting the intriguing possibility that they share some regulatory pathways which instead is absent in leukaemia cell lines that display a different set of isoforms [49]. Many nonsense mutations in the coding region are associated with exon skipping. Thus, a double point mutation in exon 7 of the neurofibromatosis 1 (*NF1*) gene disrupts the consensus binding sites for the SR factors, SRSF2 (SC35) and SRSF1, and promotes exon skipping. [50]. The *NF1* gene shows one of the highest mutation rate among genes associated with human disorders; simply on the bases of the DNA sequence, it was predicted that 37% of these genomic mutations could determine splicing defects [37]. However, when this prediction was verified through the analysis of the mature mRNAs, it became evident that up to 50% of the *NF1* gene mutations are associated with splicing defects. The negative message emerging from this analysis is that we are still unable to infer the splicing profile simply from the primary DNA sequence and that we have to rely on RNA and even protein sequences to understand the effects of any gene mutation [37].

Similar observations have been done with a number of other tumor-associated genes such as *BRCA2*, *FHIT*, *KIT*, *MLH1*, *MDM2*, *MSH2* and *LKB1* [43, 51-56]. Notably, all these cases, and many others that are continuously described, represent an extension of the concept according to which cancer progression is due to a number of stable genetic mutations that perturb the structure, the function or the abundance of critical proteins. In this sense, splicing defects can be viewed as one of routes through which gene mutations cause tumorigenesis. However, alternative splicing of oncogenes or tumor-suppressors could also be affected by mutations in splicing regulators which implies an activity of these factors as oncoproteins or tumor suppressors, depending on their antagonistic functions on splice site selection. For example, the *SRSF6* gene is mutated in breast and colorectal cancers [55]. Interestingly, SRSF6 controls the splicing profile of several tumor-associated genes, among which CD44 [57] and KIT [58]. Also some hnRNP proteins have been classified as oncogenes. In the 90% of the human myxoid liposarcomas, the t(12;16) translocation generates a fusion between the hnRNP P2 gene, that encoded a multifunctional protein involved in transcription, splicing and mRNA export, and the CHOP gene, encoding for a CCAAT/enhancer binding protein implicated in erythropoiesis, adipocyte differentiation, growth arrest and G1-S cell cycle progression. The product of the hnRNP P2-CHOP fusion contains the amino terminal transcription activation domain of hnRNP P2 and the DNA binding domain of CHOP and its over-expression in nude mice results in tumor formation [59].

## Changes in *Trans*-Acting Factors

More interesting, from our point of view, is the observation, reported by many studies over the last 20 years, that most cancer-associated splicing alterations are not associated with nucleotide changes in the affected genes, implying modifications in the expression and/or activity of splicing regulatory factors [9, 40, 60]. Indeed, changes in the repertoire of SR factors and hnRNP proteins frequently occur in tumors and are accompanied by alterations in the relative abundance of alternative splicing products, a typical signature of cancerous cells with predictable effects on cellular behavior [27, 40, 60, 61]. As a matter of fact, cancerous cell lines show a high level of alternative splicing events that are not conserved between human and mouse and are not expressed in normal tissues [62] strengthening the idea that a change in the level of splicing regulators in cancer cells may severely impinge on gene expression programs.

A salient example to illustrate how aberrant splicing in cancer can be modulated by altered activity or expression of splicing factors is provided by CD44, a trans-membrane glycoprotein involved cell–cell and cell-matrix interactions [63]. Several CD44 isoforms are generated through variable incorporation of 10 alternative exons (v1-v10) in its proximal extra-cellular domain. Standard CD44, lacking all alternative exons, is predominantly express in normal tissues, whereas CD44 isoforms, in particular those containing variant exons v5, v6 and v7, are over-expressed in various tumors and have been implicated in tumor cell invasion and metastasis [63]. The production of different CD44 isoforms correlates with changes in the abundance of SR proteins [61, 64] and several splicing factors (including hnRNPA1, SRSF6, SRSF1, Tra-2 beta, YB-1 and Sam68) that have been shown to regulate specific variant exons [65-68]. Several lines of evidence indicate that *CD44* splicing is regulated in response to extracellular stimuli. Thus, Matter *et al.* [68] found that the RNA binding protein Sam68 (Src-associated in mitosis 68-kDa) promotes v5 inclusion in response to activation of the Ras-Raf-

Mek-Erk pathway. Although mechanistic details are incomplete, it has been suggested that the interaction of phosphorylated Sam68 with exon v5 blocks the repressive activity of hnRNP A1, either by preventing the interaction of hnRNP A1 with the ESS element (by steric hindrance), or by counteracting the inhibitory effect of hnRNP A1 bound to the ESS. Recently, Cheng and Sharp identified SRm160 as another Ras-regulated splicing factor responsible for inclusion of v5 exon in *CD44* transcripts [69]. Importantly, they also showed that silencing of SRm160 decreases cellular invasiveness, linking this splicing regulator to tumorigenesis.

Another good example of splicing modulation by signaling pathways comes from the *Fibronectin* gene (FN), an extra-cellular matrix component and a key determinant in controlling proliferation, migration, invasion and metastatic behavior of tumor cells. EDA is a FN splicing isoform generated by inclusion of a single exon (EDA exon, also known as EDI or EIIIA). This splicing isoform is poorly expressed in adult normal tissues, whereas it is present in embryos as well as during wound healing and in certain tumors. Inclusion of the EDA exon is triggered by the activation of the Ras/PI3-kinase/AKT pathway by growth factors. AKT directly phosphorylates the SR proteins SRSF7 (9G8) and SRSF1, which in turn bind to the EDA exon and promote its recognition by the splicing apparatus [24]. Activation of the same pathway by insulin regulates the activity of another SR protein, SRSF5 (SRp40), and stimulates the inclusion of an alternative exon in the protein kinase C (PKC) II pre-mRNA [70, 71]. Altogether these results support the possibility that deregulation of the Ras/PI3-kinase/AKT pathway, for instance resulting from mutations in its components, could have dramatic consequences on the splicing profile of any pre-mRNAs regulated by SRSF7, SRSF1, SRSF5 and perhaps other SR proteins. An attractive hypothesis is that exons responsive to this signaling pathway belong to a set of genes that function cooperatively to modulate the cell physiology in accordance with the biological role of the signaling molecule. The identification of these exons, therefore, will be of the utmost interest. An indication in favor of this hypothesis comes from the observation that AKT also promotes translation of EDA mRNAs bound by phosphorylated SRSF7 and SRSF1 [24]. Thus, activation of a single signal-transduction pathway controls in an integrated manner both splicing and translation of specific mRNAs and stimulates the production of specific proteins. The final effect is a drastic increase both in the speed and strength of the signaling response as measured by production of the induced protein.

Although these examples provide interesting insights into the effects of signal-transduction pathways on splicing regulation, the molecular characterization is still scanty and clarification of the complete pathway has not yet been provided. Thus, in addition to illustrate a role of signal transduction in splicing control, these cases also suggest the need of further studies to elucidate this important mechanism of gene regulation.

An increasing body of data implicates alternative splicing as a mechanism to control apoptosis or programmed cell death, an essential process in development and in maintenance of cellular homeostasis in multicellular organisms. Bcl-X is a member of the *Bcl*II family that directs mitochondrial breakdown during apoptosis. The usage of alternative 5' splice sites within exon 2 determines the production of two protein isoforms: a long antiapoptotic form (Bcl-XL) and a short apoptosis-promoting protein (Bcl-XS) [72]. Therefore, a shift in the splicing pattern of these transcripts can have deep effects on proliferative activity of cancer cells and on their response to proapoptotic therapies. Several splicing factors, including Sam68, SRSF1, hnRNP F/H and SAP155, contribute in controlling the choice between the two alternative 5'-splice sites. In particular, Sam68 over-expression promotes the production of pro-apoptotic Bcl-XS and this effect is reverted upon Sam68 phosphorylation [73]. On the contrary, hnRNP F/H proteins, by binding to a G-rich stretch element, promote the usage of the Bcl-XS - 5'splice site [74]. The ratio of Bcl-X splice variants contribute to determine the sensitivity of the cells to a wide variety of apoptotic agents and may have significance in drugs resistance and chemotherapeutic responsiveness. For example, the lipid ceramide, a mediator/regulator of apoptosis promotes the expression of the pro-apoptotic splicing variants Bcl-XS; the choice between the two alternative 5'-splice sites is controlled by a ceramide responsive element (CRCE 1) located within exon 2 and bound by SAP155 [75]. Moreover, ceramide is able to modulate the phosphorylation status of SR proteins in a phosphatase-1 (PP1)-dependent manner [75]. Interestingly, one of PP1 targets is the SR factor SRSF1, another major regulator of Bcl-X pre-mRNAs processing [76].

There are several examples of alternative splicing events that control the activity of proteins involved in cell motility and invasion, a pre-requisite for the formation of cancer metastases. This is the case of splicing isoforms of the *androgen* and *estrogen receptors* that are involved in mammary carcinomas [77, 78]. Interestingly, an isoform of estrogen receptor alpha, due to skipping of exon 3 (delta3ER), is a more potent activator of vascular endothelial growth factor than the wild-type receptor [79]. The splice variant Rac1b, which is generated by inclusion of a 57-nucleotide cassette exon, has been shown to lead to anchorage-independent cell growth. Notably, Rac1b is up-regulated in colorectal tumors at various stages of neoplastic progression, as compared to adjacent normal tissues [80]. Other examples, that clearly show the potential functional effect of aberrant splicing on tumorigenesis, are the *fibroblast growth factor receptor 2* (*FGFR2*) [81], the *fibronectin* [82] and the *survivin* [83].

Recently, we have used the *Ron* (recepteur d'origine nantais) proto-oncogene as a model to investigate the relationship between alternative splicing and tumor progression [40]. Ron, the human tyrosine kinase receptor for the macrophage stimulating protein (MSP), is a heterodimeric protein (p185-Ron) composed of α and β subunits both deriving from the processing of a common precursor. Binding to MSP stimulates the intrinsic tyrosine kinase activity of Ron and results in phosphorylation of its docking site for multiple transducer and adaptor proteins leading to the activation of signaling cascades (Fig. **3A**). Along with Met, the hepatocyte growth factor (HGF) receptor, Ron belongs to a subfamily of receptor tyrosine kinases (RTK) with unique expression patterns and biological activities. In addition to promoting cell growth and protection from apoptosis, these receptors control cell dissociation, motility, and invasion of extracellular matrices, a process known as "invasive growth" or "cell scattering" [84]. Invasive growth is physiologically relevant during development, organogenesis and tissue regeneration, but it is also important to mediate invasiveness and to promote malignant progression. Currently, six variants including RonΔ170, Δ165, Δ160, Δ155, Δ110, and Δ55 with various deletions or truncations in the extracellular or intracellular regions have been identified. All these variants are constitutively active but differ in their biochemical and biological properties (for review see [85]). Moreover, the splicing profile of the *Ron* gene is frequently altered in epithelial cancers, such as colon and breast cancers, suggesting that the production of multiple Ron isoforms could contribute to pathogenesis of these tumors [40, 86]. Over-expression of any of these isoforms increases cell motility (scatter-like activity). However, only RonΔ160 or RonΔ155 are able to induce focus formation, sustained anchorage-independent growth and the ability to form metastatic tumors in mice [86]. This oncogenic potential is channeled through the PI3-Kinase/AKT pathway [87]. Also Met transcripts undergo alternative splicing and an isoform, called Met-SM, originates from skipping of exon 14 which encodes a 47 aminoacid segment in the juxtamembrane domain. This isoform has been recently shown to play an important role in development and progression of human cancers [88]. Among the mechanisms controlling the expression of the different Ron isoforms in cancer cells, the switch from constitutive to alternative splicing plays the major role. Thus, the elucidation of the regulatory pathways controlling the splicing profile of *Ron* transcripts will shed new light on both cancer initiation and progression. We have studied in detail the alternative splicing event that leads to the production of *ΔRon* mRNA. This transcript lacks a 147-bp exon (exon 11). The encoded protein bears a 49-amino-acid in frame deletion in the extracellular domain that affects the proteolytic maturation and results in the accumulation of a single-chain pro-RonΔ165 in the cytoplasm. Moreover, the deleted protein is characterized by aberrant intracellular disulfide bridges that facilitate RonΔ165 oligomerization leading to constitutive phosphorylation and activation [89]. We have shown that the choice between inclusion and skipping of exon 11 is controlled by two adjacent regulatory elements, a silencer and an enhancer, both located in the constitutive exon 12 (Fig. **3B**) [40]. The binding of the SR protein SRSF1 to ESE stimulates skipping of exon 11 and the production of ΔRon. Thus, over-expression of SRSF1 in cells that normally express Ron triggers the production of the RonΔ165 with dramatic consequences on the cell properties. Indeed, similarly to what observed after ΔRon over-expression, an increased expression of SRSF1 profoundly affects cell morphology and triggers nuclear accumulation of β-catenin, reorganization of actin cytoskeleton, and down-regulation of E-cadherin, a tumor and invasion suppressor in human carcinomas. All these morphological and molecular changes represent hallmarks of the epithelial to mesenchymal transition (EMT), which is implicated in the metastatic spreading of human carcinomas (Fig. **3B**) [90]. Notably, knockdown of SRSF1 by RNA interference (RNAi) reduces the levels of ΔRon and concomitantly decreases cell motility. Similarly, RNAi specifically directed against *ΔRon* reduces cell motility and partially reverts the morphological changes induced by SRSF1 over-expression.

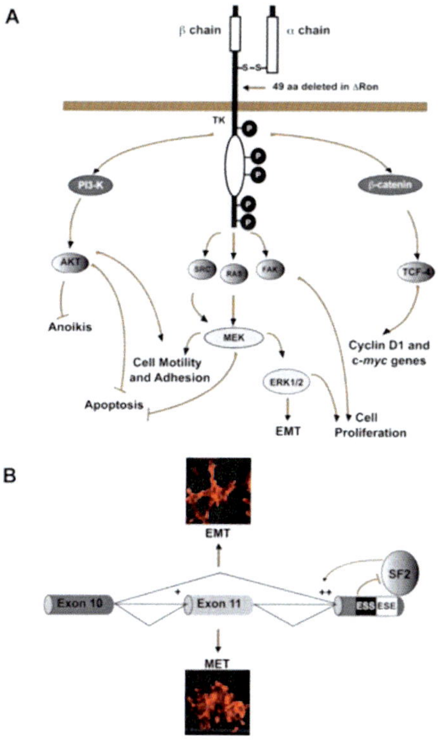


**Figure 3:** Schematic representation of the Ron receptor and of the downstream signalling pathways. (**A**) Ron is a single-pass, disulfide-linked $\alpha/\beta$ heterodimer. The $\alpha$ chain is an extracellular glycoprotein while the $\beta$ chain is a transmembrane subunit that comprises an extra-cellular sequence, a short *trans*-membrane segment, a large cytoplasmic portion with intrinsic tyrosine kinase domain (TK) and a C-terminal tail. Residues and domains important for the Ron activity are indicated. In particular, the intracellular domain includes the tyrosine kinase catalytic site (white oval) flanked by distinctive juxtamembrane and carboxy-terminal sequences. Phosphorylation of two tyrosines within the kinase domain positively regulates the enzyme activity, whereas a serine residue in the juxtamembrane domain has a negative regulatory role. Two tyrosine residues in the carboxy-terminal region, when phosphorylated, form a specific docking site for multiple signal transducers and adaptors. Activation of Ron by MSP can initiate signaling through many pathways implicated in tumor progression and metastasis, such adhesion, invasion, mobility, proliferation and inhibition of apoptosis. (**B**) The constitutively active ΔRon isoform is generated through skipping of exon 11. This event is controlled by two adjacent splicing elements, a silencer and an enhancer, located in the central part and at the 3' end of exon 12, respectively. These two regulatory elements may form a ''control cassette'' that tunes the strength of the acceptor site of intron 11 and thus the ratio between Ron and ΔRon transcripts. Splicing factor SF2 (SRSF1) directly binds to the enhancer and governs its activity. High levels of SF2 (SRSF1) increase the strength of the acceptor site of the intron 11, promote the production of ΔRon isoform and trigger morphological and molecular changes typical of the epithelial to mesenchymal transition (EMT). The primary function of the silencer could be to antagonize the enhancer and prevent exon skipping.

Given the reported up-regulation of several SR proteins, including SRSF1, during tumor progression [27, 60], it is tempting to speculate that splicing factor SRSF1 could promote the malignant transformation by inducing a ΔRon-mediated EMT. This hypothesis is consistent with a recent report showing that SRSF1 behaves as a *bona fide* proto-oncogene [60]. SRSF1 is up-regulated in a set of human tumors and, the gene coding for SRSF1 is specifically amplified in some breast tumors but not in normal breast tissue from the same patient. Limited over-expression of SRSF1, comparable to that in cancer samples, resulted in tumor formation in nude mice. Notably, several endogenous splicing targets of SRSF1, among them a novel oncogenic isoform of the mTOR substrate, S6K1, are essential for SRSF1-mediated transformation. In addition, RNA interfering (RNAi) of SRSF1 or the oncogenic S6K1 isoform, resulted in reversion of the transformed phenotype [60].

The identification of splicing regulators as a novel class of proto-oncogenes may unveil novel aspects of cancer biology and offer new opportunities for diagnosis and therapy.

Although all the examples reported above unquestionably prove that changes in the splicing activity of tumor cells is linked tumor progression, yet the molecular mechanisms responsible for these changes remain undefined. Several questions still deserve investigation. Which are the events that lead to deregulation of expression and/or activity of members of SR and hnRNP families? Is there a generalized reorganization of the splicing machinery in cancers cells or does the splicing switch result from a deregulation of a subset of splicing regulators? How do signal transduction pathways cooperate to affect splicing profiles to determine malignant conversion?

## ALTERNATIVE SPLICING SIGNATURES AS POTENTIAL DIAGNOSTIC/PROGNOSTIC INDICATORS AND PROSPECT FOR THERAPIES

In the previous sections we have discussed several examples that illustrate a causative link between alternative splicing and development of neoplasia. It is worth noticing that in many cases splicing isoforms appear to be cancer-specific [8]. Moreover, there are evidences that in different tissues a different set of genes undergoes changes of splicing profiles during tumorigenesis [91]. Thus, in kidney tumors, *Tropomyosin 1*, *Actinin α1*, *Integrin β4*, *Catenin δ1* and *Fibroblast growth factor receptor 2* change their alternative splicing patterns, whereas *N-glycanase1* does not, despite its apparent alterations in lung and uterus tumors. These findings argue against the idea that changes in splicing profiles may result from a generalized alteration of splicing machinery in cancers cells and suggest that only a specific set of splicing regulators are affected by the tumorigenesis in a tissue-specific manner. The identification of cancer-specific, tissue-specific alternatively spliced isoforms immediately incited their potential use as diagnostic, prognostic, or predictive biomarkers [8]. Although large-scale studies have yet to be carried out, initial results are promising. An important correlation between aberrant alternative splicing and tumor progression has been shown for *CD44* [63]. In particular, the isoforms containing the variant exon v6 are frequently upregulated in head and neck carcinomas or in advanced stage gastric cancers and *CD44v6* positive tumors are associated to poor prognosis [92]. Also, aberrant alternative splicing of *MDM2* transcripts were found to associate with a prognostic factor for poor survival in patients with breast cancer [93], whereas *HDMX* has been linked to soft-tissue sarcoma [94]. Furthermore, altered expression of a spliced variant of *p73*, a p53-related protein, seems to be a negative prognostic marker in patients with neuroblastoma [95].

During the past decade classical gene expression profiling performed by microarray has been a powerful tool for the discovery of new cancer biomarkers. However, most array platforms developed to date are not designed to distinguish mRNA isoforms. An important direction for the future will be the application of genome-wide screens designed for the analysis of alternative splicing signatures associates with cancer but virtually absent in normal cells. These splicing sensitive microarrays have the potential to better identify tumor sub-classes. Moreover, they do not require that cause–effect relationships between splicing profiles and cancers are identified (or even that they existed!!), but only that this associations are sufficiently consistent to be predictive. Interestingly, Li *et al.* [96] have used an approach to simultaneously quantify changes in splicing ("splicing switch") and transcript abundance (up- and down-regulation). The advantage of this "two-dimensional'' profiling strategy is that it works with partially degraded biological samples such as RNA derived from tissue blocks that have been formalin-fixed and paraffin-embedded. These authors identified a specific set of mRNA isoform biomarkers for prostate cancer using independent panels of tissue samples. In particular, they reported two variants of *AMACR* gene (which encodes a-methylacyl-CoA racemase) resulting from the alternative use of the last exon coupled with alternative polyadenylation: one isoform showed a quantitative up-regulation in prostate cancer compared with normal prostatic tissues, the other seemed to be expressed only in prostate cancer. More recently, novel microarray technology has been used to measure whole-genome exon expression in 102 normal and cancer tissue samples of different stages from colon, urinary bladder, and prostate [97]. Seven genes with tumor-specific splice variants were identified (*Alpha-actinin 1*, *Caldesmon*, *Collagen type VI alpha 3*, *Leucine-rich repeat interacting protein 2*, *Phosphatidylinositol-4 kinase catalytic beta polypeptide*, *Tropomyosin 1* and *Vinculin*). Notably, tumor-specific splicing patterns of *Alpha-actinin 1*, *caldesmon* and *Vinculin*, key cytoskeleton components, were found in all three organs suggesting that they may represent general cancer related splicing events. Moreover, *in silico* protein analysis predicts that these identified cancer-specific splice variants encode proteins with potentially altered functions, indicating that they may be involved in pathogenesis and hence represent novel potential biomarkers and drug targets [97].

Cancer-specific splice variants may not only serve as diagnostic and prognostic tumor markers, but also provide potential targets for the development of new therapeutic strategies (Fig. **4**). One promising avenue towards the development of more selective anticancer drugs consists in targeted delivery of bioactive compounds to the tumor by means of molecules (for example antibodies) that are specific for tumor-associated markers. The accessibility as well as the restricted pattern of expression of the antigene is an important criterion in the selection of a target for bio-molecular intervention. Notably, many receptors mediating cell-cell and cell-matrix interactions are regulated by alternative splicing and specific alternatively spliced variants of these molecules are associated with many human malignancies [8, 9]. Interestingly, nearly 90% of alternative splicing events affect regions located on the exposed surface of protein [98] and for many cell membrane molecules, alternative exons encode novel epitopes that are usually found in the extracellular domain of the protein [8, 9]. These epitopes seem to be ideally suited for tumor-targeting strategies. Once more, the CD44 protein, discussed in detail in the previous section, provides a remarkable example since radiolabeled antibodies directed against the CD44-v6 isoform appear to offer a promising therapeutic tool and they were used in experimental clinical trials for treatment of head and neck cancer [99]. However, despite these promising developments, solid tumors are frequently relatively resistant to antibody-based therapies [100]. This is due, in part, to the relative inaccessibility of tumor cells and to the poor penetration of antibodies into the tumor tissue. Since tumor cells are separated from the blood by endothelial cells and extracellular matrix components surrounding the vasculature, tumor uptake is highly limited by the antibody's ability to cross this layer. Importantly, therapeutic approaches directed against tumor neo-vasculature have gained significance for a number of reasons: (i) tumor vascular endothelial cells are readily accessible for drugs *via* the blood circulation; (ii) there is growing evidence that inhibition or regression of tumor vessels leads to cancer cell death since they rely on blood vessels for nutrients and oxygen to satisfy their metabolic needs [101]. An interesting example is the EDB alternatively spliced isoform of Fibronectin (FN). The EDB isoform is generated through inclusion of a single exon encoding a 91 amino acids region and it is present in neoplastic tissue blood vessels while it is virtually absent in mature/adult tissues. EDB is involved in the regulation of endothelial cell proliferation and vascular morphogenesis; an EDB–specific radiolabeled antibody was used in clinical trials for anti-angiogenic cancer treatment [102]. Recently, it has been shown that the alternatively spliced extra-domain A (EDA) of fibronectin may represent another, equally attractive, target for the antibody-based delivery of bioactive agents to the neovasculature not only of solid tumors, but also of metastatic lesions [103]. Similarly, splicing variants could represent novel leukemia-specific antigens with potential use in immunotherapeutic approaches. In Ph-positive leukemias, the t(9;22) translocation generates the Bcr/Abl fusion proteins, whose constitutive tyrosine kinase activation is responsible for the appearance of the leukemia phenotype [104]. In addition to the main *BCR/ABL* fusion transcripts, it has been reported that *BCR/ABL* transcripts, arising from alternative splicing, are also produced in a high percentage of chronic myelogenous leukemia (CML) and acute lymphoblastic leukemia (ALL) patients

[105]. In particular, *BCR/ABL* alternative transcripts involving ABL exon 4 are very attractive because the resulting fusion proteins contain a novel immunogenic epitope that is able to elicit an antigen-specific T-cell response [105].

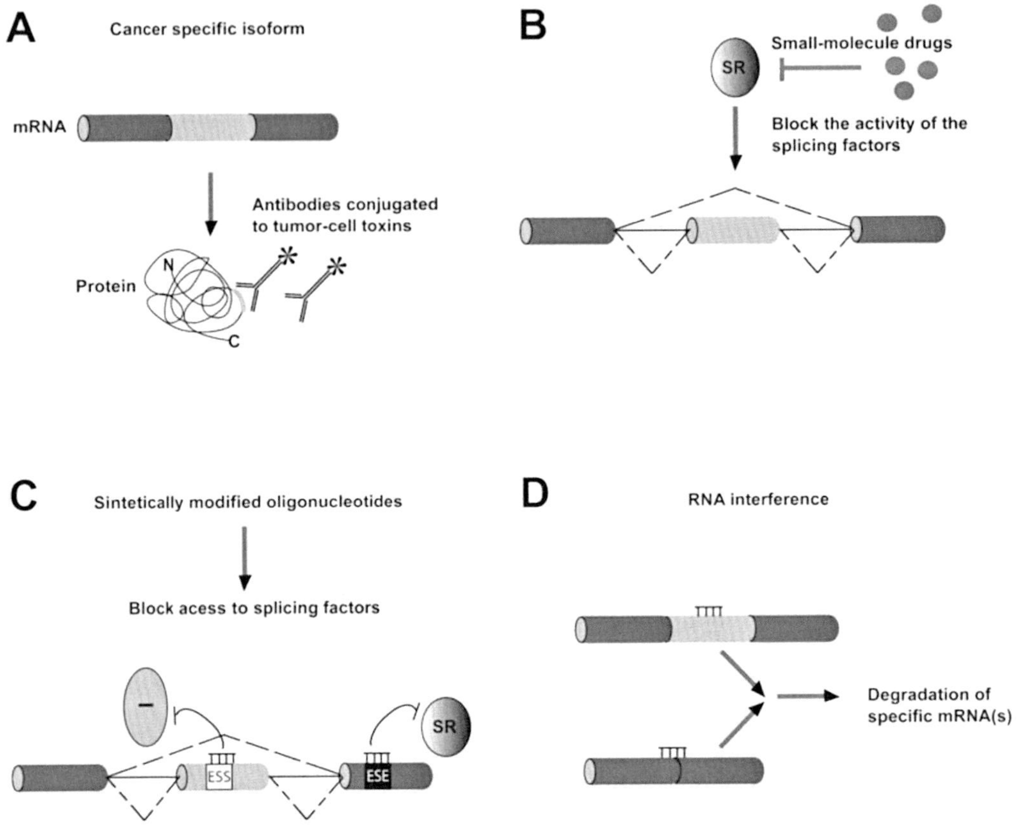


**Figure 4:** Potential use of therapeutic approaches that target alternative splicing. Various strategies are currently being used to exploit alternative splicing for treatment of cancer. (**A**) In many cases, cancer-restricted splice variants contain unique epitopes that could be used as the targets for specific antibodies conjugated to tumor-cell toxins. (**B**) Many chemical compounds have been found to affect splicing of numerous genes. Although the mechanisms by which splicing patterns are altered are still poorly understood, several compounds are able to block the activity of SR-protein kinases (SRpKs) and, consequently, reduce the phosphorylation state of SR splicing factors. (**C**) Synthetically modified oligonucleotides, such as phosphorothioate, morpholino phosphorodiamide and 2′-O-methylethyl, are able to block the interaction of the spliceosome machinery with a specific site. Moreover, they are more stable and active than regular nucleotide backbones and display low toxicity *in vivo*. (**D**) Cancer-specific mRNA transcripts that contain unique sequences can be targeted using RNAi mediated degradation.

Innovative therapeutic approaches could also be designed to target the splicing machinery. For example, post-translational modifications of splicing regulators may be relevant for tumor progression. SRPK1 is up-regulated in breast and colonic tumors compared with adjacent normal epithelium, and the levels of the kinase increase along with tumor grade [106]. Strong expression of SRPK1 protein was also evident in the majority of breast and colonic tumor cell lines [106]. Interestingly, down-regulation of SRPK1 using small interfering RNA (siRNA) affects the expression of key apoptotic factors BAX and BCL2, increases the proportion of the tumor cells (but not of nontrasformed cells) undergoing apoptosis and amplifies their sensitivity to two commonly used chemotherapeutic agents such as gemcitabine and cisplatin [106, 107]. These results suggest that pharmacological inhibition of SRPK1 activity may be effective as stand-alone agent or in combination with conventional chemotherapeutic regimens. Moreover, small molecules with SRPK-blocking activity, such as Clk1/Sty inhibitors [108] or potent inhibitors of the splicing reaction, such as indole derivatives [109], can be used to alter splicing patterns. Interestingly, recent data showed that the spliceosomal protein E (SmE) functions as suppressor of tumor cell growth in p53-independent manner [110]. Thus, considering that a large portion of human cancers are defective in p53 activity, targeting SmE in cancer gene therapy will be of particular interest.

Different therapeutic approaches are currently being explored to correct aberrant alternative splicing at the RNA level. These strategies can either alter alternative splicing patterns of specific genes or recognize particular mRNA species to elicit their degradation (for example, RNAi, microRNA and ribozymes). Correction of splicing defects through manipulation of mRNA requires the use of gene-therapy strategies, which are often connected with problems such as delivery, toxicity and immunogenicity. SiRNA-based therapy has shown great promise for many diseases such as cancer. Major targets for siRNA therapy include oncogenes and genes that are involved in angiogenesis, metastasis, cell survival, antiapoptosis and resistance to chemotherapy [111]. Moreover, the latest generations of antisense oligonucleotides that contain chemical modifications appear more stable compared to conventional nucleotide backbones. Some of these oligos prevent ribosomal assembly, and hence mRNA translation, and seem to be well tolerated in patients [112]. In addition, synthetically modified oligonucleotides targeted through sequence specific hybridization to splice sites or adjacent sequences may block inappropriate exon selection by inhibiting the binding of *trans*-acting factors. They have been successfully applied to correct aberrant splicing of *β-globin*, *CFTR*, *dystrophin*, *tau genes* [113] and to repress the generation of cancer-related splice variants of *FGFR1*, *Bcl-X*, and *MDM2*. Alone or in combination with chemotherapeutic agents, such as cisplatin, irinotecan, paclitaxel, fluorouracil they may proficiently used in therapy [114]. Finally, our lab has recently showed how cancer-associated alternatively spliced variants could be used to develop more selective anti-cancer therapeutic strategies. In the case of *Ron* gene, we have successfully used two different strategies to induce inclusion of exon 11 and reduce the expression of *ΔRon* transcript: i) a TOES (Targeted Oligonucleotide Enhancers of Splicing) oligo complementary to the first part of exon 11 and with an un-complementary tail containing an ESE element able to stimulate exon inclusion and ii) small molecules, such as indole-derived compounds (IDCs), to selectively inhibit the SRSF1 activity. Interestingly, indole derivatives can be exploited as new opportunities for pharmacological therapy since they also affect the invasive phenotype of the cells [115].

## CONCLUDING REMARKS

Following the sequencing of entire eukaryotic genomes, alternative splicing has emerged as the strategy dedicated to the interpretation of the genetic information. As widely discussed in this review, in fact, this mechanism of pre-mRNA processing accounts for a large proportion of the proteomic diversity in higher eukaryotes. While transcription regulation determines the presence and abundance of primary gene transcripts, alternative splicing is designed to manipulate the message in response to a number of stimuli. In this way it may tune protein features, such as the ability to participate to hetero-complexes or the sub-cellular localization, in order to meet the needs of the cells. Rarely are alternative splicing programs associated to "yes or no" decisions; in most of the cases splicing products co-exist in a single cell and contribute to the cell ability to cope with a number of different internal and external stimuli. From this viewpoint, therefore, this mechanism appears to be intimately connected to the complexity of metazoan organisms which, once more, underscores the necessity to understand the combinatorial rules underlying splice site choices as an obligatory step to reach a comprehensive picture of the logic of living matter.

Deciphering the biological implications of this regulatory process is a challenging task that requires the integration of a large body of data relative to the *cis*-acting sequences and *trans*-acting factors involved, the regulatory circuits and signaling pathways that modulate splice site decisions and finally the physiological and/or pathological consequences resulting from the expression of the various splicing isoforms. The attainment of this result will be certainly hastened by the recent development of new and highly sophisticated bio-informatics and system biology approaches.

An important breakthrough to fully understand how alternative splicing permeates the expression programs is expected to derive from the application of high-throughput methodologies. During the last decade, biomedical sciences have been strongly influenced by "omics": genomics, proteomics, transcriptomic and metabolomic. The application of high-throughput approaches to study splicing profiles is just at its beginning. This delay is certainly due to a number of technical problems deriving from the necessity to contemporary assess for each gene the absolute level of total transcripts and the relative abundance of any splicing isoforms. Moreover, the interpretation is significantly complicated by the necessity to understand the physiological implications resulting from a change in splicing profiles.

However, it is easy to predict that this type of analysis will provide crucial information to unveil the regulatory pathways underlying co-regulation of splicing profiles and to understand the relevance of alternative splicing in the context of the organism development. The hope is that splicing-sensitive arrays will guide the identification of circuits that, similarly to signal transduction and transcription pathways, may be causatively linked to development programs, organogenesis, body plan definition and cell identity. Another open question concerns the identification of the sub-genome that does not undergo alternative splicing events. Do these genes identify any particular critical cellular function?

Thus, the change of perspective, from detailed characterization of molecular mechanisms to global approaches, it is expected to improve our understanding of the physiological relevance of alternative splicing and to drastically increase the comprehension of important physiological and pathological conditions such as the neuronal plasticity and the complexity of cancer. We are confident that system biology approaches can help the identification of alternative splicing events that may play a critical role in tumor progression. This will offer the opportunity to develop innovative strategies for therapeutical intervention that target specific alternative splicing variants.

## CONFLICT OF INTEREST

The author(s) confirm that this chapter content has no conflict of interest.

## ACKNOWLEDGEMENTS

This work was supported by grants from the Associazione Italiana per la Ricerca sul Cancro (AIRC), Progetto di interesse CNR-MIUR "Invecchiamento" to G.B. and grants from the Associazione Italiana per la Ricerca sul Cancro (AIRC project number: 11913), the *Association for International Cancer Research* (AICR-UK) to C.G.

## REFERENCES

[1] Goodstadt, L., Ponting, C.P. Phylogenetic reconstruction of orthology, paralogy, and conserved synteny for dog and human. PLoS Comput. Biol. 2006, 2: e133.

[2] Stein, L.D., Bao, Z., Blasiar, D., Blumenthal, T., Brent, M.R., Chen, N., Chinwalla, A., Clarke, L., Clee, L., Coghlan, A., Coulson, A., D'Eustachio, P., Fitch, D.H., Fulton, L.A., Fulton, R.E, Griffiths-Jones, S., Harris, T.W., Hillier, L.W., Kamath, R., Kuwabara, P.E., Mardis, E.R., Marra, M.A., Miner, T.L., Minx, P., Mullikin, J.C., Plumb, R.W., Rogers, J., Schein, J.E., Sohrmann, M., Spieth, J., Stajich, J.E., Wei, C., Willey, D., Wilson, R.K., Durbin, R., Waterston, R.H. The genome sequence of Caenorhabditis briggsae: a platform for comparative genomics. PLoS Biol. 2003, 1: E45.

[3] Blencowe, B.J. Alternative splicing: new insights from global analyses. Cell. 2006, 126: 37-47.

[4] Graveley, B.R. Mutually exclusive splicing of the insect Dscam pre-mRNA directed by competing intronic RNA secondary structures. Cell. 2005, 123: 65-73.

[5] Wang, E.T., Sandberg, R., Luo, S., Khrebtukova, I., Zhang, L., Mayr, C., Kingsmore, S.F., Schroth, G.P., Burge, C.B. Alternative isoform regulation in human tissue transcriptomes. Nature 2008, 456: 470-476.

[6] Wang, G.S., Cooper, T.A. Splicing in disease: disruption of the splicing code and the decoding machinery. Nat. Rev. Genet. 2007, 8: 749-761.

[7] Srebrow, A., Kornblihtt, A.R. The connection between splicing and cancer. J. Cell Sci. 2006, 119: 2635-2641.

[8] Brinkman, B.M. Splice variants as cancer biomarkers. Clin. Biochem. 2004, 37: 584-594.

[9] Venables, J.P. Aberrant and alternative splicing in cancer. Cancer Res. 2004, 64: 7647-7654.

[10] Maquat, L.E., Carmichael, G.G. Quality control of mRNA function. Cell. 2001, 104: 173-176.

[11] Green, R.E., Lewis, B.P., Hillman, R.T., Blanchette, M., Lareau, L.F., Garnett, A.T., Rio, D.C., Brenner, S.E. Widespread predicted nonsense-mediated mRNA decay of alternatively-spliced transcripts of human normal and disease genes. Bioinformatics. 2003, 19 Suppl 1: i118-121.

[12][1] Ni, J.Z., Grate, L., Donohue, J.P., Preston, C., Nobida, N., O'Brien, G., Shiue, L., Clark, T.A., Blume, J.E., Ares, M. Jr. Ultraconserved elements are associated with homeostatic control of splicing regulators by alternative splicing and nonsense-mediated decay. Genes Dev. 2007, 21: 708-718.

[13] Black, D.L.. Mechanisms of alternative pre-messenger RNA splicing. Annu. Rev. Biochem. 2003, 72: 291-336.

[14] Cartegni, L., Chew, S.L., Krainer, A.R. Listening to silence and understanding nonsense: exonic mutations that affect splicing. Nat. Rev. Genet. 2002, 3: 285-298.

[15] Reddy, A.S. Alternative splicing of pre-messenger RNAs in plants in the genomic era. Annu. Rev. Plant Biol. 2007, 58: 267-294.

[16] Hastings, M.L., Krainer, A.R. Pre-mRNA splicing in the new millennium. Curr. Opin. Cell Biol. 2001, 13: 302-309.

[17] Shen, H., Green, M.R. RS domains contact splicing signals and promote splicing by a common mechanism in yeast through humans. Genes Dev. 2006, 20: 1755-1765.

[18] Xiao, S.H., Manley, J.L. Phosphorylation-dephosphorylation differentially affects activities of splicing factor ASF/SF2. Embo J. 1998, 17: 6359-6367.

[19] Gui, J.F., Lane, W.S., Fu, X.D. A serine kinase regulates intracellular localization of splicing factors in the cell cycle. Nature. 1994, 369: 678-682.

[20] Colwill, K., Pawson, T., Andrews, B., Prasad, J., Manley, J.L., Bell, J.C., Duncan, P.I. The Clk/Sty protein kinase phosphorylates SR splicing factors and regulates their intranuclear distribution. Embo J. 1996, 15: 265-275.

[21] Hernandez, F., Perez, M., Lucas, J.J., Mata, A.M., Bhat, R., Avila, J. Glycogen synthase kinase-3 plays a crucial role in tau exon 10 splicing and intranuclear distribution of SC35. Implications for Alzheimer's disease. J. Biol. Chem. 2004, 279: 3801-3806.

[22] Ko, T.K., Kelly, E., Pines, J. CrkRS: a novel conserved Cdc2-related protein kinase that colocalises with SC35 speckles. J. Cell Sci. 2001, 114: 2591-2603.

[23] Rossi, F., Labourier, E., Forne, T., Divita, G., Derancourt, J., Riou, J.F., Antoine, E., Cathala, G., Brunel, C., Tazi, J. Specific phosphorylation of SR proteins by mammalian DNA topoisomerase I. Nature. 1996, 381: 80-82.

---

[1]Ultraconserved elements are genomic sequence elements longer than 200 bp and identical in sequence in the human, mouse, and rat genomes. These elements are frequent in alternatively spliced exons of genes encoding RNA binding proteins. They are involved in unproductive splicing events that result in the inclusion of conserved stop codon-containing exons which renders the transcripts sensitive to NMD.

[24] Blaustein, M., Pelisch, F., Tanos, T., Munoz, M.J., Wengier, D., Quadrana, L., Sanford, J.R., Muschietti, J.P., Kornblihtt, A.R., Caceres, J.F., Coso, O.A., Srebrow, A. Concerted regulation of nuclear and cytoplasmic activities of SR proteins by AKT. Nat. Struct. Mol. Biol. 2005, 12: 1037-1044.

[25] Cartegni, L., Wang, J., Zhu, Z., Zhang, M.Q., Krainer, A.R. ESEfinder: A web resource to identify exonic splicing enhancers. Nucleic Acids Res. 2003, 31: 3568-3571.

[26] Caceres, J.F., Stamm, S., Helfman, D.M., Krainer, A.R. Regulation of alternative splicing *in vivo* by overexpression of antagonistic splicing factors. Science. 1994, 265: 1706-1709.

[27] Zerbe, L.K., Pino, I., Pio, R., Cosper, P.F., Dwyer-Nield, L.D., Meyer, A.M., Port, J.D., Montuenga, L.M., Malkinson, A.M. Relative amounts of antagonistic splicing factors, hnRNP A1 and ASF/SF2, change during neoplastic lung growth: implications for pre-mRNA processing. Mol. Carcinog. 2004, 41: 187-196.

[28] Vickers, S.M., Huang, Z.Q., MacMillan-Crow, L., Greendorfer, J.S., Thompson, J.A. Ligand activation of alternatively spliced fibroblast growth factor receptor-1 modulates pancreatic adenocarcinoma cell malignancy. J. Gastrointest. Surg. 2002. 6: 546-553.

[29] Luqmani, Y.A., Mortimer, C., Yiangou, C., Johnston, C.L., Bansal, G.S., Sinnett, D., Law, M., Coombes, R.C. Expression of 2 variant forms of fibroblast growth factor receptor 1 in human breast. Int. J. Cancer. 1995, 64: 274-279.

[30] Yamaguchi, F., Saya, H., Bruner, J.M., Morrison, R.S. Differential expression of two fibroblast growth factor-receptor genes is associated with malignant progression in human astrocytomas. Proc. Natl. Acad. Sci. USA. 1994, 91: 484-488.

[31] Jin, W., Cote, G.J. Enhancer-dependent splicing of FGFR1 alpha-exon is repressed by RNA interference-mediated down-regulation of SRp55. Cancer Res. 2004, 64: 8901-8905.

[32] Caceres, J.F., Screaton, G.R., Krainer, A.R. A specific subset of SR proteins shuttles continuously between the nucleus and the cytoplasm. Genes Dev. 1998, 12: 55-66.

[33][2] Michlewski, G., Sanford, J.R., Caceres, J.F. The splicing factor SF2/ASF regulates translation initiation by enhancing phosphorylation of 4E-BP1. Mol. Cell. 2008, 30: 179-189.

[34] van der Houven van Oordt, W., Diaz-Meco, M.T., Lozano, J., Krainer, A.R., Moscat, J., Caceres, J.F. The MKK(3/6)-p38-signaling cascade alters the subcellular distribution of hnRNP A1 and modulates alternative splicing regulation. J. Cell Biol. 2000, 149: 307-316.

[35] Biamonti, G. Nuclear stress bodies: a heterochromatin affair? Nat. Rev. Mol. Cell Biol. 2004, 5: 493-498.

[36] Krawczak, M., Reiss, J., Cooper, D.N. The mutational spectrum of single base-pair substitutions in mRNA splice junctions of human genes: causes and consequences. Hum. Genet. 1992, 90: 41-54.

[37] Ars, E., Serra, E., Garcia, J., Kruyer, H., Gaona, A., Lazaro, C., Estivill, X. Mutations affecting mRNA splicing are the most common molecular defects in patients with neurofibromatosis type 1. Hum. Mol. Genet. 2000, 9: 237-47.

[38] Lamolle, G., Marin, M., Alvarez-Valin, F. Silent mutations in the gene encoding the p53 protein are preferentially located in conserved amino acid positions and splicing enhancers. Mutat. Res. 2006, 600: 102-112.


[2]These authors suggest a novel mechanism for the activation of translation initiation of a subset of mRNAs bound by splicing factor SF2/ASF. This activity is mediated by interactions of SF2/ASF with both mTOR and phosphatase PP2A.

[39] Mazoyer, S., Puget, N., Perrin-Vidoz, L., Lynch, H.T., Serova-Sinilnikova, O.M., Lenoir, G.M. A BRCA1 nonsense mutation causes exon skipping. Am. J. Hum. Genet. 1998, 62: 713-715.

[40] Ghigna, C., Giordano, S., Shen, H., Benvenuto, F., Castiglioni, F., Comoglio, P.M., Green, M.R., Riva, S., Biamonti, G. Cell motility is controlled by SF2/ASF through alternative splicing of the Ron protooncogene. Mol. Cell. 2005, 20: 881-890.

[41] Mills, A.A. p53: link to the past, bridge to the future. Genes Dev. 2005, 19: 2091-2099.

[42] ElSharawy, A., Manaster, C., Teuber, M., Rosenstiel, P., Kwiatkowski, R., Huse, K., Platzer, M., Becker, A., Nurnberg, P., Schreiber, S., Hampe, J. SNPSplicer: systematic analysis of SNP-dependent splicing in genotyped cDNAs. Hum. Mutat. 2006, 27: 1129-1134.

[43] Neklason, D.W., Solomon, C.H., Dalton, A.L., Kuwada, S.K., Burt, R.W. Intron 4 mutation in APC gene results in splice defect and attenuated FAP phenotype. Fam. Cancer. 2004, 3: 35-40.

[44] Moisio, A.L., Jarvinen, H., Peltomaki, P. Genetic and clinical characterisation of familial adenomatous polyposis: a population based study. Gut. 2002, 50: 845-850.

[45] Lane, T.F. BRCA1 and transcription. Cancer Biol. Ther. 2004. 3: 528-533.

[46] Quinn, J.E., Kennedy, R.D., Mullan, P.B., Gilmore, P.M., Carty, M., Johnston, P.G., Harkin, D.P. BRCA1 functions as a differential modulator of chemotherapy-induced apoptosis. Cancer Res. 2003, 63: 6221-6228.

[47] Pettigrew, C., Wayte, N., Lovelock, P.K., Tavtigian, S.V., Chenevix-Trench, G., Spurdle, A.B., Brown, M.A. Evolutionary conservation analysis increases the colocalization of predicted exonic splicing enhancers in the BRCA1 gene with missense sequence changes and in-frame deletions, but not polymorphisms. Breast Cancer Res. 2005, 7: R929-939.

[48] Orban, T.I., Olah, E. Emerging roles of BRCA1 alternative splicing. Mol. Pathol. 2003, 56: 191-197.

[49] Orban, T.I., Olah, E. Expression profiles of BRCA1 splice variants in asynchronous and in G1/S synchronized tumor cell lines. Biochem. Biophys. Res. Commun. 2001, 280: 32-38.

[50] Colapietro, P., Gervasini, C., Natacci, F., Rossi, L., Riva, P., Larizza, L. NF1 exon 7 skipping and sequence alterations in exonic splice enhancers (ESEs) in a neurofibromatosis 1 patient. Hum. Genet. 2003, 113: 551-554.

[51] Fackenthal, J.D., Cartegni, L., Krainer, A.R., Olopade, O.I. BRCA2 T2722R is a deleterious allele that causes exon skipping. Am. J. Hum. Genet. 2002, 71: 625-631.

[52] Venables, J.P., Burn, J. EASI--enrichment of alternatively spliced isoforms. Nucleic Acids Res. 2006, 34: e103.

[53] Lee, S.H., Kim, H.Y., Kim, T.J., Park, H.K., Kim, W.H., Woo, K.M., Cho, M.H. Aberrant splicing of FHIT transcripts in human gastric cancer cell lines. Res. Commun. Mol. Pathol. Pharmacol. 2002, 112: 39-49.

[54] Hastings, M.L., Resta, N., Traum, D., Stella, A., Guanti, G., Krainer, A.R. An LKB1 AT-AC intron mutation causes Peutz-Jeghers syndrome *via* splicing at noncanonical cryptic splice sites. Nat. Struct. Mol. Biol. 2005, 12: 54-59.

[55] Sjoblom, T., Jones, S., Wood, L.D., Parsons, D.W., Lin, J., Barber, T.D., Mandelker, D., Leary, R.J., Ptak, J., Silliman, N., Szabo, S., Buckhaults, P., Farrell, C., Meeh, P., Markowitz, S.D., Willis, J., Dawson, D., Willson, J.K., Gazdar, A.F., Hartigan, J., Wu, L., Liu, C., Parmigiani, G., Park, B.H., Bachman, K.E., Papadopoulos, N., Vogelstein, B.,

Kinzler, K.W., Velculescu, V.E. The consensus coding sequences of human breast and colorectal cancers. Science. 2006, 314: 268-274.

[56] Bartel, F., Taubert, H., Harris, L.C. Alternative and aberrant splicing of MDM2 mRNA in human cancer. Cancer Cell. 2002, 2: 9-15.

[57][3] Filippov, V., Filippova, M., Duerksen-Hughes, P.J. The early response to DNA damage can lead to activation of alternative splicing activity resulting in CD44 splice pattern changes. Cancer Res. 2007, 67: 7621-7630.

[58] Chen, L.L., Sabripour, M., Wu, E.F., Prieto, V.G., Fuller, G.N. Frazier, M.L. A mutation-created novel intra-exonic pre-mRNA splice site causes constitutive activation of KIT in human gastrointestinal stromal tumors. Oncogene. 2005, 24: 4271-4280.

[59] Zinszner, H., Albalat, R., Ron, D. A novel effector domain from the RNA-binding protein TLS or EWS is required for oncogenic transformation by CHOP. Genes Dev. 1994, 8: 2513-2526.

[60][4] Karni, R., de Stanchina, E., Lowe, S.W., Sinha, R., Mu, D., Krainer, A.R. The gene encoding the splicing factor SF2/ASF is a proto-oncogene. Nat. Struct. Mol. Biol. 2007, 14: 185-193.

[61] Ghigna, C., Moroni, M., Porta, C., Riva, S., Biamonti, G. Altered expression of heterogenous nuclear ribonucleoproteins and SR factors in human colon adenocarcinomas. Cancer Res. 1998, 58: 5818-5824.

[62] Kan, Z., Garrett-Engele, P.W., Johnson, J.M., Castle, J.C. Evolutionarily conserved and diverged alternative splicing events show different expression and functional profiles. Nucleic Acids Res. 2005, 33: 5659-5666.

[63] Naor, D., Nedvetzki, S., Golan, I., Melnik, L., Faitelson, Y. CD44 in cancer. Crit. Rev. Clin. Lab. Sci. 2002, 39: 527-579.

[64] Stickeler, E., Kittrell, F., Medina, D., Berget, S.M. Stage-specific changes in SR splicing factors and alternative splicing in mammary tumorigenesis. Oncogene. 1999, 18: 3574-3582.

[65] Watermann, D.O., Tang, Y., Zur Hausen, A., Jager, M., Stamm, S., Stickeler, E. Splicing factor Tra2-beta1 is specifically induced in breast cancer and regulates alternative splicing of the CD44 gene. Cancer Res. 2006, 66: 4774-4780.

[66] Pind, M.T., Watson, P.H. SR protein expression and CD44 splicing pattern in human breast tumours. Breast Cancer Res. Treat. 2003, 79: 75-82.

[67] Galiana-Arnoux, D., Lejeune, F., Gesnel, M.C., Stevenin, J., Breathnach, R., Del Gatto-Konczak, F. The CD44 alternative v9 exon contains a splicing enhancer responsive to the SR proteins 9G8, ASF/SF2, and SRp20. J. Biol. Chem. 2003, 278: 32943-32953.

[68] Matter, N., Herrlich, P., Konig, H. Signal-dependent regulation of splicing *via* phosphorylation of Sam68. Nature. 2002, 420: 691-695.

[69] Cheng, C., Sharp, P.A. Regulation of CD44 alternative splicing by SRm160 and its potential role in tumor cell invasion. Mol. Cell Biol. 2006, 26: 362-370.

---

[3]This study demonstrates that the early response to DNA damage changes the expression levels of several SR proteins and, as a consequence, the alternative splicing profile of CD44 receptor. The results provide strong evidence that alternative splicing may play a role in regulating the stress response in the absence of the classical p53-dependent pathways.

[4] This study demonstrates that the early response to DNA damage changes the expression levels of several SR proteins and, as a consequence, the alternative splicing profile of CD44 receptor. The results provide strong evidence that alternative splicing may play a role in regulating the stress response in the absence of the classical p53-dependent pathways.

[70] Patel, N.A., Kaneko, S., Apostolatos, H.S., Bae, S.S., Watson, J.E., Davidowitz, K. K., Chappell, D.S., Birnbaum, M.J., Cheng, J.Q., Cooper, D.R. Molecular and genetic studies imply Akt-mediated signaling promotes protein kinase CbetaII alternative splicing *via* phosphorylation of serine/arginine-rich splicing factor SRp40. J. Biol. Chem. 2005, 280: 14302-14309.

[71] Li, H., Weinstein, I.B. Protein kinase C beta enhances growth and expression of cyclin D1 in human breast cancer cells. Cancer Res. 2006, 66: 11399-11408.

[72] Akgul, C., Moulding, D.A., Edwards, S.W. Alternative splicing of Bcl-2-related genes: functional consequences and potential therapeutic applications. Cell. Mol. Life Sci. 2004, 61: 2189-2199.

[73] Paronetto, M.P., Achsel, T., Massiello, A., Chalfant, C.E., Sette, C. The RNA-binding protein Sam68 modulates the alternative splicing of Bcl-x. J. Cell Biol. 2007, 176: 929-939.

[74] Dominguez, C., Allain, F.H. NMR structure of the three quasi RNA recognition motifs (qRRMs) of human hnRNP F and interaction studies with Bcl-x G-tract RNA: a novel mode of RNA recognition. Nucleic Acids Res. 2006, 34: 3634-3645.

[75] Massiello, A., Roesser, J.R., Chalfant, C.E. SAP155 Binds to ceramide-responsive RNA *cis*-element 1 and regulates the alternative 5' splice site selection of Bcl-x pre-mRNA. Faseb J. 2006, 20: 1680-1682.

[76] Massiello, A., Chalfant, C.E. SRp30a (ASF/SF2) regulates the alternative splicing of caspase-9 pre-mRNA and is required for ceramide-responsiveness. J. Lipid Res. 2006, 47: 892-897.

[77] Zhu, X., Daffada, A.A., Chan, C.M., Dowsett, M. Identification of an exon 3 deletion splice variant androgen receptor mRNA in human breast cancer. Int. J. Cancer. 1997, 72: 574-580.

[78] Gallacchi, P., Schoumacher, F., Eppenberger-Castori, S., Von Landenberg, E.M., Kueng, W., Eppenberger, U., Mueller, H. Increased expression of estrogen-receptor exon-5-deletion variant in relapse tissues of human breast cancer. Int. J. Cancer. 1998. 79: 44-48.

[79] Koduri, S., Goldhar, A.S., Vonderhaar, B.K. Activation of vascular endothelial growth factor (VEGF) by the ER-alpha variant, ERDelta3. Breast Cancer Res. Treat. 2006, 95: 37-43.

[80] Jordan, P., Brazao, R., Boavida, M.G., Gespach, C., Chastre, E. Cloning of a novel human Rac1b splice variant with increased expression in colorectal tumors. Oncogene. 1999, 18: 6835-6839.

[81] Oltean, S., Sorg, B.S., Albrecht, T., Bonano, V.I., Brazas, R.M., Dewhirst, M.W., Garcia-Blanco, M.A. Alternative inclusion of fibroblast growth factor receptor 2 exon IIIc in Dunning prostate tumors reveals unexpected epithelial mesenchymal plasticity. Proc. Natl. Acad. Sci. USA. 2006, 103: 14116-14121.

[82] Castellani, P., Viale, G., Dorcaratto, A., Nicolo, G., Kaczmarek, J., Querze, G., Zardi, L. The fibronectin isoform containing the ED-B oncofetal domain: a marker of angiogenesis. Int. J. Cancer. 1994, 59: 612-618.

[83] Ryan, B., O'Donovan, N., Browne, B., O'Shea, C., Crown, J., Hill, A.D., McDermott, E., O'Higgins, N., Duffy, M.J. Expression of survivin and its splice variants survivin-2B and survivin-DeltaEx3 in breast cancer. Br. J. Cancer. 2005, 92: 120-124.

[84] Trusolino, L., Comoglio, P.M. Scatter-factor and semaphorin receptors: cell signalling for invasive growth. Nat. Rev. Cancer. 2002, 2: 289-300.

[85] Lu, Y., Yao, H.P., Wang, M.H. Multiple variants of the RON receptor tyrosine kinase: biochemical properties, tumorigenic activities, and potential drug targets. Cancer Lett. 2007, 257: 157-164.

[86] Zhou, Y.Q., He, C., Chen, Y.Q., Wang, D., Wang, M.H. Altered expression of the RON receptor tyrosine kinase in primary human colorectal adenocarcinomas: generation of different splicing RON variants and their oncogenic potential. Oncogene. 2003, 22: 186-197.

[87] Xu, X.M., Zhou, Y.Q., Wang, M.H. Mechanisms of cytoplasmic {beta}-catenin accumulation and its involvement in tumorigenic activities mediated by oncogenic splicing variant of the receptor originated from Nantes tyrosine kinase. J. Biol. Chem. 2005, 280: 25087-25094.

[88] Lee, J.H., Gao, C.F., Lee, C.C., Kim, M.D., Vande Woude, G.F. An alternatively spliced form of Met receptor is tumorigenic. Exp. Mol. Med. 2006, 38: 565-573.

[89] Collesi, C., Santoro, M.M., Gaudino, G., Comoglio, P.M. A splicing variant of the RON transcript induces constitutive tyrosine kinase activity and an invasive phenotype. Mol. Cell. Biol. 1996, 16: 5518-5526.

[90] Thiery, J.P. Epithelial-mesenchymal transitions in tumour progression. Nat. Rev. Cancer. 2002, 2: 442-454.

[91] Okumura, M., Kondo, S., Ogata, M., Kanemoto, S., Murakami, T., Yanagida, K., Saito, A., Imaizumi, K. Candidates for tumor-specific alternative splicing. Biochem. Biophys. Res. Commun. 2005, 334: 23-29.

[92] Stauder, R., Eisterer, W., Thaler, J., Gunthert, U. CD44 variant isoforms in non-Hodgkin's lymphoma: a new independent prognostic factor. Blood. 1995, 85: 2885-2899.

[93] Lukas, J., Gao, D.Q., Keshmeshian, M., Wen, W.H., Tsao-Wei, D., Rosenberg, S., Press, M.F. Alternative and aberrant messenger RNA splicing of the mdm2 oncogene in invasive breast cancer. Cancer Res. 2001, 61: 3212-3219.

[94] Bartel, F., Schulz, J., Bohnke, A., Blumke, K., Kappler, M., Bache, M., Schmidt, H., Wurl, P., Taubert, H., Hauptmann, S. Significance of HDMX-S (or MDM4) mRNA splice variant overexpression and HDMX gene amplification on primary soft tissue sarcoma prognosis. Int. J. Cancer. 2005, 117: 469-475.

[95] Romani, M., Tonini, G.P., Banelli, B., Allemanni, G., Mazzocco, K., Scaruffi, P., Boni, L., Ponzoni, M., Pagnan, G., Raffaghello, L., Ferrini, S., Croce, M., Casciano, I. Biological and clinical role of p73 in neuroblastoma. Cancer Lett. 2003, 197: 111-117.

[96] Li, H.R., Wang-Rodriguez, J., Nair, T.M., Yeakley, J.M., Kwon, Y.S., Bibikova, M., Zheng, C., Zhou, L., Zhang, K., Downs, T., Fu, X.D., Fan, J.B. Two-dimensional transcriptome profiling: identification of messenger RNA isoform signatures in prostate cancer from archived paraffin-embedded cancer specimens. Cancer Res. 2006, 66: 4079-4088.

[97] Thorsen, K., Sorensen, K.D., Brems-Eskildsen, A.S., Modin, C., Gaustadnes, M., Hein, A.M., Kruhoffer, M., Laurberg, S., Borre, M., Wang, K., Brunak, S., Krainer, A.R., Torring, N., Dyrskjot, L., Andersen, C.L., Orntoft, T.F. Alternative splicing in colon, bladder, and prostate cancer identified by exon-array analysis. Mol. Cell. Proteomics. 2008, 7:1214-1224.

[98] Wang, P., Yan, B., Guo, J.T., Hicks, C., Xu, Y. Structural genomics analysis of alternative splicing and application to isoform structure modeling. Proc. Natl. Acad. Sci. USA. 2005, 102: 18920-18925.

[99] Borjesson, P.K., Postema, E.J., Roos, J.C., Colnot, D.R., Marres, H.A., van Schie, M.H., Stehle, G., de Bree, R., Snow, G.B., Oyen, W.J., van Dongen, G.A. Phase I therapy study with (186)Re-labeled humanized monoclonal antibody BIWA 4 (bivatuzumab) in patients with head and neck squamous cell carcinoma. Clin. Cancer Res. 2003, 9: 3961S-3972S.

[100] Jain, R.K. Delivery of molecular and cellular medicine to solid tumors. J. Control Release. 1998, 53: 49-67.

[101] Carmeliet, P. Angiogenesis in life, disease and medicine. Nature. 2005, 438: 932-936.

[102] Berndorff, D., Borkowski, S., Moosmayer, D., Viti, F., Muller-Tiemann, B., Sieger, S., Friebe, M., Hilger, C.S., Zardi, L., Neri, D., Dinkelborg, L.M. Imaging of tumor angiogenesis using 99mTc-labeled human recombinant anti-ED-B fibronectin antibody fragments. J. Nucl. Med. 2006, 47: 1707-1716.

[103][5] Rybak, J.N., Roesli, C., Kaspar, M., Villa, A., Neri, D. The extra-domain A of fibronectin is a vascular marker of solid tumors and metastases. Cancer Res. 2007, 67: 10948-10957.

[104] Deininger, M.W., Goldman, J.M., Melo, J.V. The molecular biology of chronic myeloid leukemia. Blood. 2000, 96: 3343-3356.

[105] Volpe, G., Cignetti, A., Panuzzo, C., Kuka, M., Vitaggio, K., Brancaccio, M., Perrone, G., Rinaldi, M., Prato, G., Fava, M., Geuna, M., Pautasso, M., Casnici, C., Signori, E., Tonon, G., Tarone, G., Marelli, O., Fazio, V.M., Saglio, G. Alternative BCR/ABL splice variants in Philadelphia chromosome-positive leukemias result in novel tumor-specific fusion proteins that may represent potential targets for immunotherapy approaches. Cancer Res. 2007, 67: 5300-5307.

[106] Hayes, G.M., Carrigan, P.E., Miller, L.J. Serine-arginine protein kinase 1 overexpression is associated with tumorigenic imbalance in mitogen-activated protein kinase pathways in breast, colonic, and pancreatic carcinomas. Cancer Res. 2007, 67: 2072-2080.

[107] Hayes, G.M., Carrigan, P.E., Beck, A.M., Miller, L.J. Targeting the RNA splicing machinery as a novel treatment strategy for pancreatic carcinoma. Cancer Res. 2006, 66: 3819-3827.

[108] Muraki, M., Ohkawara, B., Hosoya, T., Onogi, H., Koizumi, J., Koizumi, T., Sumi, K., Yomoda, J., Murray, M.V., Kimura, H., Furuichi, K., Shibuya, H., Krainer, A.R., Suzuki, M., Hagiwara, M. Manipulation of alternative splicing by a newly developed inhibitor of Clks. J. Biol. Chem. 2004, 279: 24246-24254.

[109] Soret, J., Bakkour, N., Maire, S., Durand, S., Zekri, L., Gabut, M., Fic, W., Divita, G., Rivalle, C., Dauzonne, D., Nguyen, C.H., Jeanteur, P., Tazi, J. Selective modification of alternative splicing by indole derivatives that target serine-arginine-rich protein splicing factors. Proc. Natl. Acad. Sci USA. 2005, 102: 8764-8769.

[110] Li, Z., Putzer, B.M. Spliceosomal protein E regulates neoplastic cell growth by modulating expression of Cyclin E/CDK2 and G2/M checkpoint proteins. J. Cell. Mol. Med. 2008,

[111] Huang, C., Li, M., Chen, C., Yao, Q. Small interfering RNA therapy in cancer: mechanism, potential targets, and clinical applications. Expert Opin. Ther. Targets. 2008, 12: 637-645.

[112] Devi, G.R., Beer, T.M., Corless, C.L., Arora, V., Weller, D.L., Iversen, P.L. *In vivo* bioavailability and pharmacokinetics of a c-MYC antisense phosphorodiamidate morpholino oligomer, AVI-4126, in solid tumors. Clin Cancer Res. 2005, 11: 3930-3938.

---

[5]This study demonstrates that the alternatively spliced extra-domain A (EDA) of the fibronectin gene may represent an attractive marker for development of antibody-based strategies targeting neovasculature not only of solid tumors, but also of metastatic lesions.

[113] Sazani, P., Kole, R. Therapeutic potential of antisense oligonucleotides as modulators of alternative splicing. J. Clin. Invest. 2003, 112: 481-486.
[114] Bartel, F., Harris, L.C., Wurl, P., Taubert, H. MDM2 and its splice variant messenger RNAs: expression in tumors and down-regulation using antisense oligonucleotides. Mol. Cancer Res.
[115] Ghigna, C., De Toledo, M., Bonomi, S., Valacca, C., Gallo, S., Apicella, M., Eperon, I., Tazi, J., Biamonti, G. Pro-metastatic splicing of Ron proto-oncogene mRNA can be reversed: therapeutic potential of bifunctional oligonucleotides and indole derivatives. RNA Biology 2010, 7: 495-503.

CHAPTER 6

# Role of HMGB Proteins in Coordinating Ribosomal Protein and Ribosomal RNA

**LiJuan Xiao[§] and Anne Grove[*]**

*Department of Biological Sciences, Louisiana State University, Baton Rouge, LA 70803, USA*

**Abstract:** Cells grow in response to nutrients or growth factors, whose presence is detected and communicated by elaborate signaling pathways. Protein kinases play crucial roles in processes such as cell cycle progression and gene expression, and misregulation of such pathways has been correlated with various diseased states. Signals intended to promote cell growth converge on ribosome biogenesis, as the ability to produce cellular proteins is intimately tied to cell growth. Part of the response to growth signals is therefore the coordinate expression of genes encoding ribosomal RNA (rRNA) and ribosomal proteins (RP). A key player in regulating cell growth is the Target of Rapamycin (TOR) kinase, one of the gatekeepers that prevent cell cycle progression from G1 to S under conditions of nutritional stress. TOR is structurally and functionally conserved in all eukaryotes. Under favorable growth conditions, TOR is active and cells maintain a robust rate of ribosome biogenesis, translation initiation and nutrient import. Under stress conditions, TOR signaling is suppressed, leading to cell cycle arrest, while the failure of TOR to respond appropriately to environmental or nutritional signals leads to uncontrolled cell growth. Emerging evidence from *Saccharomyces cerevisiae* indicates that High Mobility Group (HMGB) proteins, non-sequence-specific chromosomal proteins, participate in mediating responses to growth signals. As HMGB proteins are distinguished by their ability to alter DNA topology, they frequently function in the assembly of higher-order nucleoprotein complexes. We review here recent evidence, which suggests that HMGB proteins may function to coordinate TOR-dependent regulation of rRNA and RP gene expression.

**Keywords:** Rapamycin, TORC1, HMO1, high mobility group, yeast, RP gene, rDNA.

***Address correspondence to Anne Grove:** Department of Biological Sciences, Louisiana State University, Baton Rouge, LA 70803, USA; Tel: 225-578-5148; Fax: 225-578-8790; E-mail: agrove@lsu.edu

[§]Current address: Developmental and Regenerative Biology, Mount Sinai School of Medicine, New York, NY, USA

**Christian Neri (Ed)**

## INTRODUCTION

Eukaryotic cell growth is regulated by intricate signaling pathways in response to nutrient levels, environmental stress and the presence of growth factors. Several protein kinases contribute to such signaling, including the target of rapamycin (TOR) kinase, a highly conserved serine/threonine kinase that plays a central role in the control of cell growth (reviewed in [1-3]). TOR is structurally and functionally conserved in all eukaryotes, and in mammalian cells, its dysregulation has been linked to many forms of cancer (reviewed in [4-7]). When growth conditions are favorable, TOR is active; this results in cells supporting ribosome biogenesis, translation initiation and nutrient import. Under stress conditions, TOR signaling is suppressed, resulting in cell cycle arrest. The failure of TOR to respond appropriately to environmental factors or nutritional signals leads to uncontrolled cell growth, as TOR signaling controls many aspects of cellular metabolism, including initiation of translation and transcription.

Up to 80% of the transcriptional machinery in *Saccharomyces cerevisiae* is dedicated to synthesis of ribosomal components - rRNA by RNA polymerase (Pol) I, ribosomal proteins by Pol II and 5S ribosomal RNA and tRNA by Pol III - while in the proliferating mammalian cell, as much as 50% of nuclear transcription is committed to this goal. What has also become increasingly evident is that ribosome biogenesis is an important determinant of cell growth; to maintain steady growth in response to favorable conditions, about 2000 ribosomes need to be synthesized per minute. While the role of TOR in regulating the synthesis of ribosomal components is well established, the mechanism by which all three polymerases are coordinately regulated in response to appropriate stimuli remains to be fully elucidated. As reviewed here, emerging evidence points to a role for high mobility group (HMGB) proteins in mediating such coordinated regulation in response to TOR signaling.

## TOR COMPLEXES

Both yeast and mammals contain two functionally and structurally distinct TOR complexes, TOR complex 1 (TORC1) and TOR complex 2 (TORC2), each composed of TOR as well as subunits that determine downstream substrates [8].

TORC1 controls cell growth *via* a rapamycin-sensitive signaling pathway [9-13] while TORC2 controls the organization of the actin cytoskeleton through a rapamycin-insensitive signaling pathway [14-18]. While higher eukaryotes encode a single TOR homolog, yeast encodes two TOR proteins, TOR1 and TOR2, either of which may become components of TORC1. TOR1 and TOR2 are therefore functionally redundant as components of TORC1, which responds to rapamycin by generating cellular responses akin to those induced by stress or starvation, including down-regulation of translation, inhibition of ribosome biogenesis, autophagy, and specific changes in gene expression. In contrast, yeast TOR2 is uniquely required for mediating the cell cycle-dependent polarization of the actin cytoskeleton and is the kinase component of TORC2 [15, 19, 20].

The ~280 kDa TOR proteins share 40-60% sequence identity and contain several domains. The N-terminal domain contains a number of HEAT motifs (for Huntington, EF3, A subunit of PP2A, TOR1), required for proper subcellular localization [21]. A central FAT domain (for FRAP, an earlier name for mammalian TOR, ATM, TRAP) may mediate protein-protein interactions [21, 22]. The C-terminal domain harbors the serine/threonine protein kinase activity. TOR proteins belong to the phosphatidyl inositol kinase-related kinase (PIKK) family, as their catalytic domain resembles that of the phosphatidyl inositol 3-kinases (PI3K). The kinase domain is immediately preceded by the FKBP12-rapamycin binding domain (FRB). Both TORC1 and TORC2 are multimeric complexes (Fig. **1**); [16, 23, 24]).

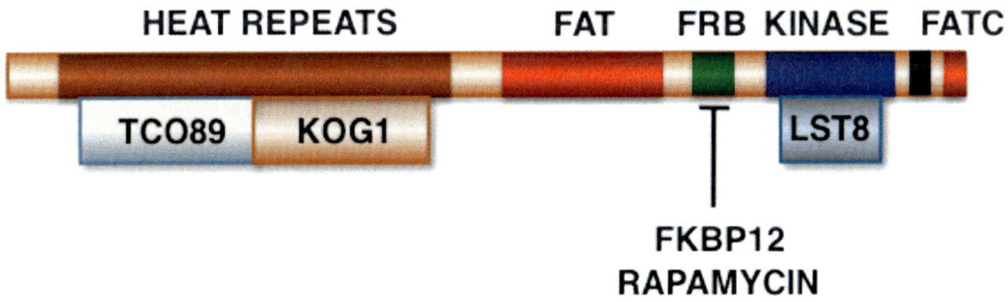


**Figure 1:** Domain organization of TOR kinases. Functional domains depicted include the N-terminal HEAT repeats, the central FAT domain, the kinase domain, and the C-terminal FRB domain to which FKBP12-rapamycin associates. Proteins associating with TORC1 include TCO89, KOG1, and LST1, of which only TCO89 has no apparent mammalian homolog.

The response to rapamycin requires that the drug first combine with the intracellular receptor FKBP12 [9, 25]. FKBP12-rapamycin binds TORC1 and functions as an allosteric inhibitor. Indeed, TOR was originally identified on the basis of mutations that confer resistance to rapamycin (clinically known as sirolimus), which is a macrocyclic lactone produced by *Streptomyces hygroscopicus* as an antifungal agent. Treatment of yeast cells with rapamycin phenocopies nitrogen starvation or inactivation of TOR by leading to down-regulation of protein synthesis and upregulation of macroautophagy (delivery of cytoplasmic contents to lysosomes and vacuoles). The mechanism by which FKBP12-rapamycin inhibits TORC1 is not known but has been proposed to include blockage of substrate access or dissociation of TOR from associated proteins [26, 27]. Proteins associated with TOR in yeast TORC1 have been identified as KOG1, TCO89, and LST8, of which only TCO89 has no obvious mammalian homolog (Fig. **1**); KOG1 deficiency in yeast resembles the phenotype of rapamycin treatment, suggesting that KOG1 is a positive regulator of TOR, an inference also reached with regard to LST8 [10, 28, 29]. Knockdown of the mammalian KOG1 homolog (Raptor), which interacts with downstream targets of TOR, likewise emulates treatment with rapamycin [26, 30, 31].

## TARGETS OF TORC1

In unicellular eukaryotes such as yeast, it is mainly nutrient levels and environmental stress such as hypoxia and DNA damage that regulate TORC1 signaling, while mammalian TORC1 also responds to environmental cues such as growth hormones and mitogens. Primary targets of TORC1 include components of the translation machinery (reviewed in [3]). Such downstream targets include the general translation initiation factor 4E-BP which functions to regulate availability of the initiation factor eIF4E, essential for cap-dependent initiation of translation [32-34]. Another well-established target of the TORC1 complex is the S6 protein kinase S6K1, which among other substrates phosphorylates ribosomal protein S6, an event that controls elongation factor 2 kinase, but probably not translation of mRNAs with a 5'-oligopyrimidine tract as originally suggested [35-38]. Thus, TORC1-mediated regulation of translation occurs by dual pathways.

## Regulation of RNA Polymerase I-Mediated Transcription

In addition to roles in regulation of translation, TORC1 regulates transcription of all genes whose products are involved in ribosome biogenesis. Production of rRNA accounts for at least 60% of total yeast transcriptional activity during normal growth, so it is a highly energy- and nutrient-consuming process. To conserve resources, cells must therefore limit the production of new ribosomes during nutrient starvation. Accordingly, both synthesis and the subsequent processing of precursor rRNA is regulated by the TOR signaling pathway in both mammalian and yeast cells [39-41].

rDNA transcription by RNA Pol I is confined to the nucleolus, where rDNA genes are arranged as hundreds of head-to-tail repeats. The primary rRNA transcript is then processed into the mature rRNAs in the nucleolus, which is also the site of ribosome biogenesis. Rapamycin and nutrient starvation causes rapid delocalization of RNA Pol I from the nucleolus, suggesting a possible mechanism for the regulation of rDNA transcription by TOR [42]. These authors also found that rDNA chromatin undergoes dramatic remodeling and becomes condensed in the presence of rapamycin. Notably, the rapamycin-mediated inhibition of rDNA transcription was blocked by a mutation in RPD3, a histone deacetylase subunit of the RPD3-SIN3 complex. Since RPD3–SIN3 histone deacetylase is important for regulation of chromatin condensation in yeast, and since both RPD3 and SIN3 were found to be rapamycin sensitive in a genomic rapamycin-sensitivity screen, these findings suggest that RPD3-mediated histone H4 deacetylation at rDNA chromatin may be a critical step in rDNA transcriptional regulation [43, 44]. Further supporting a role for chromatin remodeling in TOR-mediated rDNA gene activity is that RSC9, a component of the chromatin remodeling complex RSC (Remodels the Structure of Chromatin), is differentially localized to genomic DNA on TOR inactivation [45]. That RPD3 was also reported not to be required for rapamycin-mediated inhibition of Pol I points to mechanisms beyond chromatin remodeling [46].

Like Pol II and Pol III, Pol I requires auxiliary factors to mediate promoter recognition; for mammalian Pol I, preinitiation complex formation requires the upstream binding factor UBF and the promoter selectivity factor TIF-IB/SL1

(TIF-IB in mouse; SL1 in humans) [47-49]. Yeast does not encode a UBF homolog, instead the HMGB protein HMO1 has been reported to be a component of the Pol I machinery [50], and yeast Pol I transcription also requires the multifunctional complex UAF that binds upstream of the core promoter [51, 52]. UBF contains several HMG boxes, which result in DNA wrapping about the UBF dimer to bring the core promoter and upstream control element UCE into proximity, creating an "enhancesome" [53, 54]. It is thought that this DNA conformation is required for productive interactions between UBF and TIF-IB/SL1 that manifest in recruitment of Pol I by UBF and stabilization of TIF-IB/SL1 [55]. UBF contacts Pol I directly through interaction with PAF53, the mammalian homolog of yeast RPA49, and TIF-IB/SL1 interacts with TIF-IA, the mammalian homolog of yeast Rrn3p [56]. TIF-1A/Rrn3p is an important regulatory factor that associates with initiation-competent subpopulations of Pol I [57, 58].

In both yeast and mammals, TOR controls transcription by RNA Pol I *via* the transcription factor TIF-IA/Rrn3p, as inactivation of TOR impairs formation of the transcription initiation complex [40, 58-60]. Using chromatin immunoprecipitation assays, it was found that rapamycin treatment causes a decrease in the level of Rrn3p-Pol I complex at the yeast rDNA promoter and coding region, similar to what occurs in stationary phase, suggesting that the reduced transcriptional activity of individual active genes in stationary phase is mediated by the TOR signaling system through Rrn3p-dependent polymerase recruitment [59]. More recently, it was shown that TOR inactivation leads to proteasome-dependent degradation of Rrn3p, which correlates with decreased RNA Pol I occupancy at the 35S rDNA promoter [61].

Changes in UBF phosphorylation are key to modulating rDNA activity during cell cycle progression; for example, phosphorylation of the C-terminal activation domain of UBF by casein kinase II (CKII) facilitates the interaction between UBF and TIF-IB/SL1 [62]. Stimulation of rDNA transcription by mTOR requires both activation of S6K1 and phosphorylation of the UBF C-terminal domain (but not directly by S6K1; [63]). By contrast, rapamycin treatment leads to the rapid dephosphorylation of the UBF C-terminal domain, which significantly reduces its ability to interact with SL1.

In addition to signaling functions, including the control of cellular localization of specific transcription factors as discussed below, TOR itself is dynamically distributed in the cytoplasm and nucleus in both yeast and mammalian cells, and it has been shown to associate directly with rDNA promoters. TOR1 nuclear localization is nutrient-dependent and rapamycin sensitive, and starvation or treatment with rapamycin causes TOR1 to exit from the nucleus. TOR1 nuclear localization is critical for 35S rRNA synthesis, but not for the expression of ribosomal protein genes [64, 65]. Thus, effects of TOR signaling on rRNA gene activity appear to be exerted at the levels of both chromatin remodeling and transcription factor activity, perhaps governed by the cellular localization of TOR, as well as by direct association of TOR with RNA Pol I promoters.

**Regulation of RNA Polymerase III-Transcribed Genes**

RNA Pol III is responsible for the production of 5S rRNA and tRNA, gene products involved in ribosome assembly or translation, as well as a variety of small non-coding RNAs. Concerted regulation of 5S rDNA and tDNA genes with that of other rDNA genes would therefore be expected; indeed, TORC1 regulates transcription of RNA Pol III, but only when localized to the nucleus [66]. In this situation, TOR localizes to 5S rDNA chromatin [65, 66], and mammalian TOR was also seen to associate with tDNA genes [66, 67]. These results point to a direct role for TOR in regulating synthesis of RNA Pol III-transcribed gene products required for ribosome assembly and translation.

In addition, TORC1 also modulates RNA Pol III transcription by phosphorylation of the repressor Maf1, which is an evolutionarily conserved negative regulator that functions at the downstream end of stress signaling pathways. In yeast, repression of RNA Pol III transcription in response to inactivation of TORC1 requires Maf1, which becomes dephosphorylated, resulting in its translocation to the nucleus [68, 69]; its association with the polymerase effects a conformational change that impairs promoter binding [70]. In yeast, Sch9, a homolog of mammalian S6K1 and a substrate for TORC1, is involved in Maf1 phosphorylation, while mammalian TORC1 directly phosphorylates Maf1 at gene promoters, thus activating transcription [69, 71].

## TORC1-Mediated Regulation of Ribosomal Protein and Ribosomal Biogenesis Gene Expression

Ribosomal protein (RP) genes in eukaryotes are also regulated in response to growth stimuli and environmental stress. Indeed, it has been suggested that the reduction in ribosomal protein production that occurs after short-term inactivation of TORC1 is the primary reason for impaired ribosome production [72]. Global transcription profiling revealed that RP gene transcription is drastically repressed in response to rapamycin [39, 73]. As for rRNA gene activity, chromatin remodeling has also been implicated in regulation of RP gene expression in response to TOR signaling. For example, Rohde and Cardenas found that expression of RP genes coincides with recruitment of the ESA1 histone acetyl transferase to RP gene promoters, while inhibition of TOR with rapamycin releases ESA1 from RP gene promoters and leads to histone H4 deacetylation. Genetic and biochemical evidence demonstrated that the chromatin deacetylation complex RPD3-SIN3, also implicated in regulating rDNA gene activity, contributes to the repression of RP gene expression in response to TOR inhibition by rapamycin or nutrient limitation [74, 75].

A number of transcription factors have been implicated in regulation of RP gene expression. Genome-wide detection of transcription factor binding revealed that the Forkhead-like transcription factor Fhl1, originally isolated as a multicopy suppressor of RNA Pol III mutations, is found at yeast RP gene promoters [76, 77]. Subsequent studies found that Fhl1 may function either as a repressor or activator of RP gene transcription, depending on whether it associates with the corepressor Crf1 (Co-repressor with Fhl1) or the coactivator Ifh1 (Interacts with Forkhead). This decision is made based on the subcellular localization of Crf1, which is in turn determined by its phosphorylation state. TOR controls the subcellular localization of protein kinase A (PKA) and the PKA-regulated kinase YAK1 for which Crf1 is a substrate [78-80]; in actively growing cells, TORC1 negatively regulates the activity of YAK1, keeping Crf1 in the cytoplasm, while inhibition of TORC1 results in activation of YAK1, which in turn phosphorylates Crf1. Phosphorylated Crf1 translocates to the nucleus where it competes with Ifh1 for binding to Fhl1, which is constitutively bound at RP gene promoters [79]. The essential protein Ifh1 is recruited to RP gene promoters through the forkhead-associated domain of Fhl1 [78, 80, 81]; Fhl1 and Ifh1 were shown to associate almost exclusively with RP promoters, and the level of Ifh1 associated with RP promoters suggested to determine the level of transcription [80]. Accordingly, Ifh1 is thought to be an essential regulator of RP gene transcription (Fig. **2**). That additional factors must contribute to TOR-mediated regulation of RP gene activity was indicated by the observation that Crf1 may not be essential for rapamycin-dependent down-regulation of RP gene activity, depending on strain background, and that simply tethering of Fhl1 to RP gene promoters is insufficient to cause transcriptional activation, in spite of the fact that such tethered Fhl1 does recruit Ifh1 [82].

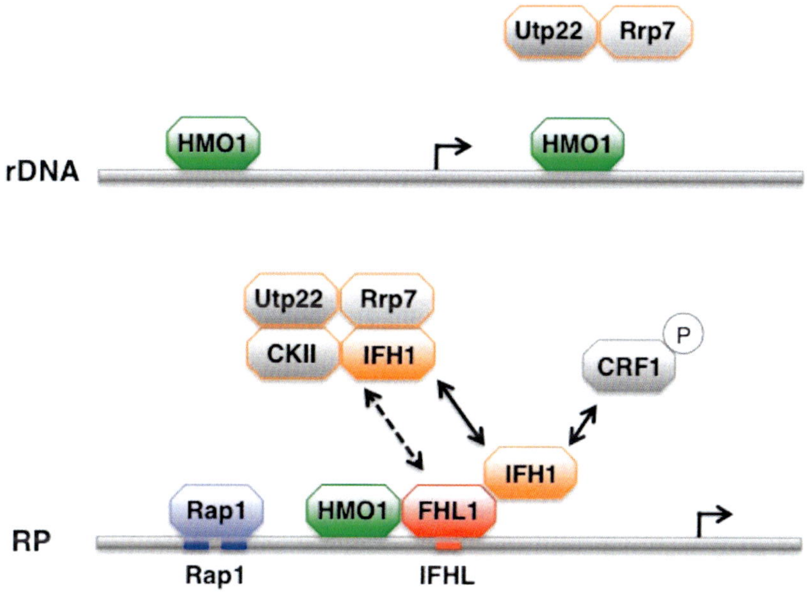


**Figure 2:** HMO1 associates with both rRNA and RP gene promoters to coordinate gene expression in response to TORC1 activity. On rDNA, HMO1 associates with both promoter and coding regions of transcriptionally active rDNA. Proposed functions include a role akin to that reported for mammalian UBP in transcriptional initiation, maintenance of transcriptionally active chromatin states, and a role in coordinating transcriptional elongation with pre-rRNA processing, to which Utp22 and Rrp7 contribute. On most RP gene promoters, Rap1 sites are further upstream than on other Rap1-dependent genes, and association of Rap1 appears to be required for the subsequent cooperative binding of HMO1 and Fhl1 at IFHL sites. When TORC1 is active, the coactivator Ifh1 binds Fhl1 to activate RP gene transcription. When TORC1 is inactive, phosphorylated Crf1 translocates to the nucleus to compete with Ifh1 for binding to Fhl1 and repress transcription. Ifh1 is found in the "CURI" complex also containing Utp22, Rrp7, and CKII, with which Fhl1 also associates. HMO1 dissociates from both classes of genes on inactivation of TORC1.

Although yeast Fhl1 has a domain that is related to the DNA binding domain of the Drosophila Forkhead protein, Fhl1 was reported not to bind DNA *in vitro* [83]. Thus, the binding of Fhl1 to an RP promoter is likely to be either indirect or facilitated by other proteins. One such candidate for Fhl1 recruitment is the repressor-activator protein 1 (Rap1), long known to be involved in the transcription of RP genes; most RP genes contain Rap1 binding sites at their promoters [84, 85]. However, Rap1 binds to many sites in the yeast genome and acts as an activator at some and as a repressor at others, as its name implies. Further, not all RP gene promoters contain Rap1 binding sites [86], suggesting that additional protein(s) contribute to Fhl1 recruitment.

The complexity of the TOR pathway is further underscored by the involvement of yet another transcription factor, Sfp1, a key cell size regulator, in regulating RP gene expression in response to nutrients and stress [87]. Non-RP genes that encode proteins involved in ribosome biogenesis (the Ribi regulon) also respond to TOR signaling, and Sfp1 has been reported to activate the Ribi regulon more rapidly than the RP regulon [88]. Under optimal growth conditions, Sfp1 is localized to the nucleus, bound to the promoters of RP genes where it activates RP gene expression. In response to inhibition of TOR signaling, Sfp1 is released from RP gene promoters, and down-regulation of RP gene transcription occurs concomitant with Sfp1 leaving the nucleus. That Sfp1 remains localized to the nucleus in a strain constitutively expressing high PKA activity in spite of rapamycin treatment suggests that Sfp1 may be regulated by the branch of the TOR pathway that depends on PKA, similar to the regulation of Crf1 localization [79, 87]. In addition, TORC1 can directly interact with and phosphorylate Sfp1, suggesting that TORC1 and Sfp1 may act directly to promote RP and Ribi gene transcription [89]. Analysis of genetic interactions further suggested that Sfp1 influences RP gene transcription *via* Fhl1 and Ifh1, and that the association of Ifh1 and Fhl1 with RP gene promoters was reduced by approximately four-fold and two-fold, respectively, in *sfp1*Δ cells [88]. As for the role of Crf1 in communicating downstream signals of the TOR pathway, the role of Sfp1 may be strain-dependent, as a subsequent genome-wide ChIP analyses suggested only limited association of Sfp1 with RP genes [90]. Further, nutrient deprivation was shown not to affect TORC1 association with and phosphorylation of Sfp1, but to result in both Tor1 and Sfp1 leaving the nucleus [64, 87-89]; in contrast, rapamycin treatment also leads to disruption of the TORC1-Sfp1 interaction, suggesting that rapamycin may in part regulate TORC1 at the level of substrate binding [89].

Another effector kinase implicated in maximal RP and Ribi gene expression is Sch9 [88]. Sch9 is a substrate for TORC1, but Sfp1 negatively regulates TORC1-mediated phosphorylation of Sch9, indicating the existence of a feedback loop controlling regulation of RP and Ribi gene activity by Sfp1 and Sch9 [89]. Sch9 appears to activate RP and Ribi gene expression by a different pathway that involves its phosphorylation of three transcription factors, Stb3, Dot6, and Tod6, factors implicated in recruiting the RPD3L histone deacetylase to RP and Ribi gene promoters when dephosphorylated [91]. Thus, while different transcription factors have been variously implicated in mediating responses to TOR signaling, suggesting the existence of more than one mechanism by which such signals are converted to changes in gene activity, signaling through Fhl1 appears to be a primary pathway.

## Coordination of rRNA and RP Gene Expression in Response to TOR Signaling

As summarized above, an effect of TOR inactivation on rRNA and RP gene expression has been well established, although the molecular basis for the response remains to be fully elucidated. Notably, recent evidence has furnished a link that may provide a clue to the mechanism by which transcriptional activity of the different RNA polymerase machineries may be coordinated in response to TOR signaling. This link comes in the form of the yeast HMGB homolog HMO1. HMGB proteins are conserved architectural proteins that bind DNA without sequence specificity, but with preference for DNA with structural distortions. As for mammalian HMGB homologs, HMO1 is composed of two HMG boxes, box A and box B. Box A has low affinity for DNA but is required for DNA structural specificity and for DNA bending, whereas box B is required for high affinity DNA binding and the C-terminal tail mediates in-phase DNA bending [92-94]. Genetic analysis showed that HMO1 is important for maximal transcription of rDNA [50]. While the molecular basis for the involvement of HMO1 in Pol I transcription is unknown, it is noteworthy that HMO1 overexpression suppresses an *rpa49Δ* mutation, as RPA49 corresponds to the mammalian PAF53 which interacts directly with the HMG-box-containing transcription factor UBF; as noted above, yeast does not encode a UBF homolog [50, 55, 56]. It is therefore tempting to speculate that HMO1 may serve to configure the yeast Pol I promoter for optimal transcriptional activity, akin to the enhancesome structure proposed for UBF [53, 54, 94].

A seminal study by Hall and co-workers demonstrated that HMO1 associates throughout the rDNA locus [95]. Their study also confirmed that HMO1 is required for maximal transcription of rRNA genes and demonstrated a role for HMO1 in rRNA maturation. Considering the localization of HMO1 throughout the rDNA locus, its reported role in rRNA processing, and recent evidence linking RNA Pol I transcriptional elongation with rRNA maturation, it is conceivable that HMO1 participates in coordinating these activities [96]. Importantly, HMO1 was also found to associate with most RP gene promoters and to do so in a Rap1-dependent fashion [95]. While Rap1 appears to be required for association of HMO1 with RP gene promoters, it is not sufficient, as evidenced by its localization to other genomic loci to which HMO1 does not associate. Notably, HMO1 association is reduced in an *fhl1Δ* strain, while inactivation of *HMO1* completely eliminates association of Fhl1 and Ifh1 with RP genes [95]. These data indicate that HMO1 and Fhl1 bind cooperatively to RP gene promoters, and that Rap1 may promote this association.

Using a global genetic approach, Berger *et al.* subsequently reported that HMO1 specifically interacts with the Pol I-transcribed region of the rDNA locus, pointing to a role in elongation, and confirmed its association with a subset of RP gene promoters [97]. In addition, the presence of HMO1 has been shown to correlate with transcriptionally active chromatin states [98]. This is consistent with an independent report indicating that HMO1 associates with the 35S rRNA gene in an RNA Pol I-dependent manner [90]. Notably, HMO1 was seen to dissociate from both rDNA and some RP gene promoters upon rapamycin treatment, and HMO1 deletion was shown to reduce the TORC1-dependent repression of RP gene expression [97]. These results clearly link HMO1 to the rapamycin-dependent TOR pathway and suggest that HMO1 is involved in coordinating rDNA transcription by Pol I and RP gene expression by Pol II in response to TOR signaling. Reinforcing this inference is the observation that expression of the *HMO1* gene is likewise regulated by TORC1 signaling, and that this regulation occurs by a mechanism that involves the association of both HMO1 and Fhl1 with the *HMO1* promoter [99].

## CONCLUSIONS

The molecular mechanisms underlying the role of HMO1 in coordinating rRNA and RP gene transcription in response to TOR-mediated signals remain to be determined. In the case of rDNA, HMO1 has been proposed to serve roles akin to those fulfilled by UBF in mammalian transcriptional initiation as well as in linking transcriptional elongation with pre-rRNA processing. Regardless of its specific function(s), it is clear that HMO1 abandons the rDNA locus when TORC1 is inhibited [97]. For UBF, dephosphorylation in response to rapamycin treatment is accompanied by reduced recruitment of Pol I [63]. Whether HMO1 is likewise subject to covalent modification in a rapamycin-dependent fashion remains to be determined.

At RP gene promoters, several transcription factors have been implicated in mediating downstream effects of TOR signaling. While the association of Sfp1 with RP gene promoters appears somewhat uncertain, a consensus is emerging based on consistent reports from several groups (Fig. **2**). At most RP gene promoters, HMO1 appears to associate in a Rap1-dependent fashion, followed or accompanied by Fhl1; however, at the promoters to which HMO1 only associates modestly, Fhl1 binding appears independent of HMO1 [90]. The transcription factor Ifh1 is essential, consistent with its reported role as a co-activator of RP gene transcription; if Ifh1 is absent, the co-repressor Crf1 would also encounter no competition for binding to Fhl1, thus potentially abolishing even basal levels of RP transcription. Depending on genetic background, a rapamycin-mediated down-regulation of RP gene activity may still be seen when Crf1 is inactivated; however, the basis for this observation remains unknown and has been suggested to involve more than one gene product [82]. One possible clue derives from the observation that Ifh1 is found to a significant extent in a complex named CURI, composed of casein kinase II (CKII), Utp22, Rrp7 and Ifh1), a complex with which Fhl1 also loosely associates [100]. Notably, both Utp22 and Rrp7 have been implicated in pre-rRNA processing, a process in which HMO1 is also involved. It is possible that TOR-dependent sequestration of Ifh1 in the CURI complex limits its availability as an RP gene activator, causing down-regulation of RP gene expression on inhibition of the TOR pathway. Activation of rDNA transcription may lead to disruption of CURI as Utp22 and Rrp7 engage in pre-rRNA processing, thus releasing Ifh1 to activate RP gene expression [100]. Whether HMO1 participates in this proposed scenario has not been addressed; however, HMO1 has been reported to interact directly with Utp22 [101], consistent with both proteins participating in pre-rRNA processing, suggesting the possibility that CURI may be recruited to both rRNA and RP genes by interaction with HMO1. That *HMO1* expression is attenuated on inactivation of TORC1 would ensure reduced HMO1 occupancy at RP and rDNA genes and perhaps affect recruitment of the CURI complex.

As HMO1 is non-sequence-specific in its DNA binding [92-94], it is likely recruited to RP gene promoters by association with sequence-specific factors; that Fhl1 fails to bind DNA *in vitro*, even in the presence of Rap1 [83], combined with the co-dependence of HMO1 and Fhl1 for binding to RP genes [95], suggests that HMO1 (or another factor, perhaps Sfp1) may be required to stabilize it on the DNA. HMO1 was reported either to have no effect or to cause a mild activation of RP gene transcription [95, 97]; such effects may be due to its interaction with TFIID, which has been reported to affect occupancy of TFIID at certain RP gene promoters as well as to participate in start site selection of a subset of Pol II genes [102]. As for rDNA, the mechanism by which HMO1 leaves the RP gene promoters following inhibition of the TOR pathway is not known.

## CONFLICT OF INTEREST

The author(s) confirm that this chapter content has no conflict of interest.

## ACKNOWLEDGEMENTS

Support of research in the authors' laboratory from the National Science Foundation (MCB-0744240 and MCB-1051610 to A.G.) and the Louisiana Governor's Biotechnology Initiative is gratefully acknowledged.

## DISCLOSURE

This manuscript is an extended and updated version of the paper titled "Coordination of Ribosomal Protein and Ribosomal RNA Gene Expression in Response to TOR Signaling", published in Current Genomics, 2009 May; 10(3): 198-205.

## ABBREVIATIONS

| | | |
|---|---|---|
| TOR | = | Target of Rapamycin |
| TORC1 | = | TOR Complex 1 |
| RP | = | Ribosomal Protein |
| HMG | = | High Mobility Group |
| Fhl1 | = | Forkhead Like |
| Ifh1 | = | Interacts with Forkhead |
| Crf1 | = | Corepressor with Forkhead |
| Pol | = | Polymerase |
| UBF | = | Upstream Binding Factor |

## REFERENCES

[1] Proud, C.G. mTOR Signalling in Health and Disease. *Biochem. Soc. Trans.* **2011**, 39: 431-436.

[2] Fingar, D.C., Blenis, J. Target of rapamycin (TOR): an integrator of nutrient and growth factor signals and coordinator of cell growth and cell cycle progression. *Oncogene* **2004**, 23: 3151-3171.

[3] Hay, N., Sonenberg, N. Upstream and downstream of mTOR. *Genes Dev.* **2004**, 18: 1926-1945.

[4] Bjornsti, M.A., Houghton, P.J. Lost in translation: dysregulation of cap-dependent translation and cancer. *Cancer Cell* **2004**, 5: 519-523.

[5] Reiling, J.H., Sabatini, D.M. Stress and mTORture signaling. *Oncogene* **2006**, 25: 6373-6383.

[6] Vivanco, I., Sawyers, C.L. The phosphatidylinositol 3-Kinase AKT pathway in human cancer. *Nat. Rev. Cancer* **2002**, 2: 489-501.

[7] Dazert, E., Hall, M.N. mTOR signaling in disease. *Curr. Opin. Cell Biol.* **2011**, 23: 744-755.

[8] Wullschleger, S., Loewith, R., Hall, M.N. TOR signaling in growth and metabolism. *Cell* **2006**, 124: 471-484.

[9] Heitman, J., Movva, N.R., Hall, M.N. Targets for cell cycle arrest by the immunosuppressant rapamycin in yeast. *Science* **1991**, 253: 905–909.

[10] Loewith, R., Jacinto, E., Wullschleger, S., Lorberg, A., Crespo, J.L., Bonenfant, D., Oppliger, W., Jenoe, P., Hall, M.N. Two TOR complexes, only one of which is rapamycin sensitive, have distinct roles in cell growth control. *Mol. Cell* **2002**, 10: 457-68.

[11] Reinke, A., Anderson, S., McCaffery, J.M., Yates, J. 3rd, Aronova, S., Chu, S., Fairclough, S., Iverson, C., Wedaman, K.P., Powers, T. TOR complex 1 includes a novel component, Tco89p (YPL180w), and cooperates with Ssd1p to maintain cellular integrity in *Saccharomyces cerevisiae. J. Biol. Chem.* **2004**, 279: 14752–14762.

[12] Ungar, L., Harari, Y., Toren, A., Kupiec, M. Tor complex 1 controls telomere length by affecting the level of Ku. *Curr. Biol.* **2011**, 21: 2115-2120.

[13] Bergkessel, M., Whitworth, G.B., Guthrie, C. Diverse environmental stresses elicit distinct responses at the level of pre-mRNA processing in yeast. *RNA* **2011**, 17: 1461-1478.

[14] Crespo, J.L., Hall, M.N. Elucidating TOR signaling and rapamycin action: lessons from *Saccharomyces cerevisiae. Microbiol. Mol. Biol. Rev.* **2002**, 66: 579–591.

[15] Schmidt, A., Kunz, J., Hall, M.N. TOR2 is required for organization of the actin cytoskeleton in yeast. *Proc. Natl. Acad. Sci. USA* **1996**, 93: 13780–13785.

[16] Wullschleger, S., Loewith, R., Oppliger, W., Hall, M.N. Molecular organization of TOR complex 2. *J. Biol. Chem.* **2005**, 280: 30697–30704.

[17] Helliwell, S.B., Wagner, P., Kunz, J., Deuter-Reinhard, M., Henriquez, R., Hall, M.N. TOR1 and TOR2 are structurally and functionally similar but not identical phosphatidylinositol kinase homologues in yeast. *Mol. Biol. Cell* **1994**, 5: 105-118.

[18] Zinzalla, V., Stracka, D., Oppliger, W., Hall, M.N. Activation of mTORC2 by association with the ribosome. *Cell* **2011**, 144: 757-768.

[19] Sarbassov, D.D., Ali, S.M., Sengupta, S., Sheen, J.H., Hsu, P.P., Bagley, A.F., Markhard, A.L., Sabatini, D.M. Prolonged rapamycin treatment inhibits mTORC2 assembly and Akt/PKB. *Mol. Cell* **2006**, 22: 159-168.

[20] Zheng, X.F., Florentino, D., Chen, J., Crabtree, G.R., Schreiber, S.L. TOR kinase domains are required for two distinct functions, only one of which is inhibited by rapamycin. *Cell* **1995**, 82: 121-130.

[21] Kunz, J., Schneider, U., Howald, I., Schmidt, A., Hall, M.N. HEAT repeats mediate plasma membrane localization of Tor2p in yeast. *J. Biol. Chem.* **2000**, 275: 37011-37020.

[22] Alarcon, C.M., Heitman, J., Cardenas, M.E. Protein kinase activity and identification of a toxic effector domain of the target of rapamycin TOR proteins in yeast. *Mol. Biol. Cell* **1999**, 10: 2531-2546.

[23] Takahara, T., Hara, K., Yonezawa, K., Sorimachi, H., Maeda, T. Nutrient-dependent multimerization of the mammalian target of rapamycin through the N-terminal HEAT repeat region. *J. Biol. Chem.* **2006**, 281: 28605-28614.

[24] Zhang, Y., Billington, C.J. Jr., Pan, D., Neufeld, T.P. Drosophila target of rapamycin kinase functions as a multimer. *Genetics* **2006**, 172: 355-362.

[25] Schreiber, S.L. Chemistry and biology of the immunophilins and their immunosuppressive ligands. *Science* **1991**, 251: 283-287.

[26] Kim, D.H., Sarbassov, D.D., Ali, S.M., King, J.E., Latek, R.R., Erdjument-Bromage, H., Tempst, P., Sabatini, D.M. mTOR interacts with raptor to form a nutrient-sensitive complex that signals to the cell growth machinery. *Cell* **2002**, 110: 163-175.

[27] Oshiro, N., Yoshino, K., Hidayat, S., Tokunaga, C., Hara, K., Eguchi, S., Avruch, J., Yonezawa, K. Dissociation of raptor from mTOR is a mechanism of rapamycin-induced inhibition of mTOR function. *Genes Cells* **2004**, 9: 359-366.

[28] Chen E.J., Kaiser, C.A. LST8 negatively regulates amino acid biosynthesis as a component of the TOR pathway, *J. Cell Biol.* **2003**, 161: 333–347.

[29] Adami, A., García-Alvarez, B., Arias-Palomo, E., Barford, D., Llorca, O. Structure of TOR and its complex with KOG1. *Mol. Cell* **2007**, 27: 509-516.

[30] Hara, K., Maruki, Y., Long, X., Yoshino, K., Oshiro, N., Hidayat, S., Tokunaga, C., Avruch, J., Yonezawa, K. Raptor, a binding partner of target of rapamycin (TOR), mediates TOR action. *Cell* **2002**, 110: 177-189.

[31] Sarbassov, D.D., Ali, S.M., Kim, D.H., Guertin, D.A. Latek, R.R., Erdjument-Bromage, H., Tempst, P., Sabatini, D.M. Rictor, a novel binding partner of mTOR, defines a rapamycin-insensitive and raptor-independent pathway that regulates the cytoskeleton. *Curr. Biol.* **2004**, 14: 1296-1302.

[32] Khaleghpour, K., Pyronnet, S., Gingras, A.C., Sonenberg, N. Translational homeostasis: eukaryotic translation initiation factor 4E control of 4E-binding protein 1 and p70 S6 kinase activities. *Mol. Cell. Biol.* **1999**, 19: 4302-4310.

[33] Miron, M., Lasko, P., Sonenberg, N. Signaling from Akt to FRAP/TOR targets both 4E-BP and S6K in *Drosophila melanogaster. Mol. Cell. Biol.* **2003**, 23, 9117-9126.

[34] Teleman, A.A., Chen, Y.W., Cohen, S.M. 4E-BP functions as a metabolic brake used under stress conditions but not during normal growth. *Genes Dev.* **2005**, 19: 1844-1848.

[35] Wang, X., Li, W., Williams, M., Terada, N., Alessi, D.R., Proud, C.G. Regulation of elongation factor 2 kinase by p90RSK1 and p70 S6 kinase, *EMBO J.* **2001**, 20: 4370–4379.

[36] Pende, M., Um, S.H., Mieulet, V., Sticker, M., Goss, V.L., Mestan, J., Mueller, M., Fumagalli, S., Kozma, S.C., Thomas, G. S6K1(-/-)/S6K2(-/-) mice exhibit perinatal lethality and rapamycin-sensitive 5'-terminal oligopyrimidine mRNA translation and reveal a mitogen-activated protein kinase-dependent S6 kinase pathway, *Mol. Cell. Biol.* **2004**, 24: 3112–3124.

[37] Ruvinsky, I., Sharon, N., Lerer, T., Cohen, H., Stolovich-Rain, M., Nir, T., Dor, Y., Zisman, P., Meyuhas, O. Ribosomal protein S6 phosphorylation is a determinant of cell size and glucose homeostasis, *Genes Dev.* **2005**, 19: 2199–2211.

[38] Magnuson, B., Ekim, B., Fingar, D.C. Regulation and function of ribosomal protein S6 kinase (S6K) within mTOR signalling networks. *Biochem. J.* **2012**, 441: 1-21.

[39] Powers, T., Walter, P. Regulation of ribosome biogenesis by the rapamycin-sensitive TOR-signaling pathway in *Saccharomyces cerevisiae. Mol. Biol. Cell* **1999**, 10: 987-1000.

[40] Mayer, C., Zhao, J., Yuan, X., I. Grummt, I. mTOR-dependent activation of the transcription factor TIF-IA links rRNA synthesis to nutrient availability, *Genes Dev.* **2004**, 18: 423–434.

[41] Wei, Y., Zheng, X.F.S. Sch9 partially mediates TORC1 signaling to control ribosomal RNA synthesis. *Cell Cycle* **2009**, 8: 4085-4090.

[42] Tsang, C.K., Bertram, P.G., Ai, W., Drenan, R., Zheng, X.F. Chromatin-mediated regulation of nucleolar structure and RNA Pol I localization by TOR. *EMBO J.* **2003**, 22: 6045-6056.

[43] Kasten, M.M., Dorland, S., Stillman, D.J. A large protein complex containing the yeast Sin3p and Rpd3p transcriptional regulators. *Mol. Cell. Biol.* **1997**, 17: 4852–4858.

[44] Chan, T.F., Carvalho, J., Riles, L., Zheng, X.F.S. A chemical genomics approach toward understanding the global functions of TOR. *Proc. Natl. Acad. Sci. USA* **2000**, 97: 13227–13232.

[45] Damelin, M., Simon, I., Moy, T.I., Wilson, B., Komili, S., Tempst, P., Roth, F.P., Young, R.A., Cairns, B.R., Silver, P.A. The genome-wide localization of Rsc9, a component of the RSC chromatin-remodeling complex, changes in response to stress. *Mol. Cell.* **2002**, 9: 563-73.

[46] Oakes, M.L., Siddiqi, I., French, S.L., Vu, L., Sato, M., Aris, J.P., Beyer, A.L., Nomura, M. Role of histone deacetylase Rpd3 in regulating rRNA gene transcription and nucleolar structure in yeast. *Mol. Cell. Biol.* **2006**, 26: 3889-3901.

[47] Jantzen, H.M., Admon, A., Bell, S.P., Tjian, R. Nucleolar transcription factor hUBF contains a DNA-binding motif with homology to HMG proteins. *Nature* **1990**, 344: 830-836.

[48] Clos, J., Buttgereit, D., Grummt, I. A purified transcription factor (TIF-IB) binds to essential sequences of the mouse rDNA promoter. *Proc. Natl. Acad. Sci. USA* **1986**, 83: 604-608.

[49] Learned, R.M., Cordes, S., Tjian, R. Purification and characterization of a transcription factor that confers promoter specificity to human RNA polymerase I. *Mol. Cell. Biol.* **1985**, 5: 1358-1369.

[50] Gadal, O., Labarre, S., Boschiero, C., Thuriaux, P. Hmo1, an HMG-box protein, belongs to the yeast ribosomal DNA transcription system. *EMBO J.* **2002**, 21: 5498-5507.

[51] Keys, D.A., Lee, B.S., Dodd, J.A., Nguyen, T.T., Vu, L., Fantino, E., Burson, L.M., Nogi, Y., Nomura, M. Multiprotein transcription factor UAF interacts with the upstream element of the yeast RNA polymerase I promoter and forms a stable preinitiation complex. *Genes Dev.* **1996**, 10: 887-903.

[52] Siddiqi, I.N., Dodd, J.A., Vu, L., Eliason, K., Oakes, M.L., Keener, J., Moore, R., Young, M.K., Nomura, M. Transcription of chromosomal rRNA genes by both RNA polymerase I and II in yeast uaf30 mutants lacking the 30 kDa subunit of transcription factor UAF. *EMBO J.* **2001**, 20: 4512-4521.

[53] Stefanovsky, V.Y., Pelletier, G., Bazett-Jones, D.P., Crane-Robinson, C., Moss, T. DNA looping in the RNA polymerase I enhancesome is the result of non-cooperative in-phase bending by two UBF molecules. *Nucleic Acids Res.* **2001**, 29: 3241-3247.

[54] Putnam, C.D., Copenhaver, G.P., Denton, M.L., Pikaard, C.S. The RNA polymerase I transactivator upstream binding factor requires its dimerization domain and high-mobility-group (HMG) box 1 to bend, wrap, and positively supercoil enhancer DNA. *Mol. Cell. Biol.* **1994**, 14: 6476-6488.

[55] Kuhn, A., Grummt, I. Dual role of the nucleolar transcription factor UBF: *trans*-activator and antirepressor. *Proc. Natl. Acad. Sci. USA* **1992**, 89: 7340-7344.

[56] Bodem, J., Dobreva, G., Hoffmann-Rohrer, U., Iben, S., Zentgraf, H., Delius, H., Vingron, M., Grummt I. TIF-IA, the factor mediating growth-dependent control of ribosomal RNA synthesis, is the mammalian homolog of yeast Rrn3p. *EMBO Rep.* **2000**, 1: 171-175.

[57] Miller, G., Panov, K.I., Friedrich, J.K., Trinkle-Mulcahy, L., Lamond, A.I., Zomerdijk, J.C. hRRN3 is essential in the SL1-mediated recruitment of RNA Polymerase I to rRNA gene promoters. *EMBO J.* **2001**, 20:1373-1382.
[58] Laferté, A., Favry, E., Sentenac, A., Riva, M., Carles, C., Chédin, S. The transcriptional activity of RNA polymerase I is a key determinant for the level of all ribosome components. *Genes Dev.* **2006**, 20: 2030-2040.
[59] Claypool, J.A., French, S.L., Johzuka, K., Eliason, K., Vu, L., Dodd, J.A., Beyer, A.L., Nomura, M. Tor pathway regulates Rrn3p-dependent recruitment of yeast RNA polymerase I to the promoter but does not participate in alteration of the number of active genes. *Mol. Biol. Cell* **2004**, 15: 946-956.
[60] Cavanaugh, A.H., Hirschler-Laszkiewicz, I., Hu, Q., Dundr, M., Smink, T., Misteli, T., Rothblum, L.I. Rrn3 phosphorylation is a regulatory checkpoint for ribosome biogenesis. *J. Biol. Chem.* **2002**, 277: 27423-27432.
[61] Philippi, A., Steinbauer, R., Reiter, A., Fath, S., Leger-Silvestre, I., Milkereit, P., Griesenbeck, J., Tschochner, H. TOR-dependent reduction in the expression level of Rrn3p lowers the activity of the yeast RNA Pol I machinery, but does not account for the strong inhibition of rRNA production. *Nucleic Acids Res.* **2010**, 38:5315-5326.
[62] Tuan, J.C., Zhai, W., Comai, L. Recruitment of TATA-binding protein-TAFI complex SL1 to the human ribosomal DNA promoter is mediated by the carboxy-terminal activation domain of upstream binding factor (UBF) and is regulated by UBF phosphorylation. *Mol. Cell. Biol.* **1999**, 19: 2872-2879.
[63] Hannan, K.M., Brandenburger, Y., Jenkins, A., Sharkey, K., Cavanaugh, A., Rothblum, L., Moss, T., Poortinga, G., McArthur, G.A., Pearson, R.B., Hannan, R.D. mTOR-dependent regulation of ribosomal gene transcription requires S6K1 and is mediated by phosphorylation of the carboxy-terminal activation domain of the nucleolar transcription factor UBF. *Mol. Cell. Biol.* **2003**, 23: 8862-8877.
[64] Li, H., Tsang, C.K., Watkins, M., Bertram, P.G., Zheng, X.F. Nutrient regulates Tor1 nuclear localization and association with rDNA promoter. *Nature* **2006**, 442: 1058-1061.
[65] Tsang, C.K., Liu, H., Zheng, X.F. mTOR binds to the promoters of RNA polymerase I- and III-transcribed genes. *Cell Cycle* **2010**, 9: 953-957.
[66] Wei, Y., Tsang, C.K., Zheng, X.F. Mechanisms of regulation of RNA polymerase III-dependent transcription by TORC1. *EMBO J.* **2009**, 28: 2220-2230.
[67] Kantidakis, T., Ramsbottom, B.A., Birch, J.L., Dowding, S.N., White, R.J. mTOR associates with TFIIIC, is found at tRNA and 5S rRNA genes, and targets their repressor Maf1. *Proc. Natl. Acad. Sci. USA* **2010**, 107: 11823-11828.
[68] Roberts, D.N., Wilson, B., Huff, J.T., Stewart, A.J., Cairns, B.R. Dephosphorylation and genome-wide association of Maf1 with Pol III-transcribed genes during repression. *Mol. Cell* **2006**, 22: 633-644.
[69] Lee, J., Moir, R.D., Willis, I.M. Regulation of RNA polymerase III transcription involves SCH9-dependent and SCH9-independent branches of the target of rapamycin (TOR) pathway. *J. Biol. Chem.* **2009**, 284: 12604-12608.
[70] Vannini, A., Ringel, R., Kusser, A.G., Berninghausen, O., Kassavetis, G.A., Cramer, P. Molecular basis of RNA polymerase III transcription repression by Maf1. *Cell* **2010**, 143: 59-70.

[71] Michels, A.A., Robitaille, A.M., Buczynski-Ruchonnet, D., Hodroj, W., Reina, J.H., Hall, M.N., Hernandez, N. mTORC1 directly phosphorylates and regulates human MAF1. *Mol. Cell. Biol.* **2010**, 30:3749-3757.

[72] Reiter, A., Steinbauer, R., Philippi, A., Gerber, J., Tschochner, H., Milkereit, P., Griesenbeck, J. Reduction in ribosomal protein synthesis is sufficient to explain major effects on ribosome production after short-term TOR inactivation in *Saccharomyces cerevisiae*. *Mol. Cell. Biol.* **2011**, 31: 803-817.

[73] Hardwick, J.S., Kuruvilla, F.G., Tong, J.K., Shamji, A.F., Schreiber, S.L. Rapamycin-modulated transcription defines the subset of nutrient-sensitive signaling pathways directly controlled by the Tor proteins. *Proc. Natl. Acad. Sci. USA* **1999**, 96: 14866–14870.

[74] Reid, J.L., Iyer, V.R., Brown, P.O., Struhl, K. Coordinate regulation of yeast ribosomal protein genes is associated with targeted recruitment of Esa1 histone acetylase. *Mol. Cell* **2000**, 6:1297-1307.

[75] Rohde, J.R., Cardenas, M.E. The tor pathway regulates gene expression by linking nutrient sensing to histone acetylation. *Mol. Cell. Biol.* **2003**, 23:629-635.

[76] Hermann-Le Denmat, S., Werner, M., Sentenac, A., Thuriaux, P. Suppression of yeast RNA polymerase III mutations by Fhl1, a gene coding for a fork head protein involved in rRNA processing. *Mol. Cell. Biol.* **1994**, 14: 2905-2913.

[77] Bar-Joseph, Z., Gerber, G.K., Lee, T.I., Rinaldi, N.J., Yoo, J.Y., Robert, F., Gordon, D.B., Fraenkel, E., Jaakkola, T.S., Young, R.A., Gifford, D.K. Computational discovery of gene modules and regulatory networks. *Nat. Biotechnol.* **2003**, 21: 1337-1342.

[78] Schmelzle, T., Beck, T., Martin, D.E., Hall, M.N. Activation of the RAS/cyclic AMP pathway suppresses a TOR deficiency in yeast. *Mol. Cell. Biol.* **2004**, 24: 338–351.

[79] Martin, D.E., Soulard, A., Hall, M.N. TOR regulates ribosomal protein gene expression *via* PKA and the Forkhead transcription factor Fhl1. *Cell* **2004**, 119: 969-979.

[80] Wade, J.T., Hall, D.B., Struhl, K. The transcription factor Ifh1 is a key regulator of yeast ribosomal protein genes. *Nature* **2004**, 432:1054-1058.

[81] Schawalder, S.B., Kabani, M., Howald, I., Choudhury, U., Werner, M., Shore, D. Growth-regulated recruitment of the essential yeast ribosomal protein gene activator Ifh1. *Nature* **2004**, 432: 1058-1061.

[82] Zhao, Y., McIntosh, K.B., Rudra, D., Schawalder, S., Shore, D., Warner,. JR. Fine-structure analysis of ribosomal protein gene transcription. *Mol. Cell. Biol.* **2006**, 26: 4853-4862.

[83] Rudra, D., Zhao, Y., Warner, J.R. Central role of Ifh1p-Fhl1p interaction in the synthesis of yeast ribosomal proteins. *EMBO J.* **2005**, 24: 533-542.

[84] Woudt, L.P., Smit, A.B., Mager, W.H., Planta, R.J. Conserved sequence elements upstream of the gene encoding yeast ribosomal protein L25 are involved in transcription activation. *EMBO J.* **1986**, 5: 1037-1040.

[85] Piña, B., Fernández-Larrea, J., García-Reyero, N., Idrissi, F.Z. The different (sur)faces of Rap1p. *Mol. Genet. Genomics* **2003**, 268: 791-798.

[86] Zhang, J., Vingron, M., Roepcke, S. Characteristic differences between the promoters of intron-containing and intronless ribosomal protein genes in yeast. *BMC Res. Notes* **2008**, 29:109.

[87] Marion, R.M., Regev, A., Segal, E., Barash, Y., Koller, D., Friedman, N., O'Shea, E.K. Sfp1 is a stress- and nutrient-sensitive regulator of ribosomal protein gene expression. *Proc. Natl. Acad. Sci. USA* **2004**, 101: 14315-14322.

[88] Jorgensen, P., Rupes, I., Sharom, J.R., Schneper, L., Broach, J.R., Tyers, M. A dynamic transcriptional network communicates growth potential to ribosome synthesis and critical cell size. *Genes Dev.* **2004**, 18: 2491-2505.

[89] Lempiäinen, H., Uotila, A., Urban, J., Dohnal, I., Ammerer, G., Loewith, R., Shore, D. Sfp1 interaction with TORC1 and Mrs6 reveals feedback regulation on TOR signaling. *Mol. Cell* **2009**, 33: 704-716.

[90] Kasahara, K., Ohtsuki, K., Ki, S., Aoyama, K., Takahashi, H., Kobayashi, T., Shirahige, K., Kokubo, T. Assembly of regulatory factors on rRNA and ribosomal protein genes in *Saccharomyces cerevisiae*. *Mol. Cell. Biol.* **2007**, 27: 6686-6705.

[91] Huber, A., French, S.L., Tekotte, H., Yerlikaya, S., Stahl, M., Perepelkina, M.P., Tyers, M., Rougemont, J., Beyer, A.L., Loewith, R. Sch9 regulates ribosome biogenesis *via* Stb3, Dot6 and Tod6 and the histone deacetylase complex RPD3L. *EMBO J.* **2011**, 30: 3052-3064.

[92] Kamau, E., Bauerle, K.T., Grove, A. The *Saccharomyces cerevisiae* high mobility group box protein HMO1 contains two functional DNA binding domains. *J. Biol. Chem.* **2004**, 279: 55234-55240.

[93] Bauerle, K.T., Kamau, E., Grove, A. Interactions between N- and C-terminal domains of the *Saccharomyces cerevisiae* high-mobility group protein HMO1 are required for DNA bending. *Biochemistry* **2006**, 45: 3635-3645.

[94] Xiao, L., Williams, A.M., Grove, A. The C-terminal domain of yeast high mobility group protein HMO1 mediates lateral protein accretion and in-phase DNA bending. *Biochemistry* **2010**, 49: 4051-4059.

[95] Hall, D.B., Wade, J.T., Struhl, K. An HMG protein, Hmo1, associates with promoters of many ribosomal protein genes and throughout the rRNA gene locus in *Saccharomyces cerevisiae*. *Mol. Cell. Biol.* **2006**, 26: 3672-3679.

[96] Schneider, D.A., Michel, A., Sikes, M.L., Vu, L., Dodd, J.A., Salgia, S., Osheim, Y.N., Beyer, A.L., Nomura, M. Transcription elongation by RNA polymerase I is linked to efficient rRNA processing and ribosome assembly. *Mol. Cell* **2007**, 6: 217-229.

[97] Berger, A.B., Decourty, L., Badis, G., Nehrbass, U., Jacquier, A., Gadal, O. Hmo1 is required for TOR-dependent regulation of ribosomal protein gene transcription. *Mol. Cell. Biol.* **2007**, 27: 8015-8026.

[98] Wittner, M., Hamperl, S., Stöckl, U., Seufert, W., Tschochner, H., Milkereit, P., Griesenbeck, J. Establishment and maintenance of alternative chromatin states at a multicopy gene locus. *Cell* **2011**, 145: 543-554.

[99] Xiao, L., Kamau, E., Donze, D., Grove, A. Expression of yeast high mobility group protein HMO1 is regulated by TOR signaling. *Gene* **2011**, 489: 55-62.

[100] Rudra, D., Mallick, J., Zhao, Y., Warner, J.R. Potential interface between ribosomal protein production and pre-rRNA processing. *Mol. Cell. Biol.* **2007**, 27: 4815-4824.

[101] Krogan, N.J., *et al.* Global landscape of protein complexes in the yeast S*accharomyces cerevisiae*. *Nature* **2006**, 440: 637-643.

[102] Kasahara, K., Ki, S., Aoyama, K., Takahashi, H,. Kokubo, T. *Saccharomyces cerevisiae* HMO1 interacts with TFIID and participates in start site selection by RNA polymerase II. *Nucleic Acids Res.* **2008**, 36: 1343-1357.

CHAPTER 7

# Databases and Web-Tools for Gene Expression Analysis in *Arabidopsis thaliana*

**Reinhard Hehl***, **Julio C. Bolívar, Jeannette Koschmann, Yuri Brill and Lorenz Bülow**

*Technische Universität Braunschweig, Institut für Genetik, Spielmannstr. 7, D-38106 Braunschweig, Germany*

**Abstract:** The number of internet resources for gene expression analysis in *Arabidopsis thaliana* has increased tremendously during the last years. These resources permit the database-assisted identification of putative *cis*-regulatory DNA sequences, their binding proteins, and the determination of common *cis*-regulatory motifs in coregulated genes. Also, genes targeted by inhibitory RNA can be identified with databases and web-servers. DNA-binding proteins may be predicted by the type of *cis*-regulatory motif. Further questions of combinatorial control based on the interaction of DNA-binding proteins and the colocalization of *cis*-regulatory motifs can be addressed. The database-assisted spatial and temporal expression analysis of DNA-binding proteins and their target genes may help to further refine experimental approaches. Signal transduction pathways upstream of regulated genes are beginning to be accessible in databases. This review focuses on the use of the AthaMap and PathoPlant databases for gene expression regulation analysis and discusses similar and complementary online databases and web-tools. Online databases are helpful for the development of working hypothesis and for designing subsequent experiments.

**Keywords:** Bioinformatics, databases, gene expression, plants, transcription, post-transcriptional regulation, web-server.

## INTRODUCTION

*Arabidopsis thaliana* is an important model organism for plant biologists. Its small genome is completely sequenced and contains only a low amount of repetitive DNA and a high gene density [1]. Furthermore, many mutants have been characterized phenotypically and insertion mutations for nearly all genes are

***Address correspondence to Reinhard Hehl:** Technische Universität Braunschweig, Institut für Genetik, Spielmannstr. 7, D-38106 Braunschweig, Germany; Tel: +49 (0)531-391-5772; Fax: +49 (0)531-391-5765; E-mail: r.hehl@tu-braunschweig.de

**Christian Neri (Ed)**

available. An integrated information resource can be accessed at http://arabidopsis.org [2]. The large amount of information available for this model plant justifies its use for almost all basic biological questions.

All scientific questions that address developmental processes or biotic and abiotic signal response reactions focus on the understanding of gene expression regulation. There are three levels of regulation, pre-transcriptional, transcriptional and post-transcriptional. The pre-transcriptional level concerns chromatin-structure and remodelling. Transcriptional control is executed mainly by transcription factors (TFs) that recruit the transcriptional preinitiation complex to the promoter. A major aspect of post-transcriptional regulation is RNA stability affected by small RNAs.

Out of these levels, transcriptional control is the most accessible level for database-assisted analysis [3]. Transcription factors bind to short sequence motifs, and families of TFs usually bind to similar sequences. In *Arabidopsis thaliana* more than 1,500 TFs were initially identified which constitute at least 5% of all protein coding genes [4, 5]. More recently, more than 2,000 protein coding sequences comprising 60 families are predicted to be TFs [6]. Experimental data on these TFs varies significantly. While usually a few type members of each family have been extensively analysed, the function of all family members often remains unknown. The same applies for information on the binding site of these factors. For members of 25 TF families, representative binding sites have been published and were annotated to databases [7-10].

A simple concept for transcriptional control is based on the presence of the binding site or *cis*-regulatory sequence in the promoters of genes which will then be a target site of a TF that regulates expression of the gene through DNA-binding [3]. A next level of complexity is exerted by the combinatorial control of gene expression where TFs will bind after homo- or heterodimerization with other TFs [11].

Based on the occurrence or combination of *cis*-regulatory elements, predictions can be established about which TF family is involved in regulating transcription [3]. However, it is still a major challenge to predict particular TF family members that bind. For this, knowledge on a possible coexpression of a member of the putatively binding TF family with the target gene may be useful [12].

Often, single TFs are not sufficient for the regulation of gene expression [11]. To gain more insight into the complexity of expression control, it is helpful to learn if *cis*-regulatory elements recognized by known interacting TFs colocalize in their target genes [13, 14]. Furthermore, additional information on protein-protein interactions may gain insight into upstream signal transduction pathways [15, 16].

Another level of complexity that can be addressed with databases is the post-transcriptional control of gene expression. A large number of small RNAs have been cloned from *Arabidopsis thaliana* and the genomic identification of their target sequences may reveal which genes are subjected to small RNA-mediated degradation. Small RNAs have a high degree of complementarity to their target sites. This led to the establishment of bioinformatic resources such as mirBase for determining target sites and target genes [17]. Predicted gene targets and expression data derived from RNA deep sequencing experiments were incorporated into mirBase [18, 19]. Currently, miRBase contains data for 243 different *A. thaliana* miRNA genes (miRBase, Release 16). In addition to mirBase, several other *A. thaliana* or plant specific databases and web-servers for siRNA and miRNA analysis are available such as ASRP, CSRDB, PMRD, and SoMART [20-24]. Tools for computer-assisted plant miRNA target prediction are for example miRU and Tapir [25, 26]. Based on experimentally verified target genes, Target-align and psRNATarget were developed for plant miRNA target detection [27, 28]. These tools take into account the sequence identity as well as the positional conservation of mismatches between miRNAs and their target genes and consider non-perfectly binding miRNAs that lead to translational inhibition [29-32].

This review describes internet resources that are available for the study of gene expression regulation in *Arabidopsis thaliana*. It will focus on two databases, AthaMap and PathoPlant, and the questions that can be addressed with them [33, 34]. The subjects discussed are schematically shown in the flow chart in Fig. (**1**). All online resources mentioned in the text are summarized in Table **1**.

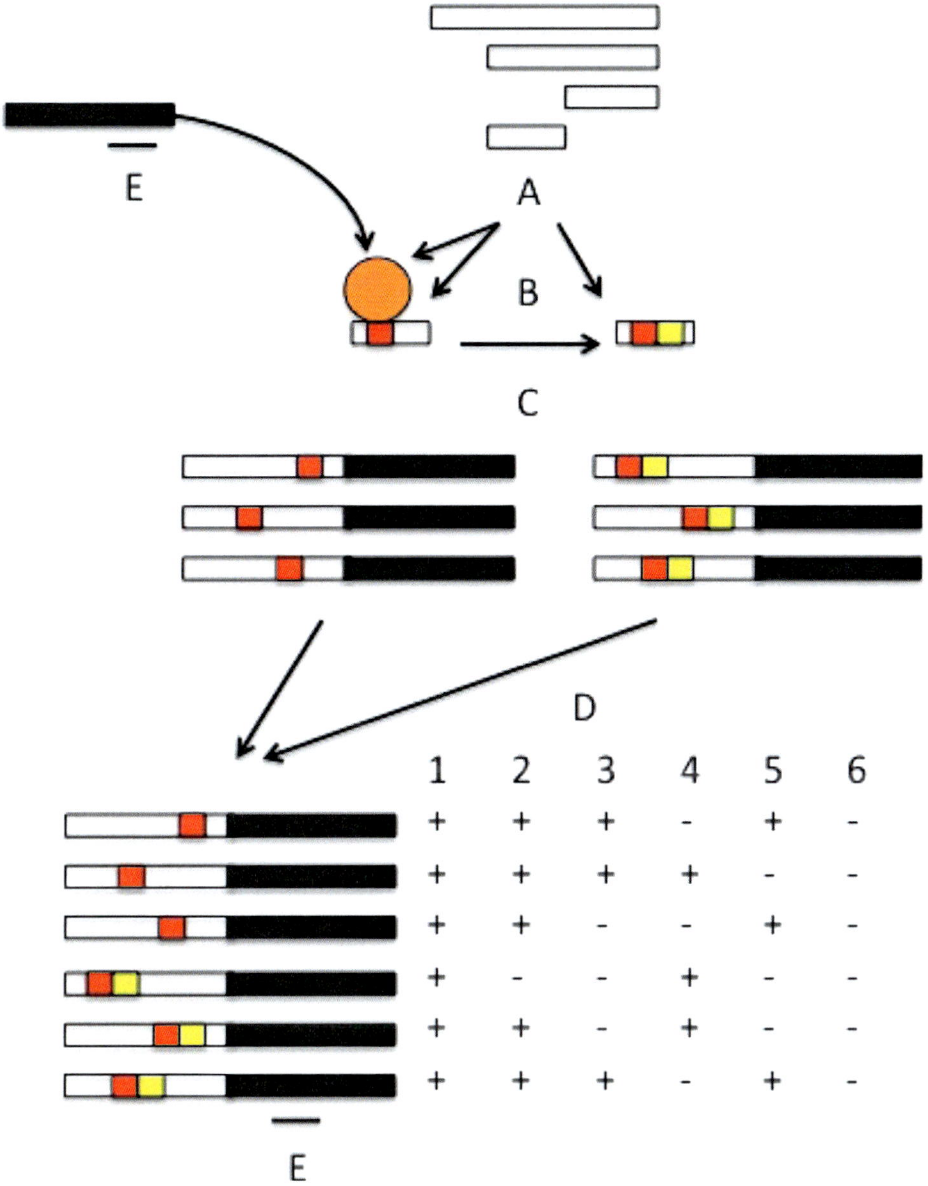


**Figure 1:** Database-based assisted anaylsis of gene expression. Internet resources permit the identification (**A**) of putative *cis*-regulatory sequences (hatched box) and binding transcription factors (circle) in an experimentally delineated regulatory region (+/-). These *cis*-regulatory sequences may be part of a known or predicted combinatorial element (**B**). Genes that harbor similar *cis*-regulatory sequences can also be determined (**C**). The identification of similar expression profiles in a set of genes with similar *cis*-regulatory sequences is illustrated for six different expression conditions (**D**). (**E**) Targets of inhibitory RNAs can also be proposed with databases. This permits the identification of potentially post-transcriptionally regulated genes.

**Table 1. Alphabetical List of Names and Links of Web Resources Mentioned in the Text**

| Name | Link |
|---|---|
| ACT | www.arabidopsis.leeds.ac.uk/ACT |
| aGFP | http://agfp.ueb.cas.cz |
| AGRIS | http://arabidopsis.med.ohio-state.edu/ |
| Arabidopsis eFP | http://www.bar.utoronto.ca/ |
| ARAMEMNON | http://aramemnon.botanik.uni-koeln.de/ |
| ASRP | http://asrp.cgrb.oregonstate.edu/ |
| AtGenExpress | http://www.weigelworld.org/resources/microarray/AtGenExpress/ |
| AthaMap | http://www.athamap.de/ |
| ATHENA | http://www.bioinformatics2.wsu.edu/cgi-bin/Athena/cgi/home.pl |
| AtProteome | http://fgcz-atproteome.unizh.ch/ |
| AtPAN | http://AtPAN.itps.ncku.edu.tw/ |
| ATTED-II | http://atted.jp/ |
| BBR | http://bbc.botany.utoronto.ca. |
| CSD.DB | http://csbdb.mpimp-golm.mpg.de/ |
| CSRDB | http://sundarlab.ucdavis.edu/smrnas/ |
| DATF | http://datf.cbi.pku.edu.cn |
| ExPlain | http://www.biobase-international.com/ |
| GEO | http://www.ncbi.nlm.nih.gov/geo |
| Genevestigator | https://www.genevestigator.com/gv/ |
| mirBASE | http://www.mirbase.org/ |
| NASCArrays | http://affy.arabidopsis.info/ |
| PatMatch | http://www.arabidopsis.org/cgi-bin/patmatch/nph-patmatch.pl |
| PathoPlant | http://www.pathoplant.de/ |
| PLACE | http://www.dna.affrc.go.jp/PLACE/ |
| PlantCARE | http://bioinformatics.psb.ugent.be/webtools/plantcare/html/ |
| PlantPAN | http://plantpan.mbc.nctu.edu.tw/index.php |
| PlnTFDB | http://plntfdb.bio.uni-potsdam.de |
| PMRD | http://bioinformatics.cau.edu.cn/PMRD/ |
| psRNATarget | http://plantgrn.noble.org/psRNATarget/ |
| RSA tools | http://rsat.ulb.ac.be/rsat |
| SMD | http://smd.stanford.edu |
| SoMART | http://somart.ist.berkeley.edu/ |
| STAMP | http://www.benoslab.pitt.edu/stamp/ |
| SUBA | http://suba.plantenergy.uwa.edu.au/index.php |
| TAIR | http://arabidopsis.org |
| TAPIR | http://bioinformatics.psb.ugent.be/webtools/tapir/ |
| Target-align | http://www.leonxie.com/targetAlign.php |
| TRANSFAC | http://www.gene-regulation.com/ |

## THE ATHAMAP DATABASE

AthaMap is a database that generates a genome-wide map of putative transcription factor binding sites (TFBS) for *Arabidopsis thaliana*. AthaMap was initially developed by matrix-based sequence searches using alignment matrices derived from several binding sites of the same TF [7]. Subsequently, AthaMap was extended with positional information for candidate TFBS based on functionally verified single TFBS [8]. Currently, the database contains 1 x $10^7$ predicted binding sites from 115 plant transcription factors [10]. Online AthaMap tools include a basic search function that requires a chromosomal position or a locus identifier (AGI). This results in a 1000 bp sequence display window with indicated putative binding sites and annotated gene structure. For matrix-based TFBS, the search function also permits the identification of highly conserved candidate binding sites. A 'Colocalization' tool permits the identification of chromosomal positions of putative combinatorial elements [13]. Combinatorial elements were also precalculated and annotated to AthaMap based on TFs that are known to interact or that contain two DNA-binding domains. A gene analysis function allows the identification of common or missing TFBS in a set of genes [9]. This function may be useful for the analysis of coregulated genes. This tool also permits the selection of TF families for analysis and a graphic display of positional information of the identified candidate TFBS. The 'Gene Identification' tool allows the identification of candidate TFBS for selected TFs. With this tool one can determine all genes that harbour putative binding sites for the selected TF. Recently, the database was extended with small RNA target sites from small RNA transcriptome datasets [35]. Furthermore, microRNA target sites were mapped genome-wide with the psRNATarget web server [28] and also annotated to AthaMap. This analysis revealed that up to 16,600 genes may be post-transcriptionally regulated by inhibitory RNAs. Two new web-tools were developed for AthaMap that permit the identification of small RNA and miRNA target sites (Bülow *et al.*, submitted). When using the 'Micro RNA Targets' tool one obtains a list of all candidate target sites for the selected miRNA in a user-selected gene region. Furthermore, the quality of the miRNA target site is represented by the psRNATarget score of each putative target site. The data content and the use of the database is described online and in the respective publications.

## THE PATHOPLANT DATABASE

PathoPlant is a database on plant-pathogen interactions and signal transduction reactions [36, 37]. The database is manually annotated with data from the literature. It currently (March 28, 2012) contains data for 99 plant species and varieties, 107 pathogens, and 638 molecules from a total of 619 references. This data represents 350 interactions and 370 reactions. *Via* a recently developed function, molecules and reactions annotated within PathoPlant can be visualized as signaling pathway maps. A map of all reactions and molecules annotated to PathoPlant can be generated as well as specific pathway maps starting from a selected molecule. Such a specific pathway map can be generated by selecting 'Specific map' from the 'Signaling pathways' menu that categorizes all molecules annotated to PathoPlant with respect to their molecular function.

A total of 173 different stimuli are currently represented by microarray data. Web-tools allow the identification of plant genes regulated by specific stimuli. Genes coregulated by up to three stimuli can be displayed as well. Furthermore, the web interface permits the submission of gene sets to be analysed for pathogen responsive gene expression. A result table lists the stimuli that act either inducing or repressing on the respective genes. This is particularly useful if sets of genes have been identified previously which harbour similar TFBS. The search in PathoPlant can be restricted to certain induction factors to identify for example strongly up- or down-regulated genes. A resulting list of coregulated genes can directly be exported to the AthaMap database for analysis of common *cis*-regulatory elements. Recently, a new web-tool designated '*in silico* expression analysis' was implemented in PathoPlant (Bolívar *et al.* manuscript in preparation). An *in silico* expression analysis can be performed in order to validate short DNA sequences as *cis*-elements responsive to different biotic and abiotic stimuli. The potential *cis*-sequence can be submitted to perform a genome-wide *Arabidopsis thaliana* promoter screening. The gene sets obtained are used to calculate mean induction factors for every *Arabidopsis thaliana* microarray experiment stored within PathoPlant. These mean values are normalized according to overall expression values of each stimulus. This results in a ranked list of microarray experiments according to their mean induction factors. The most probable stimuli to which the potential *cis*-element is responsive can be identified by looking at the highest-ranked stimuli.

The AthaMap and PathoPlant databases have been used extensively for the analysis of user-generated experimental data to further refine experimental approaches. In the future it is envisioned that both databases may also be used for database-assisted data mining. The data will then be subjected to experimental verification.

## WHICH TRANSCRIPTION FACTORS CAN BIND TO REGULATORY DNA SEQUENCE(S)?

The experimental analysis of gene expression may still include a classic deletion and reporter gene approach to identify a regulatory region in a gene. This may delineate a region that is essential for gene expression. A next step aims to predict the specific *cis*-regulatory sequence and/or transcription factor(s) that regulate(s) gene expression through this region (Fig. **1A**). For this purpose the region of interest can be displayed in AthaMap and putative TFBS that occur in this region can be predicted [7]. This allows the development of a hypothesis which TFBS may be relevant for gene expression. For example, if a drought responsive element is detected in a drought responsive gene, further experiments may focus on this element.

If the regulatory region is found in a promoter region, also other *A. thaliana* specific databases such as AGRIS, ATHENA, ATTED-II, PlantPAN and AtPAN (Table **1**) can be employed to predict *cis*-regulatory sequences [38-41]. Furthermore, the delineated regulatory sequence can also be submitted to the PLACE, PlantCARE, and TRANSFAC databases (Table **1**) to display either TFBS or *cis*-regulatory sequences that have been experimentally described before but for which no binding TF has been predicted [42-44].

## IDENTIFICATON OF COMBINATORIAL REGULATORY ELEMENTS

In many cases a single TFBS does not cause a specific expression unless in combination with other regulatory elements. A database-assisted approach could answer the question if a regulatory sequence is part of a known combinatorial element (Fig. **1B**). Such combinatorial elements will also be displayed in AthaMap if they were previously annotated [13]. Furthermore, if several *cis*-regulatory elements or TFBS occur in the experimentally defined regulatory region, it can be experimentally investigated if more than one is required for gene expression. Recently, a genome-wide screen for combinatorial elements was performed for *Arabidopsis* with *cis*-sequences from the PLACE database [45].

A different approach may start with identifying putative interacting proteins using interactome data and databases [15, 16, 46]. If predicted interacting proteins are TFs, the gene under investigation may be analysed for the occurrence of TFBSs for their TFs. If no binding site for this TF has previously been published, appropriate experiments to establish the TFBS for this factor may be carried out.

## DO OTHER GENES HARBOUR THE PREVIOUSLY PREDICTED TFBS?

If TFBS predicted by database-assisted analysis were experimentally confirmed to be relevant for gene expression, it may be interesting to find out if other genes harbour this combination of TFBS in a similar arrangement (Fig. **1C**). For this a web-tool to identify combinatorial elements in the AthaMap database can be employed [13]. Alternatively it can be checked if the combinatorial elements have been described before and occur in other genes [45]. With AthaMap it is possible to select two TFs that are relevant for gene expression and to obtain a list of genes that harbour these TFBS in a user-defined arrangement. If only the occurrence of TFBS of one TF in other genes is of interest, the ‚Gene Identification' web-tool of AthaMap can be employed [10]. Alternatively, ATHENA's data mining tool permits the identification of genes that harbour selected TFBS (Table **1**). Furthermore, another tool to identify genomic positions of previously determined *cis*-regulatory sequences is PatMatch, available at TAIR [47]. This is particularly useful if the putative *cis*-regulatory sequence has not been described before.

## IDENTIFICATION OF COREGULATED GENES WITH SIMILAR *CIS*-REGULATORY SEQUENCES

Another problem that can be addressed with web-resources is the identification of conditions under which genes with similar TFBS are coregulated (Fig. **1D**). A list of genes identified with AthaMap that harbour TFBS under investigation can be exported to the PathoPlant database to investigate if the genes are coregulated [9, 37]. PathoPlant incorporates microarray expression data for pathogen, abiotic stress, and signal substance regulated genes. If additional gene expression profiles are of interest other resources can also be used for this analysis.

Several web-based services harbour gene expression data from *Arabidopsis thaliana* microarray experiments and allow recovery of information for individual genes or gene sets (Table **1**). These are, for example, TAIR [2, 48], AtGenExpress [49, 50], NASCArrays tools [51], Stanford Microarray Database (SMD) [52, 53], Botany Array Resource now called Bio Array Resource (BBR) [54], Gene Expression Omnibus (GEO) [55], and Genevestigator [12]. Arabidopsis Co-expression Tool (ACT) [56, 57], BBR [54], the Comprehensive Systems-Biology Database (CSB.DB) [58], Genevestigator [12], and ATTED-II [59] allow comparative gene analysis to detect clusters of genes with similar expression patterns across selected or the complete set of stimuli. These tools start with a given gene of interest to determine similarities in expression patterns to other genes.

## IDENTIFICATION OF SIMILAR *CIS*-REGULATORY SEQUENCES IN COREGULATED GENES

A different approach to identify *cis*-regulatory elements starts with the identification of coregulated genes (Fig. **1D**). For example, PathoPlant allows the identification of coregulated genes and the export of the gene list to AthaMap for subsequent TFBS identification using the Gene Analysis tool [9, 37]. The commercially available ExPlain™ Analysis Platform from Biobase GmbH will identify combinatorial patterns of TFBS in a set of user-provided *Arabidopsis* genes. The analysis platform uses a matrix-based approach to identify TFBS in promoter regions and analyses the results with respect to specific patterns of TFBS that are overrepresented compared to a set of control genes [60].

Considering the fact that not all TFBS of predicted TFs are known, it may also be of interest to find patterns of conserved sequence motifs in a set of coregulated genes. For this purpose several online tools are available (Table **1**). For example the Regulatory Analysis Tools (RSA tools) offer the possibility to submit a set of promoter sequences to identify common motifs [61, 62]. A similar way of analysis is offered by Promomer at the Bio Array Resource [54]. Promomer is a web tool to discover over-represented sequence motifs in regulatory regions from sets of *A. thaliana* genes. The tools available online may also be downloaded and implemented locally. For example the Binding-site Estimation Suite of Tools (BEST) includes four commonly used motif-finding programs: AlignACE, BioProspector, CONSENSUS and MEME and the optimization program BioOptimizer [63].

Since these programs often yield sequence motifs derived from similar sequences that are conserved in a set of sequences, it may be of interest to find out if the motifs have been identified previously as regulatory sequences or TFBS. For this, STAMP may be used to query databases of known motifs with new motifs derived from similar sequences [64]. Since motifs from many plant-specific databases such as PLACE, PlantCARE, AGRIS, and AthaMap are available at STAMP it may be established that an identified motif has similarities to a known regulatory sequence or TFBS.

## ARE GENES OF INTEREST POST-TRANSCRIPTIONALLY REGULATED?

Recently, AthaMap was complemented with putative target sites of small RNAs [35]. With this data it is possible to propose if the gene(s) under investigation may be targets of inhibitory RNAs (Fig. **1E**). When using the web-tools in AthaMap it is now possible to identify those genes potentially regulated by inhibitory RNAs among the genes in a result table. By using the web-tools 'Small RNA Targets' it is also possible to obtain positional information on putative small RNA target sites for selected small RNA transcriptome datasets. Similarly, the 'MicroRNA Targets' web-tool permits the identification of positional information for miRNA target sites in the *A. thaliana* genome.

For identification of miRNA target sites also Target-align or the psRNATarget web-servers can be used (Table **1**). In this case, the small RNA sequence needs to be submitted or selected and also the sequence to be analysed for target sites needs to be provided [27, 28].

## DISCUSSION AND FUTURE DEVELOPMENTS

Many internet resources are available for gene expression analysis in *Arabidopsis thaliana* (Table **1**). It is usually very important to use more than one resource if possible. First, each database has a different level of curation and contains different data. For example, putative *cis*-regulatory sequences or TFBS can be determined with consensus sequences or can be detected with alignment matrices. They may either represent putative TFBS or sequence motifs conserved in coregulated genes. It is important to note that most of the putative *cis*-regulatory sequences or TFBS are probably not functional. This illustrates that database-assisted analysis is only a tool to refine or design experiments. Ideally a future map of TFBS contains also data from chromatin immunoprecipitations (CHIP) that reveal functional TFBS *in vivo* [65]. Unfortunately, the precise location of a TFBS within a relatively large DNA fragment obtained by CHIP is mostly not determined precisely. Therefore, the information is mainly useful to predict target genes of TFs but not their target sites. Furthermore, not all genes are transcriptionally regulated. It is estimated that, depending on the system analysed, as much as 50% of all genes may be post-transcriptionally regulated [66]. To learn which Arabidopsis genes are targets for small RNA-mediated mRNA degradation *in vivo*, the Arabidopsis Small RNA Project (ASRP) database can be employed [67].

With the application of next generation sequencing approaches, the analysis of gene expression is undergoing a revolutionary development [68]. With these developments we are not far from learning which genes are transcribed at a cellular level in time and space. The integration of such information in databases like Arabidopsis eFP Browser [69] and Arabidopsis Gene Family Profiler (aGFP) [70] will visualise the tissue- and state-specific expression of each gene during plant development. Ideally, not only transcriptome data will be integrated into databases but also proteome data [71]. Protein synthesis is under translational control and, most importantly, proteins may move between cells [72, 73]. There are already several protein databases. For example SUBA, a SUBcellular location database for Arabidopsis proteins (Table **1**) comprises 10 distinct subcellular locations, >6743 non-redundant proteins and represents the proteins encoded in the transcripts responsible for 51% of Arabidopsis expressed sequence tags [74].

Another example is ARAMEMNON, a database for Arabidopsis integral membrane proteins [75]. The AtProteome database comprises data of the high-density, organ-specific proteome map for *Arabidopsis thaliana* [76].

An integrated gene expression database of the future will contain data on functional *in vivo* TFBS, transcription factor and target gene expression and on cellular/subcellular localization of transcripts and proteins.

## CONFLICT OF INTEREST

The author(s) confirm that this chapter content has no conflict of interest.

## ACKNOWLEDGEMENTS

The work on the AthaMap and PathoPlant databases was supported by grants from the German Federal Ministry for Education and Research (BMBF) and by the KWS SAAT AG, Einbeck, Germany. We are grateful to Pascal Bartling, Sarah Pohl, and Malte Ringleb for data annotation to PathoPlant.

## DISCLOSURE

This is an updated report of the earlier review article by R. Hehl and L. Bülow, "Internet resources for gene expression analysis in *Arabidopsis thaliana*" Current Genomics, 2008, 9: 375-380.

## ABBREVIATIONS

| | | |
|---|---|---|
| TF | = | Transcription factor |
| TFBS | = | Transcription factor binding site |

## REFERENCES

[1] The Arabidopsis Genome Initiative. Analysis of the genome sequence of the flowering plant *Arabidopsis thaliana*. *Nature,* **2000**, *408,* 796-815.

[2] Swarbreck, D.; Wilks, C.; Lamesch, P.; Berardini, T. Z.; Garcia-Hernandez, M.; Foerster, H.; Li, D.; Meyer, T.; Muller, R.; Ploetz, L.; *et al.* The Arabidopsis Information Resource (TAIR): gene structure and function annotation. *Nucleic Acids Res,* **2008**, *36*, D1009-14.

[3] Hehl, R.; Wingender, E. Database-assisted promoter analysis. *Trends Plant Sci,* **2001**, *6*, 251-5.

[4] Riechmann, J. L.; Heard, J.; Martin, G.; Reuber, L.; Jiang, C.; Keddie, J.; Adam, L.; Pineda, O.; Ratcliffe, O. J.; Samaha, R, R.; *et al.* Arabidopsis transcription factors: genome-wide comparative analysis among eukaryotes. *Science,* **2000**, *290*, 2105-10.

[5] Guo, A.; He, K.; Liu, D.; Bai, S.; Gu, X.; Wei, L.; Luo, J. DATF: a database of Arabidopsis transcription factors. *Bioinformatics,* **2005**, *21,* 2568-9.

[6] Riano-Pachon, D. M.; Ruzicic, S.; Dreyer, I.; Müller-Röber, B. PlnTFDB: an integrative plant transcription factor database. *BMC Bioinformatics* 2007, 8:42.

[7] Steffens, N. O.; Galuschka, C.; Schindler, M.; Bülow, L.; Hehl, R. AthaMap: an online resource for *in silico* transcription factor binding sites in the *Arabidopsis thaliana* genome. *Nucleic Acids Res,* **2004**, *32,* D368-72.

[8] Bülow, L.; Steffens, N. O.; Galuschka, C.; Schindler, M.; Hehl, R. AthaMap: from *in silico* data to real transcription factor binding sites. *In Silico Biol,* **2006**, *6*, 243-52.

[9] Galuschka, C.; Schindler, M.; Bülow, L.; Hehl, R. AthaMap web-tools for the analysis and identification of co-regulated genes. *Nucleic Acids Res,* **2007**, *35,* D857-62.

[10] Bülow, L.; Brill, Y.; Hehl, R. AthaMap-assisted transcription factor target gene identification in *Arabidopsis thaliana. Database (Oxford),* **2010**, *2010,* baq034.

[11] Singh, K. B. Transcriptional regulation in plants: the importance of combinatorial control. *Plant Physiol,* **1998**, *118,* 1111-20.

[12] Zimmermann, P.; Hirsch-Hoffmann, M.; Hennig, L.; Gruissem, W. GENEVESTIGATOR. Arabidopsis microarray database and analysis toolbox. *Plant Physiol,* **2004**, *136,* 2621-32.

[13] Steffens, N. O.; Galuschka, C.; Schindler, M.; Bülow, L.; Hehl, R. AthaMap web tools for database-assisted identification of combinatorial *cis*-regulatory elements and the display of highly conserved transcription factor binding sites in *Arabidopsis thaliana. Nucleic Acids Res,* **2005**, *33*, W397-402.

[14] Ding, J.; Hu, H.; Li, X. Thousands of *cis*-regulatory sequence combinations are shared by Arabidopsis and Poplar. *Plant Physiol,* **2012**,*158,* 145-55

[15] Geisler-Lee, J.; O'Toole, N.; Ammar, R.; Provart, N. J.; Millar, A. H.; Geisler, M. A predicted interactome for Arabidopsis. *Plant Physiol,* **2007**, *145*, 317-29.

[16] Arabidopsis Interactome Mapping Consortium. Evidence for network evolution in an Arabidopsis interactome map. *Science, **2011***, *333,* 601-7.

[17] Griffiths-Jones, S. The microRNA Registry. *Nucleic Acids Res,* **2004**, *32*, D109-11.

[18] Griffiths-Jones S.; Grocock R. J.; van Dongen S.; Bateman A.; Enright A.J. miRBase: microRNA sequences, targets and gene nomenclature. *Nucleic Acids Res,* **2006**, *34*, D140-4.

[19] Kozomara A.; Griffiths-Jones S: miRBase: integrating microRNA annotation and deep-sequencing data. *Nucleic Acids Res,* **2011**, *39*, D152-7.

[20] Gustafson, A. M.; Allen, E.; Givan, S.; Smith, D.; Carrington, J. C.; Kasschau, K. D. ASRP: the Arabidopsis Small RNA Project Database. *Nucleic Acids Res,* **2005**, *33*, D637-40.

[21] Backman, T. W.; Sullivan, C. M.; Cumbie, J. S.; Miller Z. A.; Chapman, E. J.; Fahlgren, N.; Givan, S. A.; Carrington, J. C.; Kasschau, K. D. Update of ASRP: the Arabidopsis Small RNA Project database. *Nucleic Acids Res,* **2008**, *36*, D982-5.

[22] Johnson, C.; Bowman, L.; Adai, A. T.; Vance, V.; Sundaresan, V. CSRDB: a small RNA integrated database and browser resource for cereals. *Nucleic Acids Res,* **2007**, *35*, D829-33.

[23] Zhang, Z.; Yu, J.; Li, D.; Zhang, Z.; Liu, F.; Zhou, X.; Wang, T.; Ling, Y.; Su, Z. PMRD: plant microRNA database. *Nucleic Acids Res,* **2009**, *38*, D806-13.

[24] Li, F.; Orban, R.; Baker, B. SoMART, a web server for plant miRNA, tasiRNA and target gene analysis. *Plant J*, **2012**, *70*, 891-901

[25] Zhang, Y. miRU: an automated plant miRNA target prediction server. *Nucleic Acids Res,* **2005**, *33*, W701-4.

[26] Bonnet, E.; He, Y.; Billiau, K.; Van de Peer, Y. TAPIR, a web server for the prediction of plant microRNA targets, including target mimics. *Bioinformatics,* **2010**, *26*, 1566-8.

[27] Xie, F.; Zhang, B. Target-align: a tool for plant microRNA target identification. *Bioinformatics,* **2010**, *26*, 3002-3.

[28] Dai, X.; Zhao, P. X. psRNATarget: a plant small RNA target analysis server. *Nucleic Acids Res,* **2011**, *39*, W155-9.

[29] Rhoades, M. W.; Reinhart, B. J.; Lim, L. P.; Burge, C. B.; Bartel, B.; Bartel, D. P. Prediction of plant microRNA targets. *Cell,* **2002**, *110*, 513-20.

[30] Schwab, R.; Palatnik, J. F.; Riester, M.; Schommer, C.; Schmid, M.; Weigel, D. Specific effects of microRNAs on the plant transcriptome. *Dev Cell,* **2005**, *8*, 517-27.

[31] Brodersen, P.; Sakvarelidze-Achard, L.; Bruun-Rasmussen, M.; Dunoyer, P.; Yamamoto, Y. Y.; Sieburth, L.; Voinnet, O. Widespread translational inhibition by plant miRNAs and siRNAs. *Science,* **2008**, *320*, 1185-90.

[32] Montgomery, T. A.; Howell, M. D.; Cuperus, J. T.; Li, D.; Hansen, J. E.; Alexander, A. L.; Chapman, E. J.; Fahlgren, N.; Allen, E.; Carrington, J. C. Specificity of ARGONAUTE7-miR390 interaction and dual functionality in TAS3 *trans*-acting siRNA formation. *Cell,* **2008**, *133*, 128-41.

[33] Hehl, R.; Bülow, L. Internet resources for gene expression analysis in *Arabidopsis thaliana. Curr Genomics,* **2008**, *9*, 375-80.

[34] Hehl, R.; AthaMap, a database for the identification and analysis of transcription factor binding sites in the *Arabidopsis thaliana* genome. In *The handbook of plant functional genomics: concepts and protocols.* Edited by Kahl G, Meksem K: Wiley and Sons Ltd.; **2008**, 337-46.

[35] Bülow, L.; Engelmann, S.; Schindler, M.; Hehl, R. AthaMap, integrating transcriptional and post-transcriptional data. *Nucleic Acids Res,* **2009**, *37*, D983-6.

[36] Bülow, L.; Schindler, M.; Choi, C.; Hehl, R. PathoPlant: A database on plant-pathogen interactions. *In Silico Biol,* **2004**, *4*, 529–36.

[37] Bülow, L.; Schindler, M.; Hehl, R. PathoPlant: a platform for microarray expression data to analyze co-regulated genes involved in plant defense responses. *Nucleic Acids Res,* **2007**, *35*, D841-5.

[38] Davuluri, R. V.; Sun, H.; Palaniswamy, S. K.; Matthews, N.; Molina, C.; Kurtz, M.; Grotewold, E. AGRIS: Arabidopsis Gene Regulatory Information Server, an information resource of Arabidopsis *cis*-regulatory elements and transcription factors. *BMC Bioinformatics,* **2003**, *4*, 25.

[39] O'Connor, T. R.; Dyreson, C.; Wyrick, J. J. Athena: a resource for rapid visualization and systematic analysis of Arabidopsis promoter sequences. *Bioinformatics,* **2005**, *21*, 4411-3.

[40] Obayashi, T.; Kinoshita, K.; Nakai, K.; Shibaoka, M.; Hayashi, S.; Saeki, M.; Shibata, D.; Saito, K.; Ohta, H. ATTED-II: a database of co-expressed genes and *cis* elements for identifying co-regulated gene groups in Arabidopsis. *Nucleic Acids Res,* **2007**, *35*, D863-9.

[41] Chen, Y.-A.; Wen, Y.-C.; Chang, W.-C. AtPAN: an integrated system for reconstructing transcriptional regulatory networks in *Arabidopsis thaliana. BMC Genomics,* **2012**, *13*, 85

[42] Higo, K.; Ugawa, Y.; Iwamoto, M.; Korenaga, T. Plant *cis*-acting regulatory DNA elements (PLACE) database: 1999. *Nucleic Acids Res,* **1999**, *27*, 297-300.

[43] Rombauts, S.; Dehais, P.; Van Montagu, M.; Rouze, P. PlantCARE, a plant *cis*-acting regulatory element database. *Nucleic Acids Res,* **1999**, *27*, 295-6.

[44] Matys, V.; Fricke, E.; Geffers, R.; Gossling, E.; Haubrock, M.; Hehl, R.; Hornischer, K.; Karas, D.; Kel, A. E.; Kel-Margoulis, O. V, *et al.* TRANSFAC: transcriptional regulation, from patterns to profiles. *Nucleic Acids Res,* **2003**, *31*, 374-8.

[45] Ding, J.; Hu, H.; Li, X. Thousands of *cis*-regulatory sequence combinations are shared by Arabidopsis and poplar. *Plant Physiol,* **2012**, *158*, 145-55.

[46] Cui, J.; Li, P.; Li, G.; Xu, F.; Zhao, C.; Li, Y.; Yang, Z.; Wang, G.; Yu, Q.; Li, Y.; Shi, T. AtPID: *Arabidopsis thaliana* protein interactome database--an integrative platform for plant systems biology. *Nucleic Acids Res,* **2008**, *36*, D999-1008.

[47] Yan, T.; Yoo, D.; Berardini, T. Z.; Mueller, L. A.; Weems, D. C.; Weng, S.; Cherry, J. M.; Rhee, S. Y. PatMatch: a program for finding patterns in peptide and nucleotide sequences. *Nucleic Acids Res,* **2005**, *33*, W262-6.

[48] Rhee, S. Y.; Beavis, W.; Berardini, T. Z.; Chen, G.; Dixon, D.; Doyle, A.; Garcia-Hernandez, M.; Huala, E.; Lander, G.; Montoya, M, *et al.* The Arabidopsis Information Resource (TAIR): a model organism database providing a centralized, curated gateway to Arabidopsis biology, research materials and community. *Nucleic Acids Res,* **2003**, *31*, 224-8.

[49] Schmid, M.; Davison, T. S.; Henz, S. R.; Pape, U. J.; Demar, M.; Vingron, M.; Scholkopf, B.; Weigel, D.; Lohmann, J. U. A gene expression map of *Arabidopsis thaliana* development. *Nat Genet,* **2005**, *37*, 501-6.

[50] Kilian, J.; Whitehead, D.; Horak, J.; Wanke, D.; Weinl, S.; Batistic, O.; D'Angelo, C.; Bornberg-Bauer, E.; Kudla, J.; Harter, K. The AtGenExpress global stress expression data set: protocols, evaluation and model data analysis of UV-B light, drought and cold stress responses. *Plant J,* **2007**, *50*, 347-63.

[51] Craigon, D. J.; James, N.; Okyere, J.; Higgins, J.; Jotham, J.; May, S. NASCArrays: a repository for microarray data generated by NASC's transcriptomics service. *Nucleic Acids Res,* **2004**, *32*, D575-7.

[52] Gollub, J.; Ball, C. A.; Sherlock, G. The Stanford Microarray Database: a user's guide. *Methods Mol Biol,* **2006**, *338*, 191-208.

[53] Ball, C. A.; Awad, I. A.; Demeter, J.; Gollub, J.; Hebert, J. M.; Hernandez-Boussard, T.; Jin, H.; Matese, J. C.; Nitzberg, M.; Wymore, F.; *et al.* The Stanford Microarray Database accommodates additional microarray platforms and data formats. *Nucleic Acids Res,* **2005**, *33*, D580-2.

[54] Toufighi, K.; Brady, S. M.; Austin, R.; Ly, E.; Provart, N. J. The Botany Array Resource: e-Northerns, Expression Angling, and promoter analyses. *Plant J,* **2005**, *43*, 153-63.

[55] Barrett, T.; Suzek, T. O.; Troup, D. B.; Wilhite, S. E.; Ngau, W. C.; Ledoux, P.; Rudnev, D.; Lash, A. E.; Fujibuchi, W.; Edgar, R. NCBI GEO: mining millions of expression profiles--database and tools. *Nucleic Acids Res,* **2005**, *33*, D562-6.

[56] Manfield, I. W.; Jen, C. H.; Pinney, J. W.; Michalopoulos, I.; Bradford, J. R.; Gilmartin, P. M.; Westhead, D. R. Arabidopsis Co-expression Tool (ACT): web server tools for microarray-based gene expression analysis. *Nucleic Acids Res,* **2006**, *34*, W504-9.

[57] Jen, C. H.; Manfield, I. W.; Michalopoulos, I.; Pinney, J. W.; Willats, W. G.; Gilmartin, P. M.; Westhead, D. R. The Arabidopsis co-expression tool (ACT): a WWW-based tool and database for microarray-based gene expression analysis. *Plant J,* **2006**, *46*, 336-48.

[58] Steinhauser, D.; Usadel, B.; Luedemann, A.; Thimm, O.; Kopka, J. CSB.DB: a comprehensive systems-biology database. *Bioinformatics,* **2004**, 20, 3647-51.

[59] Obayashi, T.; Nishida, K.; Kasahara, K.; Kinoshita, K. ATTED-II updates: condition-specific gene coexpression to extend coexpression analyses and applications to a broad range of flowering plants. *Plant Cell Physiol,* **2011**, *52*, 213-9.

[60] Kel, A.; Konovalova, T.; Waleev, T.; Cheremushkin, E.; Kel-Margoulis, O.; Wingender, E. Composite Module Analyst: a fitness-based tool for identification of transcription factor binding site combinations. *Bioinformatics,* **2006**, *22*, 1190-7.

[61] van Helden, J. Regulatory sequence analysis tools. *Nucleic Acids Res,* **2003**, *31*, 3593-6.

[62] Thomas-Chollier, M.; Defrance, M.; Medina-Rivera, A.; Sand, O.; Herrmann, C.; Thieffry, D.; van Helden, J. RSAT 2011: regulatory sequence analysis tools. *Nucleic Acids Res,* **2011**, *39*, W86-91.

[63] Che, D.; Jensen, S.; Cai, L.; Liu, J. S. BEST: binding-site estimation suite of tools. *Bioinformatics,* **2005**, *21*, 2909-11.

[64] Mahony, S.; Benos, P. V. STAMP: a web tool for exploring DNA-binding motif similarities. *Nucleic Acids Res,* **2007**, *35*, W253-8.

[65] Gregory, B. D.; Yazaki, J.; Ecker, J. R. Utilizing tiling microarrays for whole-genome analysis in plants. *Plant J,* **2008**, *53*, 636-44.

[66] Cheadle, C.; Fan, J.; Cho-Chung, Y. S.; Werner, T.; Ray, J.; Do, L.; Gorospe, M.; Becker, K. G. Control of gene expression during T cell activation: alternate regulation of mRNA transcription and mRNA stability. *BMC Genomics,* **2005**, *6*, 75.

[67] Backman, T. W.; Sullivan, C. M.; Cumbie, J. S.; Miller, Z. A.; Chapman, E. J.; Fahlgren, N.; Givan, S. A.; Carrington, J. C.; Kasschau, K. D. Update of ASRP: the Arabidopsis Small RNA Project database. *Nucleic Acids Res,* **2008**, *36*, D982-5.

[68] Meyers, B. C.; Vu, T. H.; Tej, S. S.; Ghazal, H.; Matvienko, M.; Agrawal, V.; Ning, J.; Haudenschild, C. D. Analysis of the transcriptional complexity of *Arabidopsis thaliana* by massively parallel signature sequencing. *Nat Biotechnol,* **2004**, *22*, 1006-11.

[69] Winter, D.; Vinegar, B.; Nahal, H.; Ammar, R.; Wilson, G. V.; Provart, N. J. An "electronic fluorescent pictograph" browser for exploring and analyzing large-scale biological data sets. *PLoS One*, **2007**, *2*, e718.

[70] Dupl'akova, N.; Renak, D.; Hovanec, P.; Honysova, B.; Twell, D.; Honys, D. Arabidopsis Gene Family Profiler (aGFP)--user-oriented transcriptomic database with easy-to-use graphic interface. *BMC Plant Biol,* **2007**, *7*, 39.

[71] Baginsky, S.; Gruissem, W. *Arabidopsis thaliana* proteomics: from proteome to genome. *J Exp Bot,* **2006**, *57*, 1485-91.

[72] Kawaguchi, R.; Bailey-Serres, J. mRNA sequence features that contribute to translational regulation in Arabidopsis. *Nucleic Acids Res,* **2005**, *33*, 955-65.

[73] Dolan, L. Root patterning: SHORT ROOT on the move. *Curr Biol,* **2001**, *11*, R983-5.

[74] Heazlewood, J. L.; Verboom R, E.; Tonti-Filippini, J.; Small, I.; Millar, A. H. SUBA: the Arabidopsis Subcellular Database. *Nucleic Acids Res,* **2007**, *35*, D213-8.

[75] Schwacke, R.; Schneider, A.; van der Graaff, E.; Fischer, K.; Catoni, E.; Desimone, M.; Frommer, W. B.; Flugge U. I.; Kunze, R. ARAMEMNON, a novel database for Arabidopsis integral membrane proteins. *Plant Physiol,* **2003**, *131*, 16-26.

[76] Baerenfaller, K.; Grossmann, J.; Grobei, M. A.; Hull, R.; Hirsch-Hoffmann, M.; Yalovsky, S.; Zimmermann, P.; Grossniklaus, U.; Gruissem, W.; Baginsky, S. Genome-scale proteomics reveals *Arabidopsis thaliana* gene models and proteome dynamics. *Science,* **2008**, *320*, 938-41.

 *Advances in Genome Science, Vol. 2*, 2013, 194-220

# CHAPTER 8

# Genetical Genomics of Plants: From Genotype to Phenotype

**Ronny V.L. Joosen[1], Wilco Ligterink[1], Henk W.M. Hilhorst[1] and Joost J.B. Keurentjes[1,2,3,*]**

*[1]Laboratory of Plant Physiology, Wageningen University, Droevendaalsesteeg 1, NL-6708 PB Wageningen, The Netherlands; [2]Laboratory of Genetics, Wageningen University, Droevendaalsesteeg 1, NL-6708 PB Wageningen, The Netherlands; [3]Centre for Biosystems Genomics, Droevendaalsesteeg 1, NL-6708 PB Wageningen, The Netherlands*

**Abstract:** Natural variation provides a valuable resource to study the genetic regulation of quantitative traits. In quantitative trait locus (QTL) analyses this variation, captured in segregating mapping populations, is used to identify the genomic regions affecting these traits. The identification of the causal genes underlying QTLs is a major challenge for which the detection of gene expression differences is of major importance. By combining genetics with large scale expression profiling (*i.e.*, genetical genomics), resulting in expression QTLs (eQTLs), great progress can be made in connecting phenotypic variation to genotypic diversity. In this review we discuss examples from human, mouse, *Drosophila*, yeast and plant research to illustrate the advances in genetical genomics, with a focus on understanding the regulatory mechanisms underlying natural variation. With their tolerance to inbreeding, short generation time and ease to generate large families, plants are ideal subjects to test new concepts in genetics. The comprehensive resources which are available for *Arabidopsis* make it a favorite model plant but genetical genomics also found its way to important crop species like rice, barley and wheat. We discuss eQTL profiling with respect to *cis* and *trans* regulation and show how combined studies with other 'omics' technologies, such as metabolomics and proteomics may further augment current information on transcriptional, translational and metabolomic signaling pathways and enable reconstruction of detailed regulatory networks. The fast developments in the 'omics' area will offer great potential for genetical genomics to elucidate the genotype-phenotype relationships for both fundamental and applied research.

**Keywords:** Genetical genomics, e-QTL, network reconstruction, Arabidopsis thaliana, crop genetics.

***Address correspondence to J.J.B. Keurentjes:** Laboratory of Plant Physiology, Wageningen University, Droevendaalsesteeg 1, NL-6708 PB Wageningen, The Netherlands; Tel: +31 (0)317 483149; Fax: +31 (0)317 483146; E-mail: Joost.Keurentjes@wur.nl

**Christian Neri (Ed)**

# INTRODUCTION

Ever since the current paradigm of gene transcription preceding biological function, research on gene function has focused on expression studies. With the ever increasing availability of genomic sequences and the introduction of microarray technology, enabling the high-throughput analysis of gene expression, this has rapidly become a favorite tool for many researchers [1]. In a typical microarray experiment specific conditions or developmental stages are studied by comparing expression profiles and determining differences in gene transcription. The object of profiling can be a single genotype showing phenotypic diversity in a spatial and temporal manner, *e.g.,* in different tissues and developmental stages, or when exposed to different growing conditions (*e.g.,* [2]). In this way, large compendia of expression data have been acquired, providing ontological information of genes involved in developmental control and environmental responses [3-5].

Although much of the relationship between the temporal expression of genes and their function can be learned from these analyses, often no information can be obtained about the genetic regulation of transcription or whether expression differences are causal for or a consequence of phenotypic differences. Therefore, instead of comparing different conditions of a single genotype, equivalent samples of different genotypes varying in the trait of interest are often analyzed. These can be natural variants within a species, artificial mutants or transgenics like knockout and over expression lines. Such analyses have shown to be extremely powerful in determining directionality in biological pathways, especially in qualitative traits [6, 7]. However, it becomes extremely difficult to define a proper experimental setup when the trait of interest is complex and has a quantitative character. Geneticists are used to deal with this type of complex traits by using the power of natural variation within species.

In segregating populations, derived from crosses between distinct parents and genotyped with molecular markers, linkage is sought between variation in the trait of interest and genotypic diversity [8]. For this purpose a broad range of software tools and statistical analyses are available. The identified genomic regions explaining the observed phenotypic variation are commonly referred to as quantitative trait loci (QTLs), which can subsequently be used for marker assisted breeding purposes without further knowledge of the underlying genes [9]. Whenever the purpose is to identify the causal genes or even nucleotide polymorphisms (QTNs) underpinning a given QTL, one needs to invest in follow-up analyses for fine mapping and ultimately the cloning of a QTL [10]. This approach however, is very labor intensive and time consuming and confines the classical QTL mapping to a low throughput technique.

Like for many physiological traits, variation in gene expression often shows a quantitative distribution, hence, all the classical statistical tools and concepts for QTL mapping can be applied for its genetic dissection. Thus, subjecting expression variation to linkage analysis identifies genetic regulatory loci, and ideally genes, explaining the observed variation. Knowing the position of genes and their corresponding expression QTLs (eQTLs) renders great opportunities for dissecting quantitative traits. This was first recognized by Jansen and Nap [11] who outlined a concept, coined 'genetical genomics', in which the combination of a genotyped segregating population (*i.e.,* genetics) and genome-wide expression profiling (*i.e.,* genomics) is used to formulate hypothetic regulatory pathways and unravel complex traits in a more high-throughput manner [12]. Analogously, similar approaches can be followed for data derived from other 'omic' technologies such as proteomics (pQTLs) and metabolomics (mQTLs) [13].

The first study reporting a proof of principle of genetical genomics was performed in *Saccharomyces cerevisiae* [14]. In a relatively small population of 40 haploid segregants from a cross between a laboratory and a wild type strain, it was shown that parental differences in gene expression were highly heritable and amenable to genetic mapping. This first report was quickly followed by more comprehensive eQTL studies in higher eukaryotes [15] and has now been applied in a broad range of taxonomic kingdoms including yeast [14, 16-18], nematodes [19], insects [20, 21], plants [22-25], rodents [26-28] and humans [29-34]. All studies demonstrated the power of combining gene expression and genetic analyses to refine molecular pathways involved in complex phenotypes and to identify key driver genes thereof. Moreover, they have shown general and conserved mechanisms of expression regulation which improved our understanding of adaptive strategies and evolutionary concepts [20, 35].

In this review we will discuss the genetic architecture of gene expression regulation, embarking on recent findings in the reference plant *Arabidopsis thaliana*, the implications of genetical genomics approaches for crop species and the impact of genetic analyses of 'omics' data on the construction of regulatory networks. We will discuss future prospects and speculate on the utilization of advancing technological developments for genetic studies.

## GENETIC ARCHITECTURE OF GENE EXPRESSION VARIATION

The detection of eQTLs depends on a number of factors, which together determine the proportion of genetically regulated genes that can be observed. First, biological factors such as the assayed tissue, developmental stage or environmental conditions and the genotypic diversity present in the mapping population determine which genes are expressed and exhibit allelic variants, respectively. Second, statistical issues like population type and size, genetic map quality, measurement accuracy and the number of genes analyzed determine mapping power and detection thresholds. Because all these aspects vary between different experiments, reported fractions of regulated genes range from only a handful to over 50% of the total gene content.

### Regulation *in cis*

Given the prerequisite of allelic variation, there can be many reasons why genes are differentially expressed in genotypically diverse individuals of a species. Well-known phenomena are allelic variants of transcription factors and other regulators, *cis*-elemental variation in promoter sequences, differences in mRNA stability, copy number variation and genomic rearrangements such as translocations, insertions and deletions. The latter include gene loss and duplication, resulting in neo- and sub-functionalization. Most of these variations in DNA structure will result in eQTLs but depending on the position of the causal polymorphism, an important dissection is made in local and distant eQTLs (Fig. **1**) [36]. Local eQTLs can be the result of closely linked *trans*-acting factors but in the majority of cases result from *cis*-regulatory variation in the genes under study. By definition eQTLs acting *in cis* affect transcription initiation, rate and/or transcript stability in an allele-specific manner. In addition, *cis*-regulated genes might encode regulators affecting the expression of downstream target genes *in trans*. Although the exact proportion varies between studies the occurrence of *cis*-acting eQTLs is substantial ranging from one-third to half of the total number of eQTLs [37].

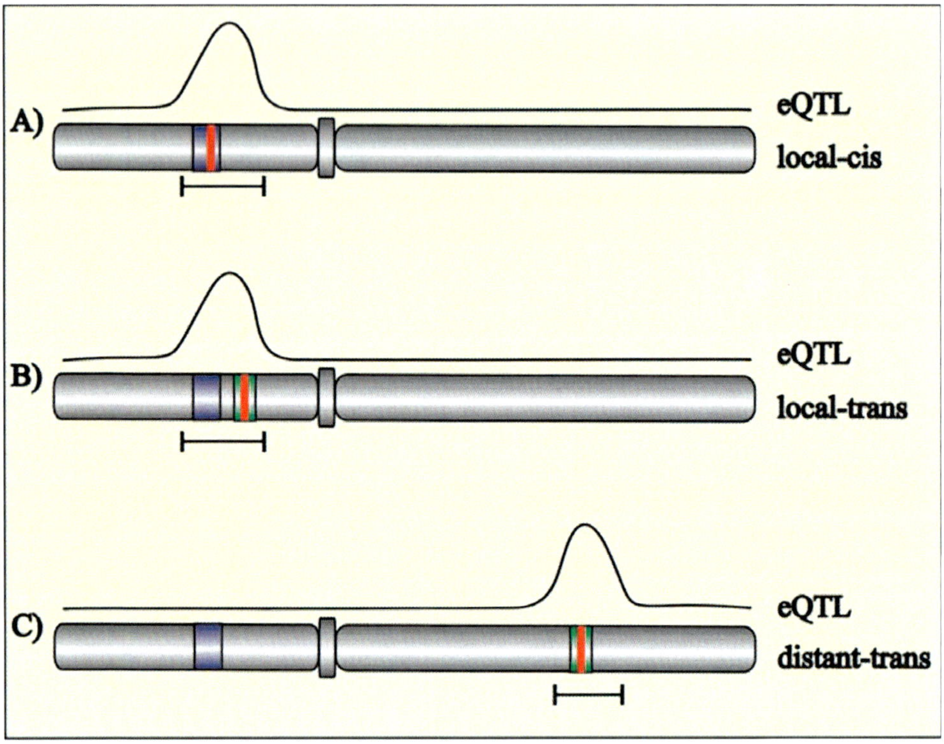


**Figure 1:** Classification of eQTLs (solid line) based on the expression of the gene under study (purple box) and location of the causal polymorphism (red bar). A) local *cis*-eQTL, result from allelic variation of the gene under study. B) local *trans*-eQTL, causal polymorphism within the eQTL confidence interval but not inside the gene under study (no allele specific expression) C) distant *trans*-eQTL, gene under study is located outside the confidence interval of its eQTL.

However, because of limitations in mapping resolution, eQTL support intervals may still contain multiple genes and as a result the classification of *cis*-eQTLs should be used with care. To discriminate true *cis*-regulatory polymorphisms from local *trans*-regulation, allele specific expression (ASE) assays can be performed [38]. In such assays a transcribed polymorphism is used to enable discrimination between the parental transcripts and test for allele specific expression in an $F_1$ hybrid. Because both parental alleles share the same genetic background in $F_1$

hybrids, and therefore are equally exposed to *trans*-acting factors, any difference in expression can only be explained by true *cis*-acting variation. Usually, ASE-assays are performed by single gene qRT-PCR approaches but the recent development of whole genome SNP-tile microarrays (*e.g.,* in Arabidopsis) enables the simultaneous testing of genome-wide ASE [39].

Although expression differences are treated as quantitative traits in mapping approaches, qualitative differences, characterized by a total lack of expression for one of the allelic variants, can also be observed. The variation in a measurable detection signal can be due to differences in hybridization efficiency, which can be confirmed with genomic DNA hybridization, or genuine loss of transcription. Hybridization efficiency differences are often caused by polymorphisms in the complementary sequences of the microarray probes or mRNA splice variation and are not necessarily accompanied by transcription differences. True transcription variation however, can be caused by strong polymorphisms in promoter regions, premature stop mutations and even the complete absence of genes in one of the parental lines [40]. Both hybridization and true transcription variation will lead to strong *cis*-eQTLs which can subsequently be used as molecular markers, allowing the construction of high-resolution maps [41, 42].

## Regulation *in trans*

The majority of differentially expressed genes will show a quantitative expression profile with complex inheritance patterns. This is because in general genes are regulated by many independent factors which can show up as *trans*-eQTLs. Because of the multiplicity of regulators and the often-observed epistasis between them, each *trans*-eQTL can have a relatively small effect. In addition, compared to the direct regulation of *cis*-eQTLs, the accumulation of stochastic variation in the expression of *trans*-regulated genes is indirectly also determined by the expression variation of one or more regulators. As a result the detected number of *trans*-eQTLs relative to the number of *cis*-eQTLs drops when the stringency for detection is increased [43].

Whereas *cis*-eQTLs are inherently associated with the gene in which they reside, a single gene can be responsible for the appearance of multiple *trans*-eQTLs throughout the genome. As a consequence the genome-wide distribution of *cis*-eQTLs is dependent on local gene density, although variation in chromatin structure can have an impact on the exposure of eQTLs. The distribution of *trans*-eQTLs however, can deviate substantially from what can be expected based on gene density. The identification of so-called hot spots, genomic regions with a high density of *trans*-eQTLs, can be explained by major regulators, *e.g.*, transcription factors, which influence the expression of many downstream genes. In Arabidopsis this was illustrated by the large number of genes mapping to the *ERECTA* locus, a gene well-known for its pleiotropic effects on many morphological and developmental traits [23]. These findings suggest that the effects of key-regulators in gene expression are progressed to the phenotypic level. This was recently confirmed in a QTL study comparing transcript, protein and metabolite data with phenotypic traits [44]. Here, only a limited number of QTL hot spots with major, system-wide effects were detected, indicating that most of the genotypic variation is phenotypically buffered. These findings support the theory of biological robustness where hotspots indicate fragilities in this genetic buffering system [45]. Until now only a few reported hotspots have been verified and the number of detected hotspots is far from consistent between different genetical genomics studies. The latter reflects differences in the analyzed populations, species and conditions used and additionally might be the consequence of different statistical procedures used to identify eQTLs [46]. Because of the difficulties in cloning QTLs and the large biological relevance of hotspots, additional sources of information are often used to reduce the number of candidate genes or even predict the causal regulator. Such methods use information on gene ontology, (co-)expression, transcription factor binding sites and targets, ChIP-Seq and protein-protein interaction [47]. Together with computational methods such as regulatory modeling this can severely reduce the number of candidate genes and prioritize remaining candidates for further experimentation (Fig. **2**).

## GENETICAL GENOMICS IN PLANTS

As discussed above many principles of genetic regulation are shared among different phylogenetic taxa. Not all species however are equally suited for large-scale experimentation. Sometimes evolutionary distances withhold translation of biological relevant findings in less conserved mechanisms, *e.g.,* in yeast and Drosophila, or long generation times, inbreeding depression and moral and ethical issues hinder experimentation, *e.g.,* in humans and other mammals. Plants, representing one of the largest kingdoms, are therefore often used to test concepts in genetic studies. The ease to generate large families from experimental crosses and the ability to store genotypes in the form of seeds or clonal propagation make plants ideal subjects to study the mechanistic basis of genetic regulation of traits.

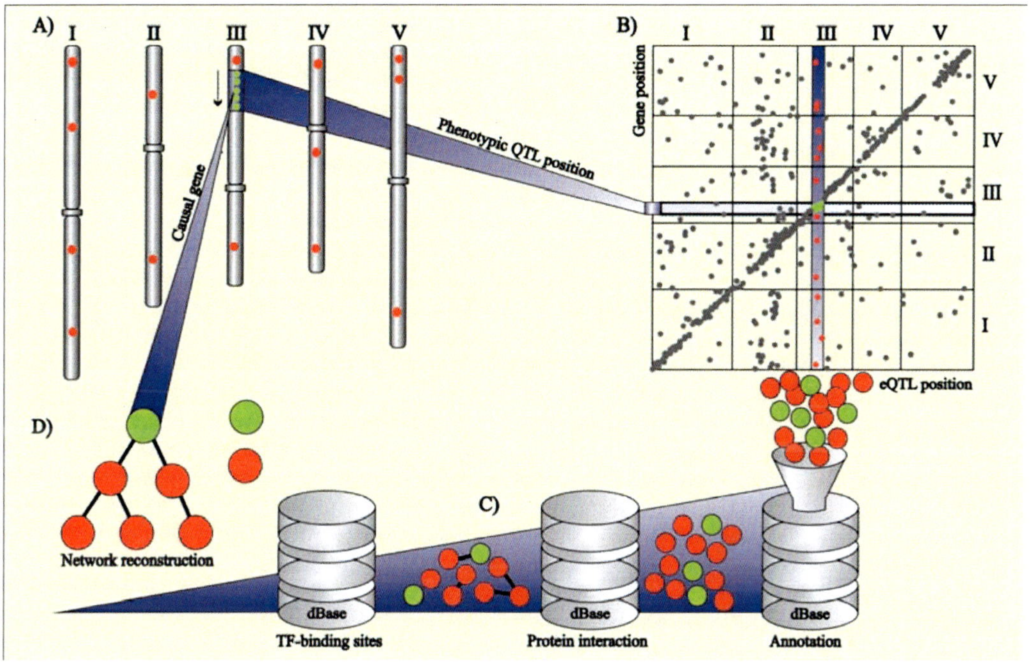


**Figure 2:** Schematic representation of candidate gene selection with a genetical genomics approach. A) Five chromosomes of Arabidopsis thaliana (I-V) with a QTL support interval for a phenotypic trait indicated on chromosome III. B) eQTL plot showing the position of genes and their corresponding eQTLs. Genes within the support interval can be causal for the observed QTL, of which *cis*-regulated genes, indicated in green, represent the strongest candidates. Genes outside the QTL support interval but regulated *in trans* by the same locus, indicated in red, might be involved in the biological process under study and represent downstream effects of the QTL. C) Available prior information of the selected genes such as gene onthology and biological interaction data can assist in limiting the number of genes to those most likely involved in the trait under study. D) Connectivity between the remaining genes is then used to construct maximum likelihood hypothetical regulatory networks which will suggest the strongest candidate regulator gene causal for the observed phenotypic QTL.

## Arabidopsis as a Reference Plant

The comprehensive resources which are available for *Arabidopsis thaliana*, such as a whole genome sequence, a large collection of natural variants and an ever-increasing number of molecular tools, made it the favorable model for genetical genomics research. As a non-obligate selfing species Arabidopsis combines the ability to cross-pollinate with high tolerance to inbreeding. Together with its short generation time and high reproductive success rate this enables the fast generation of large experimental populations such as Recombinant Inbred (RI) and Introgression Line (IL) populations. The availability and immortal character of such populations enable the accurate estimation of phenotypic values through replicated measurements and allows the testing of traits in different environments [48].

Traditionally QTL studies of 'classical' physiological traits in RIL populations are followed by mendelizing detected QTLs in near isogenic lines (NILs) for detailed analyses. By isolating QTLs from their genetic background it becomes much simpler to study their genetic effect and relate resulting phenotypes to other processes. Because it is expected that much of the phenotypic variation is the resultant of differences in gene expression and phenotypic perturbation in turn leads to transcriptional reprogramming, data mining for relationships between trait values and expression levels has become a common tool [49]. Very often mutants, knockouts or over-expression lines are used for these purposes, in which the effect of a single gene perturbation is tested on both the phenotypic and the expression level. For complex traits however, the causal genes leading to altered phenotypes are often not known and QTL analyses only identify genomic regions containing such genes. Nevertheless, using RIL populations to identify QTLs for a phenotypic trait and subsequently analyzing NILs for expression differences can be a powerful alternative to explore the functional relationship between genotype and phenotype (*e.g.,* [50, 51]). Although the regions spanned by NILs can still contain hundreds of genes, of which many may display allelic variation between accessions, the *cis*-regulated genes are strong candidates explaining phenotypic diversity. High detection stringency can limit the number of differentially expressed genes to a reasonable number of candidate genes with strong local eQTLs [50].

The availability of a whole genome sequence in Arabidopsis provides unique opportunities, especially when multiple (epistatic) phenotypic QTLs are detected. Knowing the position of genes allows the identification of strong *cis*-regulated genes collocating with phenotypic QTLs. An early eQTL study in Arabidopsis analyzed genome-wide gene expression in a limited population of only 30 individuals, mimicking shoot regeneration conditions [22]. Two of the eQTL hotspots found coincided with shoot regeneration QTLs. The most significant eQTLs within these hotspot regions showed local chromosal linkage with their corresponding genes but the majority acted distantly. These results suggest that heritable *cis*-regulated expression changes of key-regulators determine *in trans* the expression of many genes related to differences in shoot regeneration efficiency between accessions. It also indicates that a long signaling cascade may exist between the causal genotypic polymorphism and the eventual phenotype.

In contrast to the former study it is not always necessary to combine phenotypic measurements with expression analysis. Often, many genes are known to play a role in the exposure of certain traits without knowledge about the genetic regulation of these genes. Specific analysis of such genes can help to identify common regulators. In the first genome-wide eQTL study in Arabidopsis, using a complete RIL population (162 lines), this concept was used to predict possible key-regulators of flowering time and circadian rhythms [23]. The benefits of using large populations for eQTL studies became also apparent in another study where expression analyses were performed in a RIL population of 211 individuals [25]. Whereas in the majority of cases only a single QTL could be detected per differentially expressed gene in the aforementioned studies, here the expression of many genes was controlled by multiple eQTLs. Moreover, a much larger fraction of genetically regulated genes was identified with a higher proportion of *trans*-regulated genes of which the vast majority exhibited small effects.

The studies performed in Arabidopsis show that the statistical power to detect eQTLs depends largely on population size. Nonetheless, it cannot be excluded that differences in the analyzed tissues, developmental stages and populations used, such as parental variation, linkage distortion and recombination frequency, are responsible for part of the observed differences. All studies however, clearly demonstrated that variation in gene expression is for a large part genetically controlled, with much stronger effects of *cis*-eQTLs compared to *trans*-eQTLs. In general, *cis*-eQTLs also exhibit much higher heritability values and are obvious candidates to act as causal regulators of genes showing *trans*-eQTLs in the hotspots that could be detected in each of the discussed studies. The detection of regulatory loci for gene expression and the elucidation of their interaction networks might therefore provide the research community with a powerful tool to unravel the complex nature of natural variation in quantitative traits.

## Applications in Crop Species

Genetical genomics studies in Arabidopsis and other model species have shown the enormous benefits of the availability of an annotated genome sequence. However, until now full annotated genome sequence information for agronomical important species is only available for a limited number of species, including *Oryza sativa* [52, 53], *Populus trichocarpa* [54], *Vitis vinifera* [55] and papaya [56]. This relatively low number of sequenced crop species can be explained by their often immense (polyploid) genome sizes and the highly repetitive nature of many crop genomes [57]. Nevertheless, sequence efforts for many more species, are ongoing and the increasing power of next-generation sequencing will soon lead to an almost unrestricted availability of genomic sequence information. Although an annotated genome is a valuable resource for the comparison of the genomic position of genes and their respective eQTLs, for most crop species this is not feasible yet. Nonetheless, several studies in crops for which genetic maps are available have shown that comprehensive genetical genomics approaches are possible without the need for annotated genome sequences [58-63].

Illustratively, one of the first large genetical genomics experiments was performed in an economically important species, *viz. Eucalyptus* [62]. QTL analysis of transcript levels of lignin-related genes showed that their mRNA abundance is regulated by two genetic loci coinciding with QTLs for stem diameter growth. Genetic mapping of some of the candidate genes showed that most of the lignin genes are under control of a *trans* eQTL hotspot which suggests that transcription of many of the genes in this pathway are under a higher level of coordinated control. A strong *cis*-regulated gene encoding S-adenosylmethionine synthase, collocating with the growth and transcription QTLs, was presented as the possible rate limiting step in lignin biosynthesis and as such a strong candidate for the observed QTLs [62].

In some crops the required availability of genomic sequence data for large-scale classification of *cis/trans* eQTLs can be circumvented by making use of synteny with other species. In wheat, synteny with rice was used to assist the physical mapping of wheat genes [64]. A genetical genomics approach was conducted in a segregating population of 41 doubled haploid (DH) lines to study agronomic important seed quality parameters. Assuming that the most significantly different expressed genes were *cis*-regulated, a selection of genes was subjected to synteny analyses. This enabled the positioning of genes with biological relevant linkage to phenotypic traits in a species for which full genome sequence is not available yet.

In the absence of genome-wide micro-arrays, expressed sequence tag (EST) libraries allow the construction of species specific sub genome-scale microarrays. In maize, cell-wall digestibility, which is the major target for improving the feeding value of forage maize, was analyzed in a RIL population [63]. In addition forty extreme RIL lines were hybridized on a small microarray with 439 preselected candidate ESTs for cell-wall digestibility genes for which 89 eQTLs could be mapped. One eQTL hotspot collocated with a cell-wall digestibility related QTL [63]. Nowadays, RNA sequencing has become a feasible alternative for microarray hybridization. This would provide an unprecedented resolution, allowing both precise expression monitoring and polymorphism analysis [65]. The application of genetical genomics approaches can be of special interest here when the detection of eQTLs is combined with ASE assays. The thus identified *cis*-regulated genes can then be positioned on the genetic map where they may serve as candidate genes underpinning phenotypic QTLs.

An interesting alternative for species for which no (EST) sequence information is available at all, and hence no microarrays can be produced, is a gel-based cDNA-AFLP approach [66]. Here AFLP band intensities, reflecting expression differences, are profiled for a large proportion of the transcribed gene pool enabling standard eQTL analyses procedures. AFLP bands showing significant eQTLs can subsequently be sequenced to obtain the identity of the gene from which the fragment derived. Additionally, the cDNA-AFLPs can be used to construct a genetic map.

The examples given above show that genetical genomics is not necessarily restricted to model species but can be applied to any species in which experimental crosses are possible even in the absence of genomic sequence or genetic map information. The potential of combining phenotypic QTL analysis with gene expression traits is shown in a number of economically important species, *e.g.,* maize [67], barley [68], rice [69], *Populus* [58], cotton [59], jatropha [70] and sunflower [61]. The application of genetical genomics is particularly promising in breeding programs of crops that take advantage of hybrid vigour. The eQTLs involved in heterosis will segregate consistently in a $F_1$ backcross population thereby identifying valuable targets for marker assisted breeding for the best combination of alleles in the parents of the hybrid [71].

## NETWORK RECONSTRUCTION

Genetical genomics harbors the potential to dissect the genetic regulation of a specific biological process. Therefore, methods to reconstruct regulatory networks from eQTL data have obtained much attention. Prioritizing on *cis*-eQTLs that collocate with a phenotypic QTL is a valuable approach for causal gene discovery. A good example where this approach was combined with functional annotation is the identification of ELF3 as a causal gene for the shade avoidance response in *Arabidopsis* [72]. However, in many other cases little is known about the global regulation, interaction and function of genes that control a biological process. Identification of a set of genes with a *trans*-eQTL at an identical position can help to dissect genetic variation that is influencing an entire pathway and can lead to the identification of initiating polymorphisms upstream in a network [73]. Questions about the regulatory level at which *trans* polymorphisms act in the global gene expression network and what their effect is on phenotypic variation and heritability can only be addressed when eQTLs are further dissected.

With a genetical genomics approach one can use the natural genetic variation as a source of perturbations to elucidate the structure of networks. In a summation approach eQTLs for all genes in the analysis are simply superimposed to identify common regions which control many genes [15]. Such an approach does not require any *a priori* network information but applies subsequent Gene Set Enrichment Analysis (GSEA) using gene onthology (GO) annotation or other descriptors to test whether selected genes share a common biological function [74]. If the network under study is largely known or at least predicted, an *a priori* analysis can be performed. Here, the expression levels of individual genes in the network are converted into a common measure for the expression level of the entire network which is then used as the trait for QTL analysis. This strategy was tested in an Arabidopsis RIL population for 20 gene expression networks and resulted in statistically significant network variation for eighteen of the 20 predefined networks [75]. Combining summation, GSEA and *a priori* network analyses allows the generation of a more specific hypothesis about phenotypic effects of network eQTLs. In a study using 175 genes, selected to be involved in regulation of flowering and circadian rhythms, 83 genes showed an eQTL [23]. By combining co-expression analysis, which becomes feasible for microarray compendia of large populations, and positional information of genes and their eQTLs, it was possible to construct regulatory networks of key-regulators and their target genes, predicting unknown relationships and confirming common knowledge.

Pre-selection of known pathways can obviously hinder the elucidation of novel networks in a species, for which much effort is made to develop methods to translate eQTL data into network information using an *a posteriori* approach. As the precise balance of active components within a tightly controlled biological pathway is in part maintained by coordinately regulated gene expression, this creates possibilities to model networks by exploring co-expression of untargeted genes. To validate this hypothesis, gene expression in liver from a population of 60 mice with variation in diabetes susceptibility was analyzed [76]. The combination of correlation analysis across a genetic dimension and linkage mapping enabled the identification of regulatory networks, functional predictions for uncharacterized genes and characterization of novel members of known pathways. A similar approach in *Drosophila*, complemented with information about gene ontology, tissue specific expression and transcription factor binding sites, led to the construction of multiple interconnected networks with biological relevance for phenotypic traits [77].

Understanding the mechanisms underlying trait regulation requires the identification of specific causal polymorphisms. For this purpose sophisticated self-learning algorithms have been developed which make use of conservation, type and position of a particular SNP to prioritize causal regulators by estimating the likeliness that it plays a causal role in gene expression variation [78]. Extending such approaches might also provide the means to distinguish whether variation in gene expression or a regulatory network is the cause or a consequence of an altered phenotype, resulting in the construction of probabilistic directional networks [79]. Defining such causal networks is also known as reverse engineering, because it aims at understanding how the system works as an integrated whole instead of only defining the functionally related components.

## NEXT LEVEL NETWORKS: INTEGRATION OF OTHER 'OMICS' DATA

Although phenotypic variation can be partly explained by genetic variation in gene expression, this alone does not fully cover the possible differences in the regulatory mechanisms of an organism. Similar transcript levels of allelic gene variants can still result in varying protein levels because of variance in translational activity, protein degradation and post-translational modifications [80]. Furthermore, variation in coding sequences can alter protein function resulting in a flexible metabolome in terms of chemical structure and function [81]. Integrating 'omics' data such as gene expression, SNPs, metabolomics and proteomics in genetic studies can therefore reduce the number of candidate genes for a given QTL from hundreds to a manageable list without excluding regulatory mechanisms *a priori*. Because of the analytical complexity in analyzing large numbers of protein samples, genetical proteomics studies are limited (*e.g.*, [82]) but advances made in biochemical detection have already enabled the large-scale untargeted genetic analysis of metabolic content [83-85].

The complex relationship between different levels of regulation was illustrated in a study integrating parallel QTL analyses of the expression of genes, activity of encoded enzymes and metabolites involved in primary carbohydrate metabolism [86]. It could be shown that regulation acted on each of the intermediary levels of the path from genotype to phenotype. Although seemingly specific independent regulation could be observed for each analyzed trait, a strong interconnectivity existed between them resulting in coherent systematic differences between population individuals.

The importance of the tight regulation of such an essential component in plant development as primary metabolism was also demonstrated in an Arabidopsis RIL population where plant biomass was related to the metabolic profile [87]. Again, no relationship could be observed between individual metabolites and plant growth but a strong canonical correlation was observed between biomass and a specific combination of metabolites in central metabolism. The power of large-scale metabolomic profiling combined with detailed morphological analysis was also shown in tomato [84]. Significant QTLs could be detected for the accumulation of a large number of primary metabolites together with loci that modify yield-associated traits. With this information a correlation network revealing associations between phenotype, metabolic content and nutritional value could be generated. These studies show that analyzing phenotypic traits and metabolic profiles in a genetic mapping population has great potential for the generation of biomarkers in breeding programs.

Whereas primary metabolites are essential in central metabolism governing growth and development, plants also accumulate large amounts of secondary metabolites. These are believed to be less essential but may play an important role in the adaptation of plants to local environments. Since Arabidopsis can be found in a wide variety of habitats, variation in secondary metabolism might explain much of the evolutionary success of the species. A large untargeted screen of variation in secondary metabolic composition indeed revealed a high proportion of genetically controlled compounds [83]. The highly flexible nature of the metabolome was clearly shown by the fact that more than one-third of the compounds present in the RILs were not detected in either parent but were the result of recombination in biosynthesis pathways. The genetic information obtained from such studies is of great value for the construction of molecular biosynthesis networks, especially if they can be combined with expression data.

This strategy was applied in the genetic analysis of glucosinolate biosynthetic networks which were studied at both the transcriptional and metabolic level [88]. In all cases, variation in gene expression also affected the accumulation of metabolites but epistasis was detected more frequently for metabolic traits as compared to transcript traits. Within such an *a priori* defined framework it was possible to identify and unravel complex regulatory mechanisms like metabolic feed-back loops in which metabolic content regulated gene expression and *vice versa*.

The examples discussed here highlight the technological advances made in high-throughput characterization of the transcriptome, the proteome and metabolome which enables an integrated multidisciplinary approach to unravel the regulatory mechanisms involved in natural variation of complex traits.

## FUTURE CHALLENGES

Although much progress is being made in understanding the influence of genetic factors on a biological system we still have limited understanding of the interplay between environment and genetic factors. The discovery of molecular networks with genetical genomics approaches is often limited to a single experimental condition. An interesting concept, called generalized genetical genomics, studies controlled environmental perturbations combined with genetical genomics [89]. This generalization of genetical genomics will detect how the response to environmental changes is influenced by the genotype (*i.e.,* genotype x environment interactions). Here, spatial and temporal variation can also be regarded as different environments since specific tissues and developmental stages often determine the biological context in which regulatory networks function.

The advances in next generation sequence technology will continue to produce huge amounts of sequence data. Good examples are the human 1000 [90] and the 1001 Arabidopsis [91] genome projects which aim at resequencing over 1000 different humans and accessions respectively. However, *de novo* sequencing of economically important or phylogenetic strategically chosen species is of equal importance. The accumulation of genomic information, in combination with genetical genomics approaches, will enable the precise definition of functional important polymorphisms and their role in adaptation to changing environments and species formation. Having access to complete genome sequences also enables the generation of full genome tiling arrays for different (crop) species, which have been proven to be very useful for expression profiling [92, 93]. When used within a genetical genomics approach this offers unique features to elucidate the genetics behind the mechanistic basis of transcriptional differences. For Arabidopsis for instance, a SNPtile microarray was developed harboring tiling probes covering both strands of the genome and in addition probes for genome-wide detection of SNPs and CpG methylation [39]. A properly designed genetical genomics study using such arrays might reveal genetic variation for gene expression, alternative splicing, regulation of *cis*-natural antisense transcripts, allele specific expression and epigenetic regulation.

As a result of developments in SNP-discovery and platforms for genotyping large collections of individuals, the application of Linkage disequilibrium (LD) mapping for complex traits has become within reach. LD or association mapping detects the non-random inheritance of alleles at separate loci located on the same chromosome. In an experimental $F_2$ or RIL population the genetic variation is limited to the extent of natural variation present in the parental lines and resolution depends on the recombination frequency within and size of the population. In contrast LD mapping makes use of large collections of natural (wild) accessions or elite breeding lines, sampling a much larger fraction of the natural variation present within a species. Moreover, it benefits from the much higher frequency of recombination events accumulated during the evolutionary history of a species allowing higher resolution mapping [94]. The extent of LD varies between species and traits analyzed but the gain in resolution relative to experimental populations lies in the order of magnitudes, equally increasing the need for dense marker spacing to enable genome wide scans [95]. This high number of necessary markers has always been a big limitation for LD mapping but next generation sequencing will tremendously increase the available number of markers. Combining LD mapping with linkage mapping in structured populations can further help to distinguish true positive associations from false ones [96]. Therefore, we see great potential for phenotyping and expression profiling of LD populations to detect causal genes for natural variation and enable marker-assisted selection in breeding programs.

## CONCLUDING REMARKS

Since its introduction the concept of genetical genomics has proven to be a powerful approach to dissect genetic variation. Studies in crop species revealed major *cis*-eQTLs which collocated with important phenotypic traits and therefore will facilitate faster crop improvement. The genetical genomics studies in model species help to understand the extent of genetic variation and much effort is spent to develop statistical tools for building and elucidating causal networks. Recent developments of inexpensive high-throughput sequencing techniques and next generation tiling microarrays will soon create opportunities to extend genetical genomics to unravel the genetic variation of gene expression, alternative splicing, allele specific expression and epigenetic polymorphisms. Similarly, continuing technological developments have increased the power of both proteomic and metabolomic approaches. Integration of phenotypic, genetic, transcriptomic, proteomic and metabolomic data will enable accurate and detailed network reconstruction. This will ultimately result in the elucidation of the molecular pathways involved in complex phenotypic traits.

## CONFLICT OF INTEREST

The author(s) confirm that this chapter content has no conflict of interest.

## ACKNOWLEDGEMENTS

This work was supported by grants from the Netherlands Organization for Scientific Research (STW 10027), VENI scheme (863.08.019) and the Centre for Biosystems Genomics (CBSG, Netherlands Genomics Initiative).

## DISCLOSURE

This is an updated report of the earlier review article by R. V.L. Joosen, W. Ligterink, H. W.M. Hilhorst and J.J.B. Keurentjes "Advances in genetical genomics of plants". Curr. Genomics, 2009, 10: 540-549.

## REFERENCES


[1] Galbraith, D.W., DNA microarray analyses in higher plants. *OMICS A J. Integr. Biol.*, **2006**, *10*(4): p. 455-473.

[2] Joosen, R., Cordewener, J., Supena, E.D.J., Vorst, O., Lammers, M., Maliepaard, C., Zeilmaker, T., Miki, B., America, T., Custers, J., Boutilier, K. *Combined transcriptome and* 20 *Advances in Genome Science, Vol. 2 Joosen et al. proteome analysis identifies pathways and markers associated with the establishment of rapeseed microspore-derived embryo development.* Plant Physiology, **2007**, 144(1): p. 155- 172.

[3] Brazma, A., Parkinson, H., Sarkans, U., Shojatalab, M., Vilo, J., Abeygunawardena, N., Holloway, E., Kapushesky, M., Kemmeren, P., Lara, G.G., Oezcimen, A., Rocca-Serra, P., Sansone, S.A. ArrayExpress - A public repository for microarray gene expression data at the EBI. *Nucleic Acids Res*, **2003**, *31*(1): p. 68-71.

[4] Toufighi, K., Brady, S.M. , Austin, R., Ly, E., Provart, N.J. The botany array resource: e-Northerns, expression angling, and promoter analyses. *Plant J.*, **2005**, *43*(1): p. 153- 163.

[5] Zimmermann, P., L. Hennig, and W. Gruissem, Gene-expression analysis and network discovery using Genevestigator. *Trends in Plant Science*, **2005**, *10*(9): p. 407-409.

[6] Fujita, A., J. Sato, H.M. Garay-Malpartida, M.C. Sogayar, C.E. Farreira, and S. Miyano, Modeling nonlinear gene regulatory networks from time series gene expression data. *Journal of Bioinformatics and Computational Biology*, **2008**, *6*(5): p. 961-979.

[7] Gupta, A., C.D. Maranas, and R. Albert, Elucidation of directionality for co-expressed genes: Predicting intra-operon termination sites. *Bioinformatics*, **2006**, *22*(2): p. 209-214.

[8] Alonso-Blanco, C., M.G. Aarts, L. Bentsink, J.J.B. Keurentjes, M. Reymond, D. Vreugdenhil, and M. Koornneef, What has natural variation taught us about plant development, physiology, and adaptation? *Plant Cell*, **2009**, *21*(7): p. 1877-96.

[9] Ashikari, M. and M. Matsuoka, Identification, isolation and pyramiding of quantitative trait loci for rice breeding. *Trends Plant Sci*, **2006**, *11*(7): p. 344-50.

[10] Salvi, S. and R. Tuberosa, To clone or not to clone plant QTLs: Present and future challenges. *Trends in Plant Science*, **2005**, *10*(6): p. 297-304.

[11] Jansen, R.C. and J.P. Nap, Genetical genomics: the added value from segregation. *Trends Genet*, **2001**, *17*(7): p. 388-91.

[12] Joosen, R.V., W. Ligterink, H.W. Hilhorst, and J.J.B. Keurentjes, Advances in genetical genomics of plants. *Curr Genomics*, **2009**, *10*(8): p. 540-9.

[13] Keurentjes, J.J.B., M. Koornneef, and D. Vreugdenhil, Quantitative genetics in the age of omics. *Current Opinion in Plant Biology*, **2008**, *11*(2): p. 123-128.

[14] Brem, R.B., G. Yvert, R. Clinton, and L. Kruglyak, Genetic dissection of transcriptional regulation in budding yeast. *Science*, **2002**, *296*(5568): p. 752-755.

[15] Schadt, E.E., S.A. Monks, T.A. Drake, A.J. Lusis, N. Che, V. Colinayo, T.G. Ruff, S.B. Milligan, J.R. Lamb, G. Cavet, P.S. Linsley, M. Mao, R.B. Stoughton, and S.H. Friend, Genetics of gene expression surveyed in maize, mouse and man. *Nature*, **2003**, *422*(6929): p. 297-302.

[16] Yvert, G., R.B. Brem, J. Whittle, J.M. Akey, E. Foss, E.N. Smith, R. Mackelprang, and L. Kruglyak, *Trans*-acting regulatory variation in Saccharomyces cerevisiae and the role of transcription factors. *Nat Genet*, **2003**, *35*(1): p. 57-64.

[17] Leach, L.J., Z. Zhang, C. Lu, M.J. Kearsey, and Z. Luo, The role of *cis*-regulatory motifs and genetical control of expression in the divergence of yeast duplicate genes. *Mol Biol Evol,* **2007**, *24*(11): p. 2556-65.

[18] Bing, N. and I. Hoeschele, Genetical genomics analysis of a yeast segregant population for transcription network inference. *Genetics*, **2005**, *170*(2): p. 533-42.

[19] Li, Y., O.A. Alvarez, E.W. Gutteling, M. Tijsterman, J. Fu, J.A. Riksen, E. Hazendonk, P. Prins, R.H. Plasterk, R.C. Jansen, R. Breitling, and J.E. Kammenga, Mapping determinants Genetical Genomics of Plants *Advances in Genome Science, Vol. 2* 21 *of gene expression plasticity by genetical genomics in C. elegans.* PLoS genetics, **2006**, 2(12).

[20] Wittkopp, P.J., B.K. Haerum, and A.G. Clark, Evolutionary changes in *cis* and *trans* gene regulation. *Nature*, **2004**, *430*(6995): p. 85-88.

[21] Hsieh, W.P., G. Passador-Gurgel, E.A. Stone, and G. Gibson, Mixture modeling of transcript abundance classes in natural populations. *Genome Biology*, **2007**, *8*(6).

[22] DeCook, R., S. Lall, D. Nettleton, and S.H. Howell, Genetic regulation of gene expression during shoot development in Arabidopsis. *Genetics*, **2006**, *172*(2): p. 1155-1164.

[23] Keurentjes, J.J.B., J. Fu, I.R. Terpstra, J.M. Garcia, G. van den Ackerveken, L.B. Snoek, A.J. Peeters, D. Vreugdenhil, M. Koornneef, and R.C. Jansen, Regulatory network construction in Arabidopsis by using genome-wide gene expression quantitative trait loci. *Proc Natl Acad Sci U S A*, **2007**, *104*(5): p. 1708-13.

[24] Potokina, E., A. Druka, Z. Luo, R. Wise, R. Waugh, and M. Kearsey, Gene expression quantitative trait locus analysis of 16 000 barley genes reveals a complex pattern of genome-wide transcriptional regulation. *Plant J*, **2008**, *53*(1): p. 90-101.

[25] West, M.A.L., K. Kim, D.J. Kliebenstein, H. Van Leeuwen, R.W. Michelmore, R.W. Doerge, and D.A. St. Clair, Global eQTL mapping reveals the complex genetic architecture of transcript-level variation in Arabidopsis. *Genetics*, **2007**, *175*(3): p. 1441-1450.

[26] Bystrykh, L., E. Weersing, B. Dontje, S. Sutton, M.T. Pletcher, T. Wiltshire, A.I. Su, E. Vellenga, J. Wang, K.F. Manly, L. Lu, E.J. Chesler, R. Alberts, R.C. Jansen, R.W. Williams, M.P. Cooke, and G. De Haan, Uncovering regulatory pathways that affect hematopoietic stem cell function using 'genetical genomics'. *Nature Genetics*, **2005**, *37*(3): p. 225-232.

[27] Chesler, E.J., L. Lu, S. Shou, Y. Qu, J. Gu, J. Wang, H.C. Hsu, J.D. Mountz, N.E. Baldwin, M.A. Langston, D.W. Threadgill, K.F. Manly, and R.W. Williams, Complex trait analysis of gene expression uncovers polygenic and pleiotropic networks that modulate nervous system function. Nature Genetics, **2005**, *37*(3): p. 233-242.

[28] Hubner, N., C.A. Wallace, H. Zimdahl, E. Petretto, H. Schulz, F. Maciver, M. Mueller, O. Hummel, J. Monti, V. Zidek, A. Musilova, V. Kren, H. Causton, L. Game, G. Born, S. Schmidt, A. Müller, S.A. Cook, T.W. Kurtz, J. Whittaker, M. Pravenec, and T.J. Aitman, Integrated transcriptional profiling and linkage analysis for identification of genes underlying disease. *Nature Genetics*, **2005**, *37*(3): p. 243-253.

[29] Monks, S.A., A. Leonardson, H. Zhu, P. Cundiff, P. Pietrusiak, S. Edwards, J.W. Phillips, A. Sachs, and E.E. Schadt, Genetic inheritance of gene expression in human cell lines. *American Journal of Human Genetics*, **2004**, *75*(6): p. 1094-1105.

[30] Stranger, B.E., A.C. Nica, M.S. Forrest, A. Dimas, C.P. Bird, C. Beazley, C.E. Ingle, M. Dunning, P. Flicek, D. Koller, S. Montgomery, Tavaré, S., P. Deloukas, and E.T. Dermitzakis, Population genomics of human gene expression. *Nature Genetics*, **2007**, *39*(10): p. 1217-1224.

[31] Dixon, A.L., L. Liang, M.F. Moffatt, W. Chen, S. Heath, K.C.C. Wong, J. Taylor, E. Burnett, I. Gut, M. Farrall, G.M. Lathrop, G.R. Abecasis, and W.O.C. Cookson, A genomewide association study of global gene expression. *Nature Genetics*, **2007**, *39*(10): p. 1202- 1207.

[32] Göring, H.H.H., J.E. Curran, M.P. Johnson, T.D. Dyer, J. Charlesworth, S.A. Cole, J.B.M. Jowett, L.J. Abraham, D.L. Rainwater, A.G. Comuzzie, M.C. Mahaney, L. Almasy, J.W. MacCluer, A.H. Kissebah, G.R. Collier, E.K. Moses, and J. Blangero, *Discovery of* **22** ***Advances in Genome Science, Vol. 2 Joosen et al.*** *expression QTLs using large-scale transcriptional profiling in human lymphocytes.* Nature Genetics, **2007**, 39(10): p. 1208-1216.

[33] Myers, A.J., J.R. Gibbs, J.A. Webster, K. Rohrer, A. Zhao, L. Marlowe, M. Kaleem, D. Leung, L. Bryden, P. Nath, V.L. Zismann, K. Joshipura, M.J. Huentelman, D. Hu-Lince, K.D. Coon, D.W. Craig, J.V. Pearson, P. Holmans, C.B. Heward, E.M. Reiman, D. Stephan, and J. Hardy, A survey of genetic human cortical gene expression. *Nature Genetics*, **2007**, *39*(12): p. 1494-1499.

[34] Emilsson, V., G. Thorleifsson, B. Zhang, A.S. Leonardson, F. Zink, J. Zhu, S. Carlson, A. Helgason, G.B. Walters, S. Gunnarsdottir, M. Mouy, V. Steinthorsdottir, G.H. Eiriksdottir, G. Bjornsdottir, I. Reynisdottir, D. Gudbjartsson, A. Helgadottir, A. Jonasdottir, U. Styrkarsdottir, S. Gretarsdottir, K.P. Magnusson, H. Stefansson, R. Fossdal, K. Kristjansson, H.G. Gislason, T. Stefansson, B.G. Leifsson, U. Thorsteinsdottir, J.R. Lamb, J.R. Gulcher, M.L. Reitman, A. Kong, E.E. Schadt, and K. Stefansson, Genetics of gene expression and its effect on disease. *Nature*, **2008**, *452*(7186): p. 423-428.

[35] Mitchell-Olds, T. and J. Schmitt, Genetic mechanisms and evolutionary significance of natural variation in Arabidopsis. *Nature*, **2006**, *441*(7096): p. 947-952.

[36] Rockman, M.V. and L. Kruglyak, Genetics of global gene expression. *Nature Reviews Genetics*, **2006**, *7*(11): p. 862-872.

[37] Gibson, G. and B. Weir, The quantitative genetics of transcription. *Trends in Genetics*, **2005**, *21*(11): p. 616-623.

[38] Cowles, C.R., J.N. Hirschhorn, D. Altshuler, and E.S. Lander, Detection of regulatory variation in mouse genes. *Nat Genet*, **2002**, *32*(3): p. 432-7.

[39] Zhang, X., E.J. Richards, and J.O. Borevitz, Genetic and epigenetic dissection of *cis* regulatory variation. *Current Opinion in Plant Biology*, **2007**, *10*(2): p. 142-148.

[40] Gilad, Y. and J. Borevitz, Using DNA microarrays to study natural variation. *Current Opinion in Genetics and Development*, **2006**, *16*(6): p. 553-558.

[41] Borevitz, J.O. and J. Chory, Genomics tools for QTL analysis and gene discovery. *Current Opinion in Plant Biology*, **2004**, *7*(2): p. 132-136.

[42] West, M.A.L., H. Van Leeuwen, A. Kozik, D.J. Kliebenstein, R.W. Doerge, D.A. St. Clair, and R.W. Michelmore, High-density haplotyping with microarray-based expression and single feature polymorphism markers in Arabidopsis. *Genome Research*, **2006**, *16*(6): p. 787-795.

[43] Doss, S., E.E. Schadt, T.A. Drake, and A.J. Lusis, *Cis*-acting expression quantitative trait loci in mice. *Genome Research*, **2005**, *15*(5): p. 681-691.

[44] Fu, J., J.J.B. Keurentjes, H. Bouwmeester, T. America, F.W. Verstappen, J.L. Ward, M.H. Beale, R.C. de Vos, M. Dijkstra, R.A. Scheltema, F. Johannes, M. Koornneef, D. Vreugdenhil, R. Breitling, and R.C. Jansen, System-wide molecular evidence for phenotypic buffering in Arabidopsis. *Nat Genet*, **2009**, *41*(2): p. 166-7.

[45] Kitano, H., Biological robustness. *Nat Rev Genet*, **2004**, *5*(11): p. 826-37.
[46] Breitling, R., Y. Li, B.M. Tesson, J. Fu, C. Wu, T. Wiltshire, A. Gerrits, L.V. Bystrykh, G. De Haan, A.I. Su, and R.C. Jansen, Genetical genomics: Spotlight on QTL hotspots. *PLoS Genetics*, **2008**, *4*(10).
[47] Zhu, J., B. Zhang, E.N. Smith, B. Drees, R.B. Brem, L. Kruglyak, R.E. Bumgarner, and E.E. Schadt, Integrating large-scale functional genomic data to dissect the complexity of yeast regulatory networks. *Nat Genet*, **2008**, *40*(7): p. 854-61.
[48] Paran, I. and D. Zamir, Quantitative traits in plants: Beyond the QTL. *Trends in Genetics*, **2003**, *19*(6): p. 303-306.
[49] Weckwerth, W., M.E. Loureiro, K. Wenzel, and O. Fiehn, Differential metabolic networks unravel the effects of silent plant phenotypes. *Proceedings of the National Academy of Sciences of the United States of America*, **2004**, *101*(20): p. 7809-7814.
[50] Juenger, T.E., T. Wayne, S. Boles, V.V. Symonds, J. McKay, and S.J. Coughlan, Natural genetic variation in whole-genome expression in Arabidopsis thaliana: The impact of physiological QTL introgression. *Molecular Ecology*, **2006**, *15*(5): p. 1351-1365.
[51] Juenger, T.E., J.K. McKay, N. Hausmann, J.J.B. Keurentjes, S. Sen, K.A. Stowe, T.E. Dawson, E.L. Simms, and J.H. Richards, Identification and characterization of QTL underlying wholeplant physiology in Arabidopsis thaliana: ?13C, stomatal conductance and transpiration efficiency. *Plant, Cell and Environment*, **2005**, *28*(6): p. 697-708.
[52] Goff, S.A., D. Ricke, T.H. Lan, G. Presting, R. Wang, M. Dunn, J. Glazebrook, A. Sessions, P. Oeller, H. Varma, D. Hadley, D. Hutchison, C. Martin, F. Katagiri, B.M. Lange, T. Moughamer, Y. Xia, P. Budworth, J. Zhong, T. Miguel, U. Paszkowski, S. Zhang, M. Colbert, W.L. Sun, L. Chen, B. Cooper, S. Park, T.C. Wood, L. Mao, P. Quail, R. Wing, R. Dean, Y. Yu, A. Zharkikh, R. Shen, S. Sahasrabudhe, A. Thomas, R. Cannings, A. Gutin, D. Pruss, J. Reid, S. Tavtigian, J. Mitchell, G. Eldredge, T. Scholl, R.M. Miller, S. Bhatnagar, N. Adey, T. Rubano, N. Tusneem, R. Robinson, J. Feldhaus, T. Macalma, A. Oliphant, and S. Briggs, A draft sequence of the rice genome (Oryza sativa L. ssp. japonica). *Science*, **2002**, *296*(5565): p. 92-100.
[53] Yu, J., S. Hu, J. Wang, G.K.S. Wong, S. Li, B. Liu, Y. Deng, L. Dai, Y. Zhou, X. Zhang, M. Cao, J. Liu, J. Sun, J. Tang, Y. Chen, X. Huang, W. Lin, C. Ye, W. Tong, L. Cong, J. Geng, Y. Han, L. Li, W. Li, G. Hu, J. Li, Z. Liu, Q. Qi, T. Li, X. Wang, H. Lu, T. Wu, M. Zhu, P. Ni, H. Han, W. Dong, X. Ren, X. Feng, P. Cui, X. Li, H. Wang, X. Xu, W. Zhai, Z. Xu, J. Zhang, S. He, J. Xu, K. Zhang, X. Zheng, J. Dong, W. Zeng, L. Tao, J. Ye, J. Tan, X. Chen, J. He, D. Liu, W. Tian, C. Tian, H. Xia, Q. Bao, G. Li, H. Gao, T. Cao, W. Zhao, P. Li, W. Chen, Y. Zhang, J. Hu, S. Liu, J. Yang, G. Zhang, Y. Xiong, Z. Li, L. Mao, C. Zhou, Z. Zhu, R. Chen, B. Hao, W. Zheng, S. Chen, W. Guo, M. Tao, L. Zhu, and H. Yang, A draft sequence of the rice genome (Oryza sativa L. ssp. indica). *Science*, **2002**, *296*(5565): p. 79-92.
[54] Tuskan, G.A., S. Difazio, S. Jansson, J. Bohlmann, I. Grigoriev, U. Hellsten, N. Putnam, S. Ralph, S. Rombauts, A. Salamov, J. Schein, L. Sterck, A. Aerts, R.R. Bhalerao, R.P. Bhalerao, D. Blaudez, W. Boerjan, A. Brun, A. Brunner, V. Busov, M. Campbell, J. Carlson, M. Chalot, J. Chapman, G.L. Chen, D. Cooper, P.M. Coutinho, J. Couturier, S. Covert, Q. Cronk, R. Cunningham, J. Davis, S. Degroeve, A. Dejardin, C. Depamphilis, J. Detter, B. Dirks, I. Dubchak, S. Duplessis, J. Ehlting, B. Ellis, K. Gendler, D. Goodstein, M. Gribskov, J. Grimwood, A. Groover, L. Gunter, B. Hamberger, B. Heinze, Y. Helariutta, B. Henrissat, D. Holligan, R. Holt, W. Huang, N. Islam-Faridi, S. Jones, M.

Jones-Rhoades, R. Jorgensen, C. Joshi, J. Kangasjarvi, J. Karlsson, C. Kelleher, R. Kirkpatrick, M. Kirst, A. Kohler, U. Kalluri, F. Larimer, J. Leebens-Mack, J.C. Leple, P. Locascio, Y. Lou, S. Lucas, F. Martin, B. Montanini, C. Napoli, D.R. Nelson, C. Nelson, K. Nieminen, O. Nilsson, V. Pereda, G. Peter, R. Philippe, G. Pilate, A. Poliakov, J. Razumovskaya, P. Richardson, C. Rinaldi, K. Ritland, P. Rouze, D. Ryaboy, J. Schmutz, J. Schrader, B. Segerman, H. Shin, A. Siddiqui, F. Sterky, A. Terry, C.J. Tsai, E. Uberbacher, P. Unneberg, J. Vahala, K. Wall, S. Wessler, G. Yang, T. Yin, C. Douglas, M. Marra, G. Sandberg, Y. Van de Peer and D. Rokhsar, The genome of black cottonwood, Populus trichocarpa (Torr. & Gray). *Science*, **2006**, *313*(5793): p. 1596-604.

[55] Jaillon, O., J.M. Aury, B. Noel, A. Policriti, C. Clepet, A. Casagrande, N. Choisne, S. Aubourg, N. Vitulo, C. Jubin, A. Vezzi, F. Legeai, P. Hugueney, C. Dasilva, D. Horner, E. Mica, D. Jublot, J. Poulain, C. Bruyere, A. Billault, B. Segurens, M. Gouyvenoux, E. Ugarte, F. Cattonaro, V. Anthouard, V. Vico, C. Del Fabbro, M. Alaux, G. Di Gaspero, V. Dumas, N. Felice, S. Paillard, I. Juman, M. Moroldo, S. Scalabrin, A. Canaguier, I. Le Clainche, G. Malacrida, E. Durand, G. Pesole, V. Laucou, P. Chatelet, D. Merdinoglu, M. Delledonne, M. Pezzotti, A. Lecharny, C. Scarpelli, F. Artiguenave, M.E. Pe, G. Valle, M. Morgante, M. Caboche, A.F. Adam-Blondon, J. Weissenbach, F. Quetier, and P. Wincker, The grapevine genome sequence suggests ancestral hexaploidization in major angiosperm phyla. *Nature*, **2007**, *449*(7161): p. 463-7.

[56] Ming, R., S. Hou, Y. Feng, Q. Yu, A. Dionne-Laporte, J.H. Saw, P. Senin, W. Wang, B.V. Ly, K.L.T. Lewis, S.L. Salzberg, L. Feng, M.R. Jones, R.L. Skelton, J.E. Murray, C. Chen, W. Qian, J. Shen, P. Du, M. Eustice, E. Tong, H. Tang, E. Lyons, R.E. Paull, T.P. Michael, K. Wall, D.W. Rice, H. Albert, M.L. Wang, Y.J. Zhu, M. Schatz, N. Nagarajan, R.A. Acob, P. Guan, A. Blas, C.M. Wai, C.M. Ackerman, Y. Ren, C. Liu, J. Wang, J.K. Na, E.V. Shakirov, B. Haas, J. Thimmapuram, D. Nelson, X. Wang, J.E. Bowers, A.R. Gschwend, A.L. Delcher, R. Singh, J.Y. Suzuki, S. Tripathi, K. Neupane, H. Wei, B. Irikura, M. Paidi, N. Jiang, W. Zhang, G. Presting, A. Windsor, R. Navajas-PÃ©rez, M.J. Torres, F.A. Feltus, B. Porter, Y. Li, A.M. Burroughs, M.C. Luo, L. Liu, D.A. Christopher, S.M. Mount, P.H. Moore, T. Sugimura, J. Jiang, M.A. Schuler, V. Friedman, T. Mitchell-Olds, D.E. Shippen, C.W. Depamphilis, J.D. Palmer, M. Freeling, A.H. Paterson, D. Gonsalves, L. Wang, and M. Alam, The draft genome of the transgenic tropical fruit tree papaya (Carica papaya Linnaeus). *Nature*, **2008**, *452*(7190): p. 991-996.

[57] Burke, J.M., J.C. Burger, and M.A. Chapman, Crop evolution: from genetics to genomics. *Current Opinion in Genetics and Development*, **2007**, *17*(6): p. 525-532.

[58] Street, N.R., O. Skogstrom, A. Sjodin, J. Tucker, M. Rodriguez-Acosta, P. Nilsson, S. Jansson, and G. Taylor, The genetics and genomics of the drought response in Populus. *Plant J*, **2006**, *48*(3): p. 321-41.

[59] An, C., S. Saha, J.N. Jenkins, B.E. Scheffler, T.A. Wilkins, and D.M. Stelly, Transcriptome profiling, sequence characterization, and SNP-based chromosomal assignment of the EXPANSIN genes in cotton. *Mol Genet Genomics*, **2007**, *278*(5): p. 539-53.

[60] Venu, R.C., Y. Jia, M. Gowda, M.H. Jia, C. Jantasuriyarat, E. Stahlberg, H. Li, A. Rhineheart, P. Boddhireddy, P. Singh, N. Rutger, D. Kudrna, R. Wing, J.C. Nelson, and G.L. Wang, RL-SAGE and microarray analysis of the rice transcriptome after Rhizoctonia solani infection. *Mol Genet Genomics*, **2007**, *278*(4): p. 421-31.

[61] Poormohammad Kiani, S., P. Grieu, P. Maury, T. Hewezi, L. Gentzbittel, and A. Sarrafi, Genetic variability for physiological traits under drought conditions and differential

expression of water stress-associated genes in sunflower (Helianthus annuus L.). *Theoretical and Applied Genetics*, **2007**, *114*(2): p. 193-207.

[62] Kirst, M., A.A. Myburg, J.P. De Leon, M.E. Kirst, J. Scott, and R. Sederoff, Coordinated genetic regulation of growth and lignin revealed by quantitative trait locus analysis of cDNA microarray data in an interspecific backcross of eucalyptus. *Plant Physiol*, **2004**, *135*(4): p. 2368-78.

[63] Shi, C., A. Uzarowska, M. Ouzunova, M. Landbeck, G. Wenzel, and T. Lübberstedt, Identification of candidate genes associated with cell wall digestibility and eQTL (expression quantitative trait loci) analysis in a Flint x Flint maize recombinant inbred line population. *BMC Genomics*, **2007**, *8*.

[64] Jordan, M.C., D.J. Somers, and T.W. Banks, Identifying regions of the wheat genome controlling seed development by mapping expression quantitative trait loci. *Plant Biotechnol J*, **2007**, *5*(3): p. 442-53.

[65] Majewski, J. and T. Pastinen, The study of eQTL variations by RNA-seq: from SNPs to phenotypes. *Trends Genet*, **2011**, *27*(2): p. 72-9.

[66] Vuylsteke, M., H. Van Den Daele, A. Vercauteren, M. Zabeau, and M. Kuiper, Genetic dissection of transcriptional regulation by cDNA-AFLP. *Plant Journal*, **2006**, *45*(3): p. 439- 446.

[67] Holloway, B., S. Luck, M. Beatty, J.A. Rafalski, and B. Li, Genome-wide expression quantitative trait loci (eQTL) analysis in maize. *BMC Genomics*, **2011**, *12*: p. 336.

[68] Chen, X., C.A. Hackett, R.E. Niks, P.E. Hedley, C. Booth, A. Druka, T.C. Marcel, A. Vels, M. Bayer, I. Milne, J. Morris, L. Ramsay, D. Marshall, L. Cardle, and R. Waugh, *An eQTL analysis of partial resistance to Puccinia hordei in barley.* PLoS One, **2010**, 5(1): p. e8598.

[69] Wang, J., H. Yu, W. Xie, Y. Xing, S. Yu, C. Xu, X. Li, J. Xiao, and Q. Zhang, A global analysis of QTLs for expression variations in rice shoots at the early seedling stage. *Plant J*, **2010**, *63*(6): p. 1063-74.

[70] Liu, P., C.M. Wang, L. Li, F. Sun, and G.H. Yue, Mapping QTLs for oil traits and eQTLs for oleosin genes in jatropha. *BMC Plant Biol*, **2011**, *11*: p. 132.

[71] Kirst, M., C.J. Basten, A.A. Myburg, Z.B. Zeng, and R.R. Sederoff, Genetic architecture of transcript-level variation in differentiating xylem of a eucalyptus hybrid. *Genetics*, **2005**, *169*(4): p. 2295-2303.

[72] Jimenez-Gomez, J.M., A.D. Wallace, and J.N. Maloof, Network analysis identifies ELF3 as a QTL for the shade avoidance response in Arabidopsis. *PLoS genetics*, **2010**, *6*(9).

[73] Hansen, B.G., B.A. Halkier, and D.J. Kliebenstein, Identifying the molecular basis of QTLs: eQTLs add a new dimension. *Trends in Plant Science*, **2008**, *13*(2): p. 72-77.

[74] Subramanian, A., P. Tamayo, V.K. Mootha, S. Mukherjee, B.L. Ebert, M.A. Gillette, A. Paulovich, S.L. Pomeroy, T.R. Golub, E.S. Lander, and J.P. Mesirov, Gene set enrichment analysis: a knowledge-based approach for interpreting genome-wide expression profiles. *Proc Natl Acad Sci U S A*, **2005**, *102*(43): p. 15545-50.

[75] Kliebenstein, D.J., M.A.L. West, H. van Leeuwen, O. Loudet, R.W. Doerge, and D.A. St. Clair, Identification of QTLs controlling gene expression networks defined a priori. *BMC Bioinformatics*, **2006**, *7*.

[76] Lan, H., M. Chen, J.B. Flowers, B.S. Yandell, D.S. Stapleton, C.M. Mata, E.T.K. Mui, M.T. Flowers, K.L. Schueler, K.F. Manly, R.W. Williams, C. Kendziorski, and A.D. Attie, Combined expression trait correlations and expression quantitative trait locus mapping. *PLoS Genetics*, **2006**, *2*(1): p. 51-61.

[77] Ayroles, J.F., M.A. Carbone, E.A. Stone, K.W. Jordan, R.F. Lyman, M.M. Magwire, S.M. Rollmann, L.H. Duncan, F. Lawrence, R.R.H. Anholt, and T.F.C. Mackay, Systems genetics of complex traits in Drosophila melanogaster. *Nat Genet*, **2009**, *41*(3): p. 299-307.

[78] Lee, S.I., A.M. Dudley, D. Drubin, P.A. Silver, N.J. Krogan, D. Pe'er, and D. Koller, Learning a prior on regulatory potential from eQTL data. PLoS Genetics, **2009**, *5*(1).

[79] Rockman, M.V., Reverse engineering the genotype-phenotype map with natural genetic variation. *Nature*, **2008**, *456*(7223): p. 738-744.

[80] Stylianou, I.M., J.P. Affourtit, K.R. Shockley, R.Y. Wilpan, F.A. Abdi, S. Bhardwaj, J. Rollins, G.A. Churchill, and B. Paigen, Applying gene expression, proteomics and singlenucleotide polymorphism analysis for complex trait gene identification. *Genetics*, **2008**, *178*(3): p. 1795-1805.

[81] Keurentjes, J.J.B., Genetical metabolomics: closing in on phenotypes. *Curr Opin Plant Biol*, **2009**, *12*(2): p. 223-30.

[82] Chevalier, F., O. Martin, V. Rofidal, A.D. Devauchelle, S. Barteau, N. Sommerer, and M. Rossignol, Proteomic investigation of natural variation between Arabidopsis ecotypes. *Proteomics*, **2004**, *4*(5): p. 1372-81.

[83] Keurentjes, J.J.B., J. Fu, C.H.R. De Vos, A. Lommen, R.D. Hall, R.J. Bino, L.H.W. Van Der Plas, R.C. Jansen, D. Vreugdenhil, and M. Koornneef, The genetics of plant metabolism. *Nature Genetics*, **2006**, *38*(7): p. 842-849.

[84] Schauer, N., Y. Semel, U. Roessner, A. Gur, I. Balbo, F. Carrari, T. Pleban, A. Perez-Melis, C. Bruedigam, J. Kopka, L. Willmitzer, D. Zamir, and A.R. Fernie, Comprehensive metabolic profiling and phenotyping of interspecific introgression lines for tomato improvement. *Nat Biotechnol*, **2006**, *24*(4): p. 447-54.

[85] Lisec, J., R.C. Meyer, M. Steinfath, H. Redestig, M. Becher, H. Witucka-Wall, O. Fiehn, O. Torjek, J. Selbig, T. Altmann, and L. Willmitzer, Identification of metabolic and biomass QTL in Arabidopsis thaliana in a parallel analysis of RIL and IL populations. *Plant J*, **2008**, *53*(6): p. 960-72.

[86] Keurentjes, J.J.B., R. Sulpice, Y. Gibon, M.C. Steinhauser, J. Fu, M. Koornneef, M. Stitt, and D. Vreugdenhil, Integrative analyses of genetic variation in enzyme activities of primary carbohydrate metabolism reveal distinct modes of regulation in Arabidopsis thaliana. *Genome Biology*, **2008**, *9*(8).

[87] Meyer, R.C., M. Steinfath, J. Lisec, M. Becher, H. Witucka-Wall, O. Torjek, O. Fiehn, A. Eckardt, L. Willmitzer, J. Selbig, and T. Altmann, The metabolic signature related to high plant growth rate in Arabidopsis thaliana. *Proc Natl Acad Sci U S A*, **2007**, *104*(11): p. 4759-64.

[88] Wentzell, A.M., H.C. Rowe, B.G. Hansen, C. Ticconi, B.A. Halkier, and D.J. Kliebenstein, Linking metabolic QTLs with network and *cis*-eQTLs controlling biosynthetic pathways. *PLoS Genetics*, **2007**, *3*(9): p. 1687-1701.

[89] Li, Y., R. Breitling, and R.C. Jansen, Generalizing genetical genomics: getting added value from environmental perturbation. *Trends in Genetics*, **2008**, *24*(10): p. 518-524.

[90] Siva, N., 1000 Genomes project. *Nature biotechnology*, **2008**, *26*(3): p. 256.

[91] Ossowski, S., K. Schneeberger, R.M. Clark, C. Lanz, N. Warthmann, and D. Weigel, Sequencing of natural strains of Arabidopsis thaliana with short reads. *Genome Research*, **2008**, *18*(12): p. 2024-2033.

[92] Laubinger, S., G. Zeller, S.R. Henz, T. Sachsenberg, C.K. Widmer, N. Naouar, M. Vuylsteke, B. Schölkopf, G. Rätsch, and D. Weigel, At-TAX: A whole genome tiling array

resource for developmental expression analysis and transcript identification in Arabidopsis thaliana. *Genome Biology*, **2008**, *9*(7).

[93] Matsui, A., J. Ishida, T. Morosawa, Y. Mochizuki, E. Kaminuma, T.A. Endo, M. Okamoto, E. Nambara, M. Nakajima, M. Kawashima, M. Satou, J.M. Kim, N. Kobayashi, T. Toyoda, K. Shinozaki, and M. Seki, Arabidopsis transcriptome analysis under drought, cold, high salinity and ABA treatment conditions using a tiling array. *Plant and Cell Physiology*, **2008**, *49*(8): p. 1135-1149.

[94] Buckler, E. and M. Gore, An Arabidopsis haplotype map takes root. *Nat Genet*, **2007**, *39*(9): p. 1056-1057.

[95] Sorkheh, K., L.V. Malysheva-Otto, M.G. Wirthensohn, S. Tarkesh-Esfahani, and P. Martínez-Gómez, Linkage disequilibrium, genetic association mapping and gene localization in crop plants. *Genetics and Molecular Biology*, **2008**, *31*(4): p. 805-814.

[96] Brachi, B., N. Faure, M. Horton, E. Flahauw, A. Vazquez, M. Nordborg, J. Bergelson, J. Cuguen, and F. Roux, Linkage and association mapping of Arabidopsis thaliana flowering time in nature. *PLoS Genet*, **2010**, *6*: p. e1000940.

# CHAPTER 9

# Expanded CTG Repeat Induced Molecular Alterations in Myotonic Dystrophy

**Beatriz Llamusí*[,1,2], Ariadna Bargiela[2] and Ruben Artero[2]**

*[1]Fundación Investigación Hospital Clínico Universitario/INCLIVA, Valencia, Spain; [2]Department of Genetics, University of Valencia, Doctor Moliner, 50, E46100 Burjasot, Valencia, Spain*

**Abstract:** Myotonic Dystrophy type 1 (DM1) is a multi-system disorder characterised by muscle wasting, myotonia, cardiac conduction defects, cataracts, and neuropsychological dysfunction. DM1 is caused by expansion of a CTG repeat in the 3′ untranslated region (UTR) of the *Dystrophia Myotonica Protein Kinase* (*DMPK*) gene. A body of work demonstrates that *DMPK* mRNAs containing abnormally expanded CUG repeats are toxic to several cell types. A core mechanism underlying symptoms of DM1 is that mutant *DMPK* RNA interferes with the developmentally regulated alternative splicing of defined pre-mRNAs by altering the function of important splicing regulators such as MBNL1, CELF1 and hnRNP H. However, other RNA metabolism pathways have also been found to be altered in DM1. Bidirectional transcription allows generation of poly-CAG and poly-CUG mRNAs, both of which can generate toxic polyaminoacidic proteins by Repeat Associated Non-ATG (RAN) translation. Double stranded RNAs (CAG)•(CUG) and CUG expansion hairpins might act as Dicer ribonuclease targets, thus leading to silencing of transcripts containing complementary sequences. Upon transcription, mutant *DMPK* RNA coexists as ssCUG and dsCUG forming hairpins. CELF1 accumulates in the nucleus and binds to ssCUG while MBNL1 is sequestered in hairpins, thus altering their activity as splicing factors. ssCUG also binds different types of nuclear factors such as transcription factors, contributing to gene transcription deficiencies. More recently, several microRNAs have also been found to be altered in DM1 affected muscles. In summary, mutant *DMPK* transcripts have been shown to alter gene transcription, translation and alternative splicing of specific gene transcripts, and have the ability to trigger gene-specific silencing effects and microRNA dysregulation in DM1 cells. Therapies aimed at reversing these gene expression alterations are likely to be effective ways to treat DM1.

**Keywords:** Muscleblind, myotonic dystrophy, missplicing, nucleotide expansions, RNA toxicity.

***Address correspondence to Beatriz Llamusí:** Department of Genetics, University of Valencia, Doctor Moliner, 50, E46100 Burjasot, Valencia, Spain; Tel: 34+96-3625580; Fax: 34+96-3864875; E-mail: mbeatriz.llamusi@uv.es

**Christian Neri (Ed)**

# INTRODUCTION

Myotonic Dystrophy type 1 (DM1; OMIM 160900) is a particularly complex inherited disease, both from clinical and genetic standpoints. Clinically, it affects up to 11 organ systems including the muscular, nervous, ocular, digestive, respiratory and cardiovascular systems. Characteristic disabilities are loss of muscle strength with a distal to proximal pattern, myotonia, excessive daytime sleepiness, excessive fatigue, abdominal pain as well as dysphagia [1-3]. Although rare, it is estimated that the worldwide prevalence ranges between 2.1 and 14.3 per 100,000 [4, 5], making it the most common form of muscular dystrophy in adults. Genetically, DM1 is caused by a CTG trinucleotide repeat expansion in the 3' untranslated region of the *Dystrophia Myotonica Protein Kinase* (*DMPK*) gene, which is assigned to chromosome band 19q13.2-q13.3 [6-10]. DM1 was the first dominantly inherited disease found to be caused by non-coding repeat expansions; the mutation is transcribed into RNA but not translated into protein. A second mutation in chromosome 3 was identified in patients with very similar symptoms that was classified as either Proximal Myotonic Myopathy (PROMM) or Myotonic Dystrophy type 2 (DM2) [11, 12]. DM2 was found to be associated with CCTG tetranucleotide repeat expansions in the first intron of the *CCHC type zinc finger, nucleic acid binding protein (CNBP)* gene ([11, 13] reviewed in [14]). Thus, DM1 and DM2 originate from long non-coding repeat expansions, which cause a similar chronic, slowly progressing, multisystemic disease with a dominant inheritance pattern. If the proteins encoded by *DMPK* and *CNBP* were normal, how can a dominant phenotype be explained? Recently there have been several excellent reviews which update the reader on the current state of knowledge in this area [15, 16]. In the present review we focus specifically on the levels of gene expression regulation that mutant *DMPK* affects, with particular emphasis on recent data.

## Gene Transcription Changes in DM1: *cis*-Effects

Since the discovery of the genetic mutation responsible for DM1, much effort has been devoted to precisely defining whether *DMPK* transcripts and/or protein levels change in the presence of CTG expansions. Initial reports using antisera and semiquantitative reverse transcriptase-polymerase chain reaction techniques reported decreased levels of both *DMPK* mRNA and protein in adult forms of DM1, suggesting that large CTG repeat expansions altered *DMPK* mRNA synthesis and processing [17-20]. These observations, and the detection of mutant *DMPK* transcripts retained in the nucleus of DM1 fibroblasts and muscle biopsies, supported the hypothesis of loss of *DMPK* function (*DMPK* haploinsufficiency) as the mechanism of pathogenesis of DM1 [21, 22]. However, subsequent improvements to the protocols used for detection and quantification of total cytoplasmic mRNA levels indicated that there was no significant change in *DMPK* transcription [23], and showed that the anti-DMPK antisera used in previous reports crossreacted with other proteins [24]. However, other studies revealed that *DMPK* transcription was altered in patients and that there was no correlation between CTG repeat length and DMPK protein reduction [25-27].

The functional implications of a reduction in *DMPK* expression were genetically tested by the generation of knockout mice. *DMPK*$^{-/-}$ mice develop a mild late-onset, progressive skeletal myopathy which suggested that *DMPK* might be necessary for the maintenance of skeletal muscle structure [28]. Subsequent studies showed that *DMPK*-deficient mice also exhibited cardiac conduction abnormalities [29] and altered β-agonist-induced responses as a consequence of reduced $\beta_1$-adrenergic receptor density in cardiac plasma membranes [30]. Moreover, diminished levels of DMPK produce metabolic impairment such as abnormal glucose tolerance, reduced glucose uptake and impaired insulin-dependent GLUT4 trafficking in muscle [31]. As similarly proposed for mutant *DMPK*, recent studies also suggest that *CNBP* haploinsufficiency may contribute to the DM2 disorder. *CNBP* knockout mice show phenotypes that resemble DM2, including muscle histopathology, myotonia and heart conduction abnormalities [32]. However, the fact that *DMPK*$^{-/-}$ mice showed only a mild phenotype for just some of the DM1 symptoms and that no point mutations have been associated with DM1 or DM2, strongly suggested that the multisystemic features of DM were not simply caused by *DMPK* nor *CNBP* haploinsufficiency.

The possibility that CTG and CCTG repeat expansions were altering the local structure of the chromatin, thus interfering with the expression of *DMPK* and neighbouring genes, has also been considered (Fig. **1A**). Immediately adjacent to *DMPK* is *SIX5* which encodes a homeodomain transcription factor, and *Myotonic*

*Dystrophy gene with WD repeats* (*DMDW*) which is prominently expressed in testis and brain [33-35]. The hypothesis was chiefly supported by the fact that *DMDW* expression levels were reported as decreased in repeat expansion bearing patients [34, 36]. Also, between *DMPK* and *SIX5* there are two CTCF binding sites that form an insulator element. In congenital DM1 those sites are methylated and consequently CTCF binding is avoided, thus disrupting insulator function and producing *DMPK* overexpression [37]. Abnormal methylation patterns have also been found in the CTG repeats boundaries in adult DM1 patient tissues [38]. Mutational analysis in *Drosophila* has shown that *D-Six4*, the closest *Six5* homolog in flies, is required for the normal development of muscle and the mesodermal components of the gonad. This suggested that human *Six5* could participate in muscle wasting and testicular atrophy phenotypes in DM1 [39]. However, *Six5* knockout mice only developed cataracts that lacked the distinctive iridescent opacities characteristic of cataracts from DM1 patients [40].

Cho and collaborators (2005) described that *DMPK* undergoes antisense transcription emanating from the adjacent *SIX5* regulatory region in DM1 [41]. Among mammals, antisense transcription is widespread, about 70% of the genome is transcribed in sense and antisense strands [42]. Both strands can be protein coding or not. However, the antisense strand commonly has a regulatory function as it hybridizes with its complementary RNA strand or with DNA to interfere with the transcription or translation process. As a result of *DMPK* bidirectional transcription (CAG)•(CUG) dsRNA is generated, and processing these types of RNA structures might activate RNAi pathways. Thus, gene silencing could be an underlying pathogenic mechanism in DM1 and many other expansion diseases (see below and Fig. **1B**). Moreover, recent studies have demonstrated that simultaneous forward and reverse transcription in human fibroblasts containing 800 CTG repeats increased the somatic instability of the expanded (CTG)•(CAG) repeats [43]. As a consequence, bidirectional transcription might contribute to modifications in the expansion length.

## Gene Transcription Changes in DM1: *trans*-Effects

Electron microscope examination has revealed that CUG repeat RNA forms dsRNA segments [44] that were additionally able to bind the dsRNA activated protein kinase PKR [45], and Muscleblind protein family members [46]. More recent crystallographic data have confirmed that CUG repeat RNA forms stable antiparallel double-stranded helices similar in structure to A-form RNA except for the unique U-U-mismatches [47]. The identification of this peculiar structure supported an RNA dominant mutation model in which long stable dsRNAs gain the capacity to sequester certain RNA-binding proteins that are correspondingly depleted from their normal subcellular localizations.

It has also been found in a DM1 cell model that a number of regulatory transcription factors (TFs) change their nuclear compartment distribution from active chromatin to CUG repeat-containing ribonuclear particles, a process known as RNA leaching (Fig. **1A**) [48]. Fractions of Specificity protein 1 (Sp1), Signal Transducer and Activator of Transcription (STAT1 and STAT3) and Retinoic Acid Receptor gamma (RARγ) measured in the chromatin, were dramatically reduced after three weeks of mutant *DMPK* RNA expression. Consequently, diverse genes had their expression altered, including the muscle-specific chlorine channel 1 (*CLCN1*), which causes myotonia [48]. Another example of a TF found to be altered by repeat expansions is MyoD. Levels of MyoD, a TF necessary for muscle differentiation and regeneration, were significantly reduced in myoblasts expressing the mutant 3'UTR *DMPK* RNA [49]. Changes in TF activity should alter target gene transcription. Consistent with this, gene transcription changes in DM1 have been further verified by comparing the expression profile of muscle biopsies from DM1 and DM2 patients to controls in macroarray analysis of 96 neuroscience-related genes [50]. Six genes critical for calcium and potassium metabolism and mitochondrial functions were misregulated both in DM1 and DM2 also strengthening the current view that both diseases share a common pathogenetic pathway [50].

DM1 has also been associated with overexpression of the cardiac-specific transcription factor *NK2 transcription factor related, locus 5 (Drosophila) (NKX2-5)*. Levels of *NKX2-5* increased in cardiac tissue in an inducible transgenic mouse model for RNA toxicity [51]. Moreover, overexpression of *DMPK* 3' UTR mRNA in mouse skeletal muscle also induced transcriptional activation of Nkx2-5 and its targets. In human muscles these changes were specific to DM1 and were

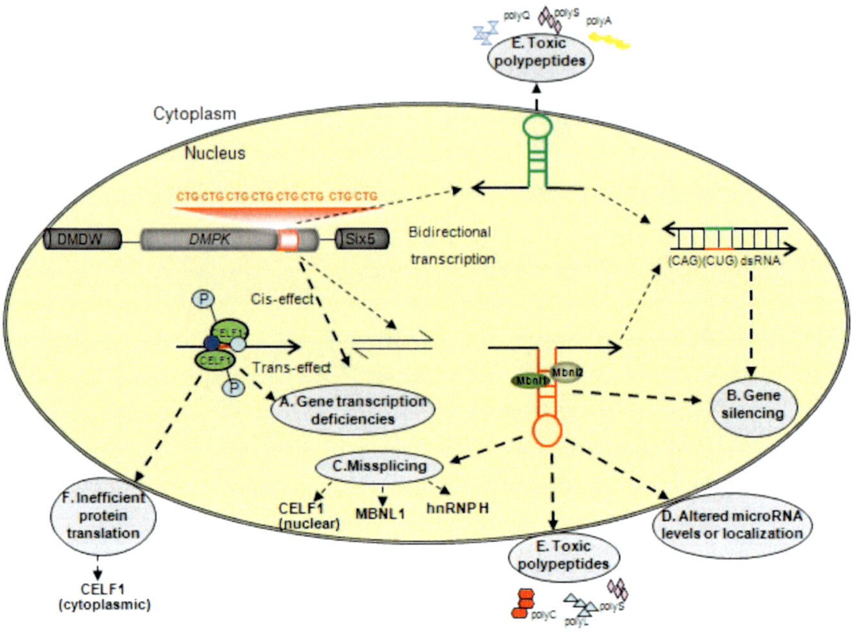


**Figure 1:** CTG trinucleotide expansions interfere with gene expression regulation at several levels. The genomic organization around CTG expansions is represented in the upper left corner of the figure, with grey cylinders denoting genes, and lines representing the intergenic DNA. The red cylinder of the *DMPK* gene represents the CTG repeat expansion. Bidirectional transcription (direction of transcription is marked by the arrow) allows generation of poly-CAG (antisense; in green) and poly-CUG mRNA (sense; in red), both of which can generate toxic homoaminoacidic proteins by RAN translation (**E. Toxic polypeptides**). CTG-induced local changes in chromatin organization may explain reduced transcription in *DMPK*, *Six5* and *DMDW* genes. Upon transcription, mutant *DMPK* RNA containing CUG repeats fold into dsRNA hairpins in the cell nucleus, some coexisting as soluble ssRNA. CELF1 binds to ssRNA regions while MBNL1 binds to hairpins forming insoluble foci. CUG hairpins may also bind different types of nuclear factors (colored circles). Transcription factor leaching may be the origin of additional transcriptional deficiencies (**A. Gene transcription deficiencies**). Aberrant MBNL1 binding to the ds(CUG) hairpins sequesters the protein. Steady-state levels of nuclear CELF1 increase by protein kinase C mediated phosphorylation. Both effects result in unbalanced cytoplasmic levels of these antagonistic alternative splicing (AS) regulators, leading to changes in the transcript isoforms originating from several pre-mRNAs in brain, heart and skeletal muscle (**C. Missplicing**). MBNL1 retention by CUG expansions also explains alterations in microRNA biogenesis leading to miR-1 downregulation. Other dysregulated microRNAs have also been reported but the cause of their altered expression is still unknown (**D. Altered microRNA levels or localization**). (CAG)(CUG) dsRNA and ds(CUG) hairpins are a substrate for the Dicer ribonuclease, which reduces levels of mutant *DMPK,* simultaneously generating CUG siRNA with the ability to silence transcripts which contain complementary CAG repeats such as those of the Huntington Disease and Spinocerebellar ataxia genes (**B. Gene silencing**). Finally, a combination of nuclear stabilization and changes in key phosphorylation events likely explain the inefficient activation of *p21, Mef2A,* and *C/EBPβ* mRNA translation by cytoplasmic CELF1 (**F. Inefficient protein translation**).

not present in other muscular dystrophies, possibly contributing to the cardiac conduction defects found in DM1 [51]. The mislocalization of SHARP (SMART/HDAC1-associated repressor protein) in the cytoplasm in DM1 myoblasts could be the cause of altered steady-state levels of a set of RNA transcripts implicated in muscle development and function [52]. Despite the still limited supporting experimental data, modification of TFs expression by leaching from chromatin or other mechanisms, provides a powerful explanation for broad gene expression changes, especially if their effects impinge on alternative splicing event altered in DM1 (see below). A general requirement for transcription factors in different tissues that express *DMPK* would account for the multisystemic and multisymptomatic nature of DM1. Moreover, a common *trans*-acting mechanism with trapping and depletion of similar TFs might contribute to the clinical analogies between DM1 and DM2.

## CUG repeat RNA Sequesters alternative Splicing Factors

The identification of RNA-binding proteins bound to CUG repeat expansions clearly suggested an RNA-gain-of function model for DM1, and provided a molecular explanation for the phenotypical similarities between DM1 and DM2 since the same or similar RNA-binding proteins could be sequestered by mutant CUG and CCUG repeat RNA [46].

### *MBNL Family Proteins*

Several studies have established that one of the reasons for the toxicity of mutant *DMPK* RNA is its interference with human Muscleblind-like 1, 2 and 3 (MBNL1-3) protein activity (see [53] for a review about MBNL proteins). MBNL proteins were directly implicated in DM1 pathogenesis when they were found to specifically bind CUG trinucleotide repeats of 20 or longer *in vitro* in HeLa cell extracts [46]. MBNL proteins are orthologous to the *Drosophila* Muscleblind proteins, which are required for terminal differentiation of muscle and photoreceptor cells [54, 55]. Co-localization studies in both cell and tissue samples from DM1 and DM2 patients showed that MBNL1-3 were retained in the cell nucleus as insoluble ribonuclear foci that included mutant transcripts [56-62]. *Drosophila* Muscleblind similarly co-localized with CUG ribonuclear foci in muscle cells in DM1 fly models [63-65]. Human MBNL proteins seem to have acquired specialized functions although all three were capable of regulating the use of alternative exons in *cardiac troponin T (cTNT)*, *insulin receptor (IR)* [66] and *myomesin 1 (MYOM1)* [67]. *MBNL1 and MBNL3* functions in skeletal muscle have been widely studied. While MBNL1 is involved in promoting muscle differentiation, MBNL3 represses it. Importantly, recent studies have reported the MBNL3 mediated mechanism of DM1 muscle differentiation repression. Expression of CUG expanded RNAs can lead to an increase in MBNL3 expression. MBNL3 disrupts Mef2 β-exon splicing which produces a less transcriptionally active Mef2 protein, thus causing decreased expression of Mef2 target proteins such as MyoD [68, 69]. In contrast, MBNL2 role in muscle development is not so clear. Apart from participating in splicing regulation [70] [71], it participates in the subcellular localization of *alpha3-integrin* transcripts [46]. In brain, it has recently been discovered that major pathological features of the DM brain result from disruption of the MBNL2-mediated developmental splicing program thus, underlining the important role of MBNL2 as splicing regulator in this tissue [72]. These results are supported by the analysis of alternative splicing events in brains of Mbnl1-null mice, which confirmed that many of the misregulated transcripts in the DM1 brain are normal in the Mbnl1-null mice [73]. Conversely, whereas Mbnl1-null mice exhibit many of the skeletal muscle features of DM1 patients, Mbnl2-null muscles are normal [72, 74, 75]. More recent studies using RNA-seq and CLIP-seq approaches have identified several hundred splicing events whose regulation depended on Mbnl function in a pattern indicating functional interchangeability between Mbnl1 and Mbnl2 in brain, heart, muscle, and myoblasts. Transcriptomic analysis of subcellular compartments have also uncovered a global role for Mbnls in regulating localization of mRNAs in both mouse and *Drosophila* cells, and Mbnl-dependent translation and protein secretion were observed for a subset of mRNAs with Mbnl-dependent localization [76].

Consistent with the involvement of MBNL proteins in the RNA gain of function mechanism, *Mbnl1* knockdown mice [74] showed DM1-like phenotypes including iridescent cataracts, myotonia and missplicing of muscle transcripts that had previously been reported as altered in DM1 patients such as insulin receptor (IR), chlorine channel 1 (CLC1) and Bin1 pre-mRNA [77, 78]. *muscleblind* mutant embryos similarly showed altered expression of muscle transcript isoforms from the *Drosophila ZASP* homologue and *alpha-actinin* genes [79]. More recently, the generation of an *Mbnl2* deficient mouse also displaying myotonia, skeletal muscle pathology consistent with human DM, and reduced expression of chloride channel mRNA in skeletal muscle, suggested that depletion of MBNL2 might also contribute to the human DM pathogenesis [71]. Conversely, overexpression of MBNL1 *in vivo* using a recombinant adeno-associated viral vector rescued disease-associated muscle myotonia and adult-splicing patterns of muscle pre-mRNAs (which would otherwise be misspliced into fetal forms), in transgenic mice expressing 250 CTG repeats in the 3′ UTR of a modified human skeletal *alpha-actin* (HSA) transcript [80] ($HSA^{LR}$ mice). MBNL1 overexpression as a valuable strategy for treating skeletal muscle features of DM has also been supported by recent studies on a transgenic MBNL1 overexpression mouse model. These studies have confirmed that early and long-term MBNL1 overexpression prevents CUG-induced myotonia, myopathy and alternative splicing misregulation in DM1 mice [81]. Gene expression patterns in $HSA^{LR}$ DM1 mice was compared with *Mbnl1* knockouts and *Clcn1* null mice, which have generalized myotonia. It was observed that most alterations present in model mice were also present in *Mbnl1* knockouts. About 24% of dysregulated transcripts in the $HSA^{LR}$ mice showed similar expression levels in $Clcn1^{-/-}$ mice. These data indicated that major changes in gene expression were a consequence of increased electrical activity due to *Clcn1* transcript missplicing, and most (around 80%) of the biggest alterations in splicing events resulted directly or indirectly (*via* myotonia) from *Mbnl1* loss of function [82, 83]. Taken together, these results support the hypothesis that loss of MBNL1 activity is a primary pathogenic event in the development of the disease.

### *CELF1*

CUG repeat expansions interfere with the activity of alternative splicing factors other than Muscleblind. CELF1, a member of the CUG-BP1 ETR-3 like factor (CELF) family, does not bind ds(CUG) hairpins nor co-localizes with ribonuclear foci, but its activity is increased in DM1 myoblasts, skeletal muscle, and heart tissues [44, 84, 85] (see [86] for a review about CELF RNA binding proteins).

This demonstrated that CELF1 was not precipitated in insoluble foci. Furthermore, there are studies that suggest that CUG expressing RNA may be present in a non-focused form with a contributing role in DM1 [87, 88].

Indeed, it has been proposed that expanded CUG RNA is in equilibrium between ssRNA and dsRNA and this equilibrium can be displaced upon CELF1 binding to ssCUG, or MBNL1 binding to dsCUG. A critical aspect of this model is that while CELF1 binding to ssCUG stabilizes and increases the function of the protein, binding to dsCUG blocks MBNL1 function through sequestration [15]. Although the molecular mechanism leading to CELF1 activation is not completely understood, inappropriate activation of the protein kinase C (PKC) pathway contributes to the pathogenic effect of noncoding CUG repeat RNA through hyperphosphorylation of nuclear CELF1, which is stabilized in this cellular compartment [89]. Transgenic mice that overexpress CELF1 in muscle and heart reproduce missplicing alterations typical of DM1, which has confirmed the involvement of this protein in the pathogenesis of the disease as well as the antagonism between MBNL1 and CELF1 in alternative splicing regulation ([90-92] and below). Results from several groups support the hypothesis that increased activity of CELF1 in DM1 is pathogenic as CELF1 regulates alternative splicing of pre-mRNA transcripts which antagonizes Muscleblind activity [79, 85, 93, 94]. In fact, some transcripts altered in DM1 brain have been found to be targets of CELF [60, 86]. Recent studies targeting CELF1 as a therapeutic approach have shown that repressing CELF1 activity rescues CELF1-dependent alternative splicing in cell culture and transgenic mouse models of DM1. However, CELF1-independent splicing remained dysregulated [95]. Lee and collaborators (2010) have found a possible new contribution of CELF1 in DM1 as it regulates decay of mRNA from a wide range of genes in muscle cells. Among these genes are myogenic transcription factors *Myod1* and *Myog* [96].

### *hnRNP H*

Another splicing regulator altered by CUG repeat expansions is heterogeneous nuclear ribonucleoprotein H (hnRNP H). The protein was identified in UV-crosslinking assays as a factor capable of binding and modulating nuclear retention of mutant *DMPK* mRNA. The specific binding of hnRNP H requires not only a CUG expansion but also a splicing branch point distal to the repeats [97]. It has been proposed that hnRNP H and CELF1 form an RNA-dependent complex which inhibits inclusion of the *insulin receptor* (*IR)* exon 11 in normal myoblasts. That fact is counteracted by the exon 11 splice enhancing activity of Muscleblind proteins [98]. Because Muscleblind proteins are required facilitators of IR exon 11 exclusion, this event is a typical misregulated splicing in DM1 [77].

### *Staufen1*

Recent studies have found that the RNA-binding protein Staufen1 is increased in skeletal muscle from DM1 mouse models and patients. Staufen1 interacts with mutant CUG mRNAs and promotes their nuclear export and translation. Overexpression of Staufen1 rescues alternative splicing of two key pre-mRNAs known to be aberrantly spliced in DM1, suggesting its increased expression represents an adaptive response to the pathology [99].

Given the involvement of the alternative splicing regulators, it is not surprising that a defining molecular alteration in DM1 is the missplicing of a defined set of muscle, brain and heart transcripts (for a compilation of splice alterations in DM1 see [100]) (Fig. **1C**). Changes in tissue-specific splice patterns have been shown to contribute to particular symptoms in DM1 patients. Myotonia has been attributed to reduced expression of *CLC-1* due to the combined effects of a decrease in transcription from the *CLC-1* gene and inclusion of the alternative exon 7a, which includes a TGA stop codon which triggers nonsense mediated decay [101]. Indeed, patch clamp analysis has confirmed that myotonia is associated with a marked reduction in chloride channel activity [88] thus suggesting that myotonia originates from an increase in transcription and generation of protein variants with no intrinsic channel activity [60, 88, 102]. Recently it has been shown that the alternative splicing of *BIN1,* which is regulated by MBNL1, contributes to muscle weakness, another predominant feature of DM. *BIN1* exon 11 splicing is skipped in DM1 patients generating an inactive form of BIN1 which lacks phosphatidylinositol 5-phosphate binding ability which disrupts membrane function, leading to T tubule alterations and muscle weakness [103]. The number of genes and alternative exons altered in DM1 is still unknown, although several efforts have been made to identify exon/intron alterations using exon arrays [82, 104]. The data available support a view in which the process of constitutive RNA processing is not affected in DM1 cells [66, 105, 106]. During embryonic development and postnatal periods in mammals, a series of changes are required at the transcriptional and post-transcriptional levels that account for the remodeling necessary for adult body development [107]. Some of these changes are based on the alternative splicing of many transcripts in these tissues, which are regulated by RNA binding proteins including members of the Muscleblind and CELF families. Activation of MBNL1 during muscle differentiation stages brings about the exclusion of fetal exons from multiple proteins, resulting in production of their adult encoding isoforms [75, 108]. Indeed, the DM pathology seems to selectively affect a defined set of pre-mRNAs that undergo such a developmentally regulated switch in their alternative splicing pattern. Failure to switch on an adult-type alternative splicing pattern, due to lack of Muscleblind function, results in the maintenance of a fetal-like splicing pattern in adult tissue [75, 108]. A *trans*-effect on the alternative splicing of many RNAs that leads to the expression of splice products that are developmentally inappropriate for a particular tissue was termed spliceopathy [100] with DM1 being the first human example.

Compared with other genetic diseases the therapeutic opportunities for DM1 are particularly favourable because elimination of toxic RNA has already been shown to reverse the cardinal features of DM1 in a transgenic mouse model. Overexpression of a normal *DMPK* 3′ UTR (five CUG repeats) as part of an inducible RNA transcript was reported to cause myotonia, cardiac conduction abnormalities, histopathology and RNA splicing defects in the absence of detectable nuclear inclusions. Five CUG repeats were apparently sufficient to cause DM1-like phenotypes without the formation of RNA or Mbnl nuclear foci. In these transgenic mice CELF1 levels were found to be increased in skeletal muscle while no change in MBNL1 was detected. Although surprising, the authors explained the DM1 phenotype by the imbalance between CELF1 (upregulated) and MBNL1 (unchanged) antagonistic activities as alternative splicing regulators, resulting in missplicing of characteristic transcripts such as *ClC-1* and *cTNT* [94].

## Dicer Mediated Gene Silencing

The double stranded nature of CUG repeat expansions *in vivo* prompted speculation about the possibility that it might act as a Dicer ribonuclease target, thus generating small interfering RNAs (siRNAs) that would silence transcripts containing complementary, or near complementary sequences (Fig. **1B**) [109]. The MBNL1 protein itself includes a run of seven alanines, encoded by a sequence with the potential to act as a CUG siRNA target. *In vitro* studies in HeLa cells have already shown that the mechanism that permits specific target RNA silencing by siRNAs and microRNAs is active in the cell nucleus [110]. Indeed, Krol and co-workers have shown that transcripts containing long hairpin structures composed of CNG repeats represent a sub-class of Dicer ribonuclease targets. In these studies, it was suggested that Dicer elicits two opposing effects in DM fibroblast cells [111]. On the one hand, the activity of the ribonuclease might downregulate the mutant *DMPK* transcript, which removes the toxic RNA from the cell and contributes to the reduction of *DMPK* expression levels. While on the other hand, the Dicer-induced short CUG repeats generated as siRNAs use the RNA interference pathway to silence the expression of transcripts containing long complementary repeats, which may add up to potentially pathogenic gene expression imbalances. Supporting these studies, more recent reports have revealed dramatically enhanced toxicity in flies that co-express CTG with CAG expanded repeats. Expression of the two transcripts led to novel pathogenesis with the generation of dcr-2 and ago2-dependent 21-nt triplet repeat-derived siRNAs which targeted the expression of CAG-containing genes, such as Ataxin-2 and TATA binding protein (TBP) [112]. In contrast, other groups have failed to demonstrate the presence of short poly(CUG) RNAs in skeletal muscle of the $HSA^{LR}$ mice model of DM1 [83]. Given the current state of knowledge, the contribution of gene silencing to DM1 is still controversial.

### Involvement of microRNAs in DM

MicroRNAs (miRNAs) are short (20-24 nt) non-coding RNAs that are involved in post-transcriptional regulation of gene expression in multicellular organisms by affecting both the stability and translation of mRNAs (reviewed in [113]). miRNAs have been shown to participate in a wide range of biological processes, including myogenesis and muscle regeneration. Recent studies have reported evidence that miRNAs may be involved in the molecular pathogenesis of DM1 (Fig. **1D**). Firstly, pull-down analysis using GST-MBNL1-6xHis as bait, identified DDX1, which participates in miRNA biogenesis [114]. Secondly, MBNL1 binds to a UGC motif located within the loop of pre-miR-1 and competes for LIN28 binding, which promotes pre-miR-1 uridylation and blocks Dicer processing thus inhibiting miR-1 biogenesis. As a consequence of MBNL1 depletion, miR-1 is increased in both DM1 and DM2-affected hearts [115]. Moreover, independent studies have described aberrant localization, or expression alterations of miRNAs in muscle biopsies from DM1 patients. miR-1 and miR-335 were found to be up-regulated, whereas miR-29b and c, and miR-33 were down-regulated in DM1 skeletal muscles. The cellular distribution of muscle specific miR-1, miR-133b and miR-206 was severely altered in human DM1 skeletal muscle biopsies [116]. In other studies, miR-206 has also been found to be over-expressed in the *vastus lateralis* in 5 out of 7 DM1 patients when compared to the control group [117]. This evidence suggests that CTG expansions alter miRNA function and the reversion of these alterations could provide new therapeutic approaches to DM. Moreover, the possibility of targeting genes in an allele-specific manner was also considered since miRNAs with CAG repeats in their seed regions may more tightly regulate mutated alleles. It was proposed by Hon and Zhang that the length of CUG repeats could correlate with miRNA repression in the case of the *DMPK* 3' UTR. Six miRNAs carrying CAG repeats in their seed regions were identified and *in silico* studies indicated that they could regulate the expression of *DMPK* by targeting the CUG repeat region in its 3' UTR [118]. However, further analysis and experimental validation of this proposal is needed. Recent studies have investigated the relevance of miRNA deregulation also in DM2, and found 11 miRNAs altered in skeletal muscle biopsies of patients supporting the role of miRNA in DM2 pathogenic mechanisms [82].

### Effects on mRNA Translation

CELF1 is a multifunctional RNA-binding protein that regulates RNA processing at several stages including activating cap-dependent and cap-independent translation, RNA stability, and splicing [96, 108, 119]. CELF1 can also promote translation of *p21* [120], *C/EBPβ* [121] and *Mef2A* [90] during normal muscle differentiation by binding to mRNAs and to translation initiation factor eIF2. However, the mechanism which made translation inefficient in DM1 muscle cells was not fully understood, in particular after confirming that the levels of CELF1 mRNA did not show any significant changes in DM1 muscle cells when compared to normal myoblasts [122]. Salisbury and co-workers [119] showed that two key phosphorylation events in CELF1 were mis-regulated during DM1 myogenesis. The serine/threonine-specific protein kinase family member Akt phosphorylates CELF1 at Ser28 in proliferating myoblasts, increasing the interaction of CELF1 with cyclin D1 mRNA, which is a strong cell proliferation promoter [119]. CELF1 was also phosphorylated at Ser302 by cyclinD3/cdk4 complexes. During normal muscle differentiation, CELF1 phosphorylation by cyclinD3/cdk4 increased CELF1 interactions with eIF2 and with cdk inhibitor p21 mRNA, encouraging formation of the translational CELF1-eIF2 complex that promotes delivery of mRNAs to polysomes [123]. In DM1 myoblasts Ser28 phosphorylation was increased whereas Ser302 phosphorylation in myotubes was diminished, resulting in an increase of cyclin D1, a decrease of p21, and decreased formation of the translational complex CELF1-eIF2 [124]. Consequently, alteration in CELF1 phosphorylation delays myogenesis. However, recent studies in primary human cell lines show that the muscle wasting typical in DM1 is due to impairment of muscle mass maintenance-regeneration, through premature apoptotic-autophagic activation, rather than altered myogenesis [90, 119, 125-127]. Moreover, it provides a reasonable explanation for lower translational levels of some mRNAs in DM1 differentiated muscle (Fig. **1F**). Consistent with these results, Eyk and co-workers (2010) observed that RNA containing CUG repeats altered expression levels of several components of the Akt/Gsk3-b signalling pathway in *Drosophila* [128]. Thus, this could be the origin of changes in CELF1 phosphorylation levels and therefore represent a potential new therapeutic target in DM1.

It has been shown that RNA transcripts containing CAG and CUG expansions can be translated in the absence of the starting codon AUG. This process is known as Repeat Associated Non-ATG translation (RAN translation) (Fig. **1E**). Bidirectional transcription in DM1 generates both polyCUG and polyCAG transcripts which can undergo RAN translation. These hairpins have been suggested to recruit initiation factors and ribosomal subunits, acting as internal ribosome entry sites (IRESs) [129], that promote initiation of translation at non-ATG codons [130]. As a result of RAN translation homopolymeric proteins carrying long polyGln tracts have been found in nuclear aggregates in DM1 mouse models, patient cardiac myocytes, and leukocytes. Neuroblastoma cells transfected with CAG repeat RNA are sensitized to apoptosis, which might be due to the generation of toxic polyGln, polyAla, and polySer polypeptides. Moreover, RAN translation can also occur in *DMPK* sense transcripts to produce polyLeu, polyCys and polyAla tracts [126]. This fact suggests that RAN translation products may be toxic to cells and may therefore have a role in DM1 pathogenesis. Although the functional relevance of RAN translation is still not clear, it is likely that DM1 pathogenesis involves not only RNA, but also protein toxicity.

## Perspectives

The toxic RNA hypothesis is now a well established model to explain the pathogenic effect of mutant *DMPK* transcripts and is being extended to additional trinucleotide repeat expansion diseases such as spinal and bulbar muscular atrophy (SBMA) [131], fragile X syndrome (FXS) [132], Hungtinton's disease [133], Huntington's Disease-Like 2 (HDL-2) [134] and a number of spinocerebellar ataxias (SCAs) (reviewed in [135]). Despite the fact that this hypothesis builds on results obtained over more than a decade of work, a number of key issues still deserve immediate attention. First, a number of studies have described the involvement of multiple proteins in DM1 pathogenesis. However, further studies are needed to build a picture of the detailed molecular mechanisms in order to explain the basis of particular disease symptoms. A wide range of proteins or transcripts have been found to be altered in DM1 cells such as protein kinase C, Akt [89, 119] and extracellular matrix proteins [82]. Whether this is a primary or secondary effect of CUG repeat RNA remains to be determined, as well as the relevance of signal transduction changes in the regulation of alternative splicing factor activity. Second, scientific work has mainly addressed the study of skeletal muscle phenotypes in DM1. However, other organs and tissues such as the cardiac and nervous system are receiving more attention as they are more relevant from a clinical stand point. Third, new therapy strategies addressed to silence mutant DMPK expression are under quick development. Antisense oligonucleotides (ASOs) have been widely studied to induce gene silencing as a therapeutic strategy in diseases that involve RNA or protein gain-of-function effects. Recent studies have shown that nuclear-retained transcripts containing expanded CUG repeats are unusually sensitive to antisense silencing. In a transgenic mouse model of DM1, systemic administration of ASOs caused a rapid knockdown of RNA expressing CUG repeats in skeletal muscle, which rescue different features of the disease [136]. Functional silencing of *DMPK* has also been tested using mirtron-mediated gene knockdown. Mirtrons are a type of microRNAs that are located in the introns of the mRNA encoding host genes. Mirtrons arise from the spliced-out introns and are known to function in gene expression. In a murine myoblast line containing a pathogenic copy of human *DMPK* an artificial mirtron of that gene corrected DM1 associated splicing abnormalities demonstrating the therapeutic potential of mirtron-mediated RNAi [137]. Finally, the generation of improved animal models for DM1 has allowed the development and evaluation of new therapies. In the last few years, DM models in *Drosophila*, mice, and *C. elegans* have been generated and tested for several potential treatments ranging from activation of the RNA interference pathways to the administration of chemical compounds [65, 94, 138-141]. Moreover, new systems for performing high-throughput drug screening in *in vivo* and *in vitro* models, and rational drug design applicable to DM1 are in development [142]. Finding an effective anti-DM1 treatment will require further exploration of the therapies already discussed in this chapter and newer experimental therapies in animal models, as well as a continued determined effort to untangle the still unexplored complexity of this multifaceted disease.

## CONFLICT OF INTEREST

The author(s) confirm that this chapter content has no conflict of interest.

## ACKNOWLEDGEMENTS

We thank members of our laboratory for insightful discussions during the writing of this chapter. B.L. was funded by a research grant from Fundació La Marató of

TV3 (100230) to R.A. A research grant from Conselleria d´Educació, Formació i Ocupació (Prometeo 2010/081) to R.A. provided funds to support A.B. A grant was also provided to R.A. from Ministero de Ciencia e Innovación ( BFU2009-10940) including funds from plane.

## DISCLOSURE

This is an update report of the earlier review article by B.Llamusi and R.Artero, "Molecular Effects of the CTG Repeats in Mutant Dystrophia Myotonica Protein Kinase Gene". Curr. Genomics, 2008 Dec; 9(8): 509-16.

## REFERENCES

[1] Romeo, V., Myotonic Dystrophy Type 1 or Steinert's disease. *Adv. Exp. Med. Biol.*, **2012**, *724*, 239-57.

[2] Gagnon, C., *et al.*, Towards an integrative approach to the management of myotonic dystrophy type 1. J. Neurol. *Neurosurg. Psychiatry.*, **2007**, *78*, 800-6.

[3] Udd, B. and R. Krahe, The myotonic dystrophies: molecular, clinical, and therapeutic challenges. *Lancet. Neurol.*, **2012**, *11*, 891-905.

[4] Mathieu, J., M. De Braekeleer, and C. Prevost, Genealogical reconstruction of myotonic dystrophy in the Saguenay-Lac-Saint-Jean area (Quebec, Canada). *Neurology.*, **1990**, *40*, 839-42.

[5] Harper, P., Myotonic dystrophy, Saunders: London, **2001**.

[6] Shutler, G., *et al.*, Physical mapping and cloning of the proximal segment of the myotonic dystrophy gene region. Genomics., 1992, 13, 518-25.

[7] Fu, Y.H., *et al.*, An unstable triplet repeat in a gene related to myotonic muscular dystrophy. Science., 1992, 255, 1256-8.

[8] Mahadevan, M., *et al.*, Myotonic dystrophy mutation: an unstable CTG repeat in the 3' untranslated region of the gene. Science., 1992, 255, 1253-5.

[9] Brook, J.D., *et al.*, Molecular basis of myotonic dystrophy: expansion of a trinucleotide (CTG) repeat at the 3' end of a transcript encoding a protein kinase family member. Cell., 1992, 69, 385.

[10] Aslanidis, C., *et al.*, Cloning of the essential myotonic dystrophy region and mapping of the putative defect. Nature., 1992, 355, 548-51.

[11] Ranum, L.P., *et al.*, Genetic mapping of a second myotonic dystrophy locus. Nat. Genet., 1998, 19, 196-8.

[12] Liquori, C.L., *et al.*, Myotonic dystrophy type 2 caused by a CCTG expansion in intron 1 of ZNF9. Science., 2001, 293, 864-7.

[13] Ranum, L.P. and J.W. Day, Dominantly inherited, non-coding microsatellite expansion disorders. Curr. Opin. Genet. Dev., 2002, 12, 266-71.

[14] Day, J.W. and L.P. Ranum, RNA pathogenesis of the myotonic dystrophies. Neuromuscul. Disord., 2005, 15, 5-16.

[15] Junghans, R.P., Dystrophia myotonia: why focus on foci? Eur. J. Hum. Genet., 2009, 17, 543-53.

[16] Sicot, G., G. Gourdon, and M. Gomes-Pereira, Myotonic dystrophy, when simple repeats reveal complex pathogenic entities: new findings and future challenges. Hum. Mol. Genet., 2011, 20, 116-23.

[17] Fu, Y.H., *et al.*, Decreased expression of myotonin-protein kinase messenger RNA and protein in adult form of myotonic dystrophy. Science., 1993, 260, 235-8.

[18] Hofmann-Radvanyi, H., *et al.*, Myotonic dystrophy: absence of CTG enlarged transcript in congenital forms, and low expression of the normal allele. Hum. Mol. Genet., 1993, 2, 1263-6.

[19] Novelli, G., *et al.*, Failure in detecting mRNA transcripts from the mutated allele in myotonic dystrophy muscle. Biochem. Mol. Biol. Int., 1993, 29, 291-7.

[20] Carango, P., *et al.*, Absence of myotonic dystrophy protein kinase (DMPK) mRNA as a result of a triplet repeat expansion in myotonic dystrophy. Genomics., 1993, 18, 340-8.

[21] Taneja, K.L., *et al.*, Foci of trinucleotide repeat transcripts in nuclei of myotonic dystrophy cells and tissues. J. Cell. Biol., 1995, 128, 995-1002.

[22] Davis, B.M., *et al.*, Expansion of a CUG trinucleotide repeat in the 3' untranslated region of myotonic dystrophy protein kinase transcripts results in nuclear retention of transcripts. Proc. Natl. Acad. Sci. U.S.A., 1997, 94, 7388-93.

[23] Eriksson, M., *et al.*, Simultaneous analysis of expression of the three myotonic dystrophy locus genes in adult skeletal muscle samples: the CTG expansion correlates inversely with DMPK and 59 expression levels, but not DMAHP levels. Hum. Mol. Genet., 1999, 8, 1053-60.

[24] Narang, M.A., *et al.*, Myotonic dystrophy (DM) protein kinase levels in congenital and adult DM patients. Eur. J. Hum. Genet., 2000, 8, 507-12.

[25] Salvatori, S., *et al.*, Decreased expression of DMPK: correlation with CTG repeat expansion and fibre type composition in myotonic dystrophy type 1. Neurol. Sci., 2005, 26, 235-42.

[26] Hamshere, M.G., *et al.*, Transcriptional abnormality in myotonic dystrophy affects DMPK but not neighboring genes. Proc. Natl. Acad. Sci. U.S.A., 1997, 94, 7394-9.

[27] Tiscornia, G. and M.S. Mahadevan, Myotonic dystrophy: the role of the CUG triplet repeats in splicing of a novel DMPK exon and altered cytoplasmic DMPK mRNA isoform ratios. Mol. Cell., 2000, 5, 959-67.

[28] Reddy, S., *et al.*, Mice lacking the myotonic dystrophy protein kinase develop a late onset progressive myopathy. Nat. Genet., 1996, 13, 325-35.

[29] Berul, C.I., *et al.*, Progressive atrioventricular conduction block in a mouse myotonic dystrophy model. J. Interv. Card. Electrophysiol., 2000, 4, 351-8.

[30] Llagostera, E., *et al.*, Altered beta-adrenergic response in mice lacking myotonic dystrophy protein kinase. Muscle Nerve., 2012, 45, 128-30.

[31] Llagostera, E., *et al.*, Role of myotonic dystrophy protein kinase (DMPK) in glucose homeostasis and muscle insulin action. PLoS ONE, 2007, 2, e1134.

[32] Chen, W., *et al.*, Haploinsuffciency for Znf9 in Znf9+/- mice is associated with multiorgan abnormalities resembling myotonic dystrophy. J. Mol. Biol., 2007, 368, 8-17.

[33] Winchester, C.L., *et al.*, Characterization of the expression of DMPK and SIX5 in the human eye and implications for pathogenesis in myotonic dystrophy. Hum. Mol. Genet., 1999, 8, 481-92.

[34] Alwazzan, M., *et al.*, Myotonic dystrophy is associated with a reduced level of RNA from the DMWD allele adjacent to the expanded repeat. Hum. Mol. Genet., 1999, 8, 1491-7.

[35] Junghans, R.P., A. Ebralidze, and B. Tiwari, Does (CUG)n repeat in DMPK mRNA 'paint' chromosome 19 to suppress distant genes to create the diverse phenotype of myotonic dystrophy?: A new hypothesis of long-range *cis* autosomal inactivation. Neurogenetics., 2001, 3, 59-67.

[36] Gennarelli, M., *et al.*, Reduction of the DM-associated homeo domain protein (DMAHP) mRNA in different brain areas of myotonic dystrophy patients. Neuromuscul. Disord., 1999, 9, 215-9.

[37] Filippova, G.N., *et al.*, CTCF-binding sites flank CTG/CAG repeats and form a methylation-sensitive insulator at the DM1 locus. Nat. Genet., 2001, 28, 335-43.

[38] Lopez Castel, A., *et al.*, Expanded CTG repeat demarcates a boundary for abnormal CpG methylation in myotonic dystrophy patient tissues. Hum. Mol. Genet., 2011, 20, 1-15.

[39] Kirby, R.J., *et al.*, Drosophila homolog of the myotonic dystrophy-associated gene, SIX5, is required for muscle and gonad development. Curr. Biol., 2001, 11, 1044-9.

[40] Klesert, T.R., *et al.*, Mice deficient in Six5 develop cataracts: implications for myotonic dystrophy. Nat. Genet., 2000, 25, 105-9.

[41] Cho, D.H., *et al.*, Antisense transcription and heterochromatin at the DM1 CTG repeats are constrained by CTCF. Mol. Cell., 2005, 20, 483-9.

[42] Katayama, S., *et al.*, Antisense transcription in the mammalian transcriptome. Science., 2005, 309, 1564-6.

[43] Nakamori, M., C.E. Pearson, and C.A. Thornton, Bidirectional transcription stimulates expansion and contraction of expanded (CTG)*(CAG) repeats. Hum. Mol. Genet., 2011, 20, 580-8.

[44] Michalowski, S., *et al.*, Visualization of double-stranded RNAs from the myotonic dystrophy protein kinase gene and interactions with CUG-binding protein. Nucleic. Acids. Res., 1999, 27, 3534-42.

[45] Tian, B., *et al.*, Expanded CUG repeat RNAs form hairpins that activate the double-stranded RNA-dependent protein kinase PKR. Rna., 2000, 6, 79-87.

[46] Miller, J.W., *et al.*, Recruitment of human muscleblind proteins to (CUG)(n) expansions associated with myotonic dystrophy. Embo J., 2000, 19, 4439-48.

[47] Mooers, B.H., J.S. Logue, and J.A. Berglund, The structural basis of myotonic dystrophy from the crystal structure of CUG repeats. Proc. Natl. Acad. Sci. U.S.A., 2005, 102, 16626-31.

[48] Ebralidze, A., *et al.*, RNA leaching of transcription factors disrupts transcription in myotonic dystrophy. Science., 2004, 303, 383-7.

[49] Amack, J.D., S.R. Reagan, and M.S. Mahadevan, Mutant DMPK 3'-UTR transcripts disrupt C2C12 myogenic differentiation by compromising MyoD. J. Cell. Biol., 2002, 159, 419-29.

[50] Botta, A., *et al.*, Gene expression analysis in myotonic dystrophy: indications for a common molecular pathogenic pathway in DM1 and DM2. Gene Expr., 2007, 13, 339-51.
[51] Yadava, R.S., *et al.*, RNA toxicity in myotonic muscular dystrophy induces NKX2-5 expression. Nat. Genet., 2008, 40, 61-8.
[52] Dansithong, W., *et al.*, RNA steady-state defects in myotonic dystrophy are linked to nuclear exclusion of SHARP. EMBO Rep., 2011, 12, 735-42.
[53] Fernandez-Costa, J.M., *et al.*, Alternative splicing regulation by Muscleblind proteins: from development to disease. Biol. Rev. Camb. Philos. Soc., 2011, 86, 947-58.
[54] Begemann, G., *et al.*, muscleblind, a gene required for photoreceptor differentiation in Drosophila, encodes novel nuclear Cys3His-type zinc-finger-containing proteins. Development., 1997, 124, 4321-31.
[55] Artero, R., *et al.*, The muscleblind gene participates in the organization of Z-bands and epidermal attachments of Drosophila muscles and is regulated by Dmef2. Dev. Biol., 1998, 195, 131-43.
[56] Fardaei, M., *et al.*, *In vivo* co-localisation of MBNL protein with DMPK expanded-repeat transcripts. Nucleic. Acids. Res., 2001, 29, 2766-71.
[57] Fardaei, M., *et al.*, Three proteins, MBNL, MBLL and MBXL, co-localize *in vivo* with nuclear foci of expanded-repeat transcripts in DM1 and DM2 cells. Hum. Mol. Genet., 2002, 11, 805-14.
[58] Mankodi, A., *et al.*, Muscleblind localizes to nuclear foci of aberrant RNA in myotonic dystrophy types 1 and 2. Hum. Mol. Genet., 2001, 10, 2165-70.
[59] Mankodi, A., *et al.*, Nuclear RNA foci in the heart in myotonic dystrophy. Circ. Res., 2005, 97, 1152-5.
[60] Jiang, H., *et al.*, Myotonic dystrophy type 1 is associated with nuclear foci of mutant RNA, sequestration of muscleblind proteins and deregulated alternative splicing in neurons. Hum. Mol. Genet., 2004, 13, 3079-88.
[61] Cardani, R., *et al.*, Muscleblind-like protein 1 nuclear sequestration is a molecular pathology marker of DM1 and DM2. Eur. J. Histochem., 2006, 50, 177-82.
[62] Wheeler, T.M., M.C. Krym, and C.A. Thornton, Ribonuclear foci at the neuromuscular junction in myotonic dystrophy type 1. Neuromuscul. Disord., 2007, 17, 242-7.
[63] Houseley, J.M., *et al.*, Myotonic dystrophy associated expanded CUG repeat muscleblind positive ribonuclear foci are not toxic to Drosophila. Hum. Mol. Genet., 2005, 14, 873-83.
[64] Haro, M.D., *et al.*, MBNL1 and CUGBP1 modify expanded CUG-induced toxicity in a Drosophila model of Myotonic Dystrophy Type 1. Hum. Mol. Genet., 2006.
[65] Garcia-Lopez, A., *et al.*, Genetic and chemical modifiers of a CUG toxicity model in Drosophila. PLoS ONE., 2008, 3, e1595.
[66] Ho, T.H., *et al.*, Muscleblind proteins regulate alternative splicing. Embo J., 2004, 23, 3103-12.
[67] Koebis, M., *et al.*, Alternative splicing of myomesin 1 gene is aberrantly regulated in myotonic dystrophy type 1. Genes Cells., 2011, 16, 961-72.
[68] Wang, G.S. and T.A. Cooper, Splicing in disease: disruption of the splicing code and the decoding machinery. Nat. Rev. Genet., 2007, 8, 749-61.
[69] Lee, K.S., *et al.*, RNA-binding protein Muscleblind-like 3 (MBNL3) disrupts myocyte enhancer factor 2 (Mef2) {beta}-exon splicing. J. Biol. Chem., 2010, 285, 33779-87.

[70] Machuca-Tzili, L.E., *et al.*, Zebrafish deficient for Muscleblind-like 2 exhibit features of myotonic dystrophy. Dis. Model. Mech., 2011, 4, 381-92.

[71] Hao, M., *et al.*, Muscleblind-like 2 (Mbnl2) -deficient mice as a model for myotonic dystrophy. Dev. Dyn., 2008, 237, 403-10.

[72] Charizanis, K., *et al.*, Muscleblind-like 2-mediated alternative splicing in the developing brain and dysregulation in myotonic dystrophy. Neuron., 2012, 75, 437-50.

[73] Suenaga, K., *et al.*, Muscleblind-like 1 knockout mice reveal novel splicing defects in the myotonic dystrophy brain. PLoS One., 2012, 7, e33218.

[74] Kanadia, R.N., *et al.*, A muscleblind knockout model for myotonic dystrophy. Science., 2003, 302, 1978-80.

[75] Lin, X., *et al.*, Failure of MBNL1-dependent postnatal splicing transitions in myotonic dystrophy. Hum. Mol. Genet., 2006.

[76] Wang, E.T., *et al.*, Transcriptome-wide regulation of pre-mRNA splicing and mRNA localization by muscleblind proteins. Cell., 2012, 150, 710-24.

[77] Savkur, R.S., A.V. Philips, and T.A. Cooper, Aberrant regulation of insulin receptor alternative splicing is associated with insulin resistance in myotonic dystrophy. Nat. Genet., 2001, 29, 40-7.

[78] Charlet, B.N., *et al.*, Loss of the muscle-specific chloride channel in type 1 myotonic dystrophy due to misregulated alternative splicing. Mol. Cell., 2002, 10, 45-53.

[79] Machuca-Tzili, L., *et al.*, Flies deficient in Muscleblind protein model features of myotonic dystrophy with altered splice forms of Z-band associated transcripts. Hum. Genet., 2006, 120, 487-499.

[80] Kanadia, R.N., *et al.*, Reversal of RNA missplicing and myotonia after muscleblind overexpression in a mouse poly(CUG) model for myotonic dystrophy. Proc. Natl. Acad. Sci. U.S.A., 2006, 103, 11748-53.

[81] Chamberlain, C.M. and L.P. Ranum, Mouse model of muscleblind-like 1 overexpression: keletal muscle effects and therapeutic promise. Hum. Mol. Genet., 2012, 21, 4645-54.

[82] Du, H., *et al.*, Aberrant alternative splicing and extracellular matrix gene expression in mouse models of myotonic dystrophy. Nat. Struct. Mol. Biol., 2010, 17, 187-93.

[83] Osborne, R.J., *et al.*, Transcriptional and post-transcriptional impact of toxic RNA in myotonic dystrophy. Hum. Mol. Genet., 2009, 18, 1471-81.

[84] Timchenko, N.A., *et al.*, RNA CUG repeats sequester CUGBP1 and alter protein levels and activity of CUGBP1. J. Biol. Chem., 2001, 276, 7820-6.

[85] Dansithong, W., *et al.*, MBNL1 is the primary determinant of focus formation and aberrant insulin receptor splicing in DM1. J. Biol. Chem., 2005, 280, 5773-80.

[86] Dasgupta, T. and A.N. Ladd, The importance of CELF control: molecular and biological roles of the CUG-BP, Elav-like family of RNA-binding proteins. Wiley Interdiscip. Rev. RNA., 2012, 3, 104-21.

[87] Mankodi, A., *et al.*, Ribonuclear inclusions in skeletal muscle in myotonic dystrophy types 1 and 2. Ann. Neurol., 2003, 54, 760-8.

[88] Lueck, J.D., *et al.*, Muscle chloride channel dysfunction in two mouse models of myotonic dystrophy. J. Gen. Physiol., 2007, 129, 79-94.

[89] Kuyumcu-Martinez, N.M., G.S. Wang, and T.A. Cooper, Increased steady-state levels of CUGBP1 in myotonic dystrophy 1 are due to PKC-mediated hyperphosphorylation. Mol. Cell., 2007, 28, 68-78.

[90] Timchenko, N.A., *et al.*, Overexpression of CUG triplet repeat-binding protein, CUGBP1, in mice inhibits myogenesis. J. Biol. Chem., 2004, 279, 13129-39.
[91] Koshelev, M., *et al.*, Heart-specific overexpression of CUGBP1 reproduces functional and molecular abnormalities of myotonic dystrophy type 1. Hum. Mol. Genet., 2010, 19, 1066-75.
[92] Ward, A.J., *et al.*, CUGBP1 overexpression in mouse skeletal muscle reproduces features of myotonic dystrophy type 1. Hum. Mol. Genet., 2010, 19, 3614-22.
[93] Philips, A.V., L.T. Timchenko, and T.A. Cooper, Disruption of splicing regulated by a CUG-binding protein in myotonic dystrophy. Science., 1998, 280, 737-41.
[94] Mahadevan, M.S., *et al.*, Reversible model of RNA toxicity and cardiac conduction defects in myotonic dystrophy. Nat. Genet., 2006, 38, 1066-70.
[95] Berger, D.S. and A.N. Ladd, Repression of nuclear CELF activity can rescue CELF-regulated alternative splicing defects in skeletal muscle models of myotonic dystrophy. PLoS Curr., 2012, 4, RRN1305.
[96] Lee, J.E., *et al.*, Systematic analysis of *cis*-elements in unstable mRNAs demonstrates that CUGBP1 is a key regulator of mRNA decay in muscle cells. PLoS One., 2010, 5, e11201.
[97] Kim, D.H., *et al.*, HnRNP H inhibits nuclear export of mRNA containing expanded CUG repeats and a distal branch point sequence. Nucleic Acids Res., 2005, 33, 13866-74.
[98] Paul, S., *et al.*, Interaction of muscleblind, CUG-BP1 and hnRNP H proteins in DM1-associated aberrant IR splicing. Embo J., 2006, 25, 4271-83.
[99] Ravel-Chapuis, A., *et al.*, The RNA-binding protein Staufen1 is increased in DM1 skeletal muscle and promotes alternative pre-mRNA splicing. J. Cell. Biol., 2012, 196, 699-712.
[100] Osborne, R.J. and C.A. Thornton, RNA-dominant diseases. Hum. Mol. Genet., 2006, 15, R162-9.
[101] Mankodi, A., *et al.*, Expanded CUG repeats trigger aberrant splicing of ClC-1 chloride channel pre-mRNA and hyperexcitability of skeletal muscle in myotonic dystrophy. Mol. Cell., 2002, 10, 35-44.
[102] Berg, J., *et al.*, Truncated ClC-1 mRNA in myotonic dystrophy exerts a dominant-negative effect on the Cl current. Neurology., 2004, 63, 2371-5.
[103] Fugier, C., *et al.*, Misregulated alternative splicing of BIN1 is associated with T tubule alterations and muscle weakness in myotonic dystrophy. Nat. Med., 2011, 17, 720-5.
[104] Yamashita, Y., *et al.*, Four parameters increase the sensitivity and specificity of the exon array analysis and disclose 25 novel aberrantly spliced exons in myotonic dystrophy. J. Hum. Genet., 2012, 57, 368-74.
[105] Ranum, L.P. and J.W. Day, Myotonic dystrophy: RNA pathogenesis comes into focus. Am. J. Hum. Genet., 2004, 74, 793-804.
[106] Ishiura, S., *et al.*, Regulation of splicing by MBNL and CELF family of RNA-binding protein. Acta Myol., 2005, 24, 74-7.
[107] Jasper, H., *et al.*, A genomic switch at the transition from cell proliferation to terminal differentiation in the Drosophila eye. Dev. Cell., 2002, 3, 511-21.
[108] Ladd, A.N., N. Charlet, and T.A. Cooper, The CELF family of RNA binding proteins is implicated in cell-specific and developmentally regulated alternative splicing. Mol. Cell. Biol., 2001, 21, 1285-96.

[109] Malinina, L., Possible involvement of the RNAi pathway in trinucleotide repeat expansion diseases. J. Biomol. Struct. Dyn., 2005, 23, 233-5.

[110] Robb, G.B., *et al.*, Specific and potent RNAi in the nucleus of human cells. Nat. Struct. Mol. Biol., 2005, 12, 133-7.

[111] Krol, J., *et al.*, Ribonuclease dicer cleaves triplet repeat hairpins into shorter repeats that silence specific targets. Mol. Cell., 2007, 25, 575-86.

[112] Yu, Z., X. Teng, and N.M. Bonini, Triplet repeat-derived siRNAs enhance RNA-mediated toxicity in a Drosophila model for myotonic dystrophy. PLoS Genet., 2011, 7, e1001340.

[113] Janga, S.C. and S. Vallabhaneni, MicroRNAs as post-transcriptional machines and their interplay with cellular networks. Adv. Exp. Med. Biol., 2011, 722, 59-74.

[114] Onishi, H., *et al.*, MBNL1 associates with YB-1 in cytoplasmic stress granules. J. Neurosci. Res., 2008, 86, 1994-2002.

[115] Rau, F., *et al.*, Misregulation of miR-1 processing is associated with heart defects in myotonic dystrophy. Nat. Struct. Mol. Biol., 2011, 18, 840-5.

[116] Perbellini, R., *et al.*, Dysregulation and cellular mislocalization of specific miRNAs in myotonic dystrophy type 1. Neuromuscul. Disord., 2011, 21, 81-8.

[117] Gambardella, S., *et al.*, Overexpression of microRNA-206 in the skeletal muscle from myotonic dystrophy type 1 patients. J. Transl. Med., 2010, 8, 48.

[118] Hon, L.S. and Z. Zhang, The roles of binding site arrangement and combinatorial targeting in microRNA repression of gene expression. Genome Biol., 2007, 8, R166.

[119] Salisbury, E., *et al.*, Ectopic expression of cyclin D3 corrects differentiation of DM1 myoblasts through activation of RNA CUG-binding protein, CUGBP1. Exp. Cell. Res., 2008, 314, 2266-78.

[120] Iakova, P., *et al.*, Competition of CUGBP1 and calreticulin for the regulation of p21 translation determines cell fate. Embo J., 2004, 23, 406-17.

[121] Karagiannides, I., *et al.*, Increased CUG triplet repeat-binding protein-1 predisposes to impaired adipogenesis with aging. J. Biol. Chem., 2006, 281, 23025-33.

[122] Nezu, Y., *et al.*, Expression of MBNL and CELF mRNA transcripts in muscles with myotonic dystrophy. Neuromuscul. Disord., 2007, 17, 306-12.

[123] Timchenko, N.A., G.L. Wang, and L.T. Timchenko, RNA CUG-binding protein 1 increases translation of 20-kDa isoform of CCAAT/enhancer-binding protein beta by interacting with the alpha and beta subunits of eukaryotic initiation translation factor 2. J. Biol. Chem., 2005, 280, 20549-57.

[124] Ho, T.H., *et al.*, Transgenic mice expressing CUG-BP1 reproduce splicing mis-regulation observed in myotonic dystrophy. Hum. Mol. Genet., 2005, 14, 1539-47.

[125] Timchenko, N.A., *et al.*, CUG repeat binding protein (CUGBP1) interacts with the 5' region of C/EBPbeta mRNA and regulates translation of C/EBPbeta isoforms. Nucleic Acids Res., 1999, 27, 4517-25.

[126] Zu, T., *et al.*, Non-ATG-initiated translation directed by microsatellite expansions. Proc. Natl. Acad. Sci. U.S.A., 2011, 108, 260-5.

[127] Loro, E., *et al.*, Normal myogenesis and increased apoptosis in myotonic dystrophy type-1 muscle cells. Cell. Death. Differ., 2010, 17, 1315-24.

[128] van Eyk, C.L., *et al.*, Perturbation of the Akt/Gsk3-beta signalling pathway is common to Drosophila expressing expanded untranslated CAG, CUG and AUUCU repeat RNAs. Hum. Mol. Genet., 2011, 20, 2783-94.

[129] Jackson, R.J., C.U. Hellen, and T.V. Pestova, The mechanism of eukaryotic translation initiation and principles of its regulation. Nat. Rev. Mol. Cell. Biol., 2010, 11, 113-27.

[130] Domier, L.L., N.K. McCoppin, and C.J. D'Arcy, Sequence requirements for translation initiation of Rhopalosiphum padi virus ORF2. Virology., 2000, 268, 264-71.

[131] La Spada, A.R., *et al.*, Androgen receptor gene mutations in X-linked spinal and bulbar muscular atrophy. Nature., 1991, 352, 77-9.

[132] Verkerk, A.J., *et al.*, Identification of a gene (FMR-1) containing a CGG repeat coincident with a breakpoint cluster region exhibiting length variation in fragile X syndrome. Cell., 1991, 65, 905-14.

[133] A novel gene containing a trinucleotide repeat that is expanded and unstable on Huntington's disease chromosomes. The Huntington's Disease Collaborative Research Group. Cell., 1993, 72, 971-83.

[134] Rudnicki, D.D., *et al.*, Huntington's disease--like 2 is associated with CUG repeat-containing RNA foci. Ann. Neurol., 2007, 61, 272-82.

[135] Verbeek, D.S. and B.P. van de Warrenburg, Genetics of the dominant ataxias. Semin. Neurol., 2011, 31, 461-9.

[136] Wheeler, T.M., *et al.*, Targeting nuclear RNA for *in vivo* correction of myotonic dystrophy. Nature., 2012, 488, 111-5.

[137] Seow, Y., C.R. Sibley, and M.J. Wood, Artificial mirtron-mediated gene knockdown: functional DMPK silencing in mammalian cells. RNA., 2012, 18, 1328-37.

[138] Gomes-Pereira, M. and D.G. Monckton, Chemical modifiers of unstable expanded simple sequence repeats: what goes up, could come down. Mutat. Res., 2006, 598, 15-34.

[139] Seznec, H., *et al.*, Transgenic mice carrying large human genomic sequences with expanded CTG repeat mimic closely the DM CTG repeat intergenerational and somatic instability. Hum. Mol. Genet., 2000, 9, 1185-94.

[140] Denovan-Wright, E.M. and B.L. Davidson, RNAi: a potential therapy for the dominantly inherited nucleotide repeat diseases. Gene Ther., 2006, 13, 525-31.

[141] Garcia-Lopez, A., *et al.*, *In vivo* discovery of a peptide that prevents CUG-RNA hairpin formation and reverses RNA toxicity in myotonic dystrophy models. Proc. Natl. Acad. Sci. U.S.A., 2011, 108, 11866-71.

[142] Garcia-Alcover, I., López Castel, A, Perez-Alonso, M., Artero, R., *In vivo* strategies for drug discovery in myotonic dystrophy disorders. Drug Discovery Today: Technologies., 2012, 306, 6.

*Advances in Genome Science, Vol. 2,* 2013, 246-296

# CHAPTER 10

# Saturation of the Human Phenome

**Mark E. Samuels***

*Centre de Recherche du CHU Ste-Justine, University of Montreal, Montreal, Canada*

**Abstract:** The phenome is the complete set of phenotypes resulting from genetic variation in populations of an organism. Saturation of a phenome implies the identification and phenotypic description of mutations in all genes in an organism, potentially constrained to those encoding proteins. The human genome is believed to contain 20-25,000 protein coding genes, but only a small fraction of these have documented mutant phenotypes, thus the human phenome is far from complete. In model organisms, genetic saturation entails the identification of multiple mutant alleles of a gene or locus, allowing a consistent description of mutational phenotypes for that gene. Saturation of several model organisms has been attempted, usually by targeting annotated coding genes with insertional transposons (*Drosophila melanogaster*, *Mus musculus*) or by sequence directed deletion (*Saccharomyces cerevisiae*) or using libraries of antisense oligonucleotide probes injected directly into animals (*Caenorhabditis elegans*, *Danio rerio*). This paper reviews the general state of the human phenome, and discusses theoretical and practical considerations toward a saturation analysis in humans. Throughout, emphasis is placed on high penetrance genetic variation, of the kind typically asociated with monogenic *versus* complex traits.

**Keywords:** Phenome, human genome, saturation mutagenesis, model organisms.

## INTRODUCTION

The phenome is the complete set of phenotypes resulting from genetic variation in populations of an organism [1-4]. The concept of phenomic saturation cannot be approached without a careful consideration of the meaning of phenotype, and the genotype-phenotype map. In the most general sense, the phenotype of an individual organism is the sum total of its physiology. Such a broad definition is however not particularly useful. In the context of genetics, phenotypes may be

***Address correspondence to Mark E. Samuels:** Centre de Recherche du CHU Ste-Justine, 3175, Côte Ste-Catherine, Montréal QC H3T 1C5, Canada; Tel: 514-345-4931 p4265; Fax: 514-345-4801;
E-mail: mark.e.samuels@umontreal.ca

**Christian Neri (Ed)**

defined more specifically as those physiological traits which vary measurably as a function of genomic sequence differences among individuals in a species population. Thus, many aspects of physiology can be ignored depending on the type of sequence differences being studied. For example, mutations in the gene encoding the low density lipoprotein receptor (LDLR) have huge impact on plasma cholesterol, atherosclerosis, and heart disease, but have no obvious or known impact on eye color or height. In contrast, mutations in the genes encoding pigment-producing enzymes in the iris may be irrelevant to heart disease. One of the advantages of forward genetics is that one begins with an unusual physiological trait observed in actual organisms (human or model animal), and proceeds to search for genetic variation that underlies that trait. In contrast, if one starts with a gene of interest, and mutates it in a directed way (or even through screening undirected libraries of mutants), one has no idea what the physiological effects might be so one is obliged to explore all possible phenotype consequences – a time consuming and difficult task with animals that cannot tell us how they are feeling! Both approaches have been employed in model studies, but human genetics in intact organisms is generally restricted to forward genetics except in some special cases of clinical trials involving somatic cell genetic manipulation (*i.e.* gene therapy).

## MONOGENIC *VERSUS* COMPLEX PHENOTYPES

To some extent, the distinction between monogenic and complex phenotypes is operational. All genetic variants in nuclear chromosomes are *de facto* transmitted in a Mendelian fashion. Thus even weakly penetrant variants, such as most single nucleotide polymorphisms (SNPs) found through large scale genome-wide association studies (GWAS) in case control cohorts, are nonetheless "Mendelian". Monogenic or Mendelian conditions are defined as being caused by variants in single genes, with the variants having a high penetrance (*i.e.* added risk). As a simple example, the clinical phenotype of cystic fibrosis (CF) has been made for many thousands of patients across all human populations. In all cases that have been studied, mutations are found in a single protein-coding gene, *CFTR*. Thus there is a clear one gene-one phenotype correlation for this condition. In contrast, oligogenic or polygenic or complex phenotypes are usually thought of as those for which combined input of many different genetic variants is required, potentially with significant environmental factors required as well (Fig. **1**).

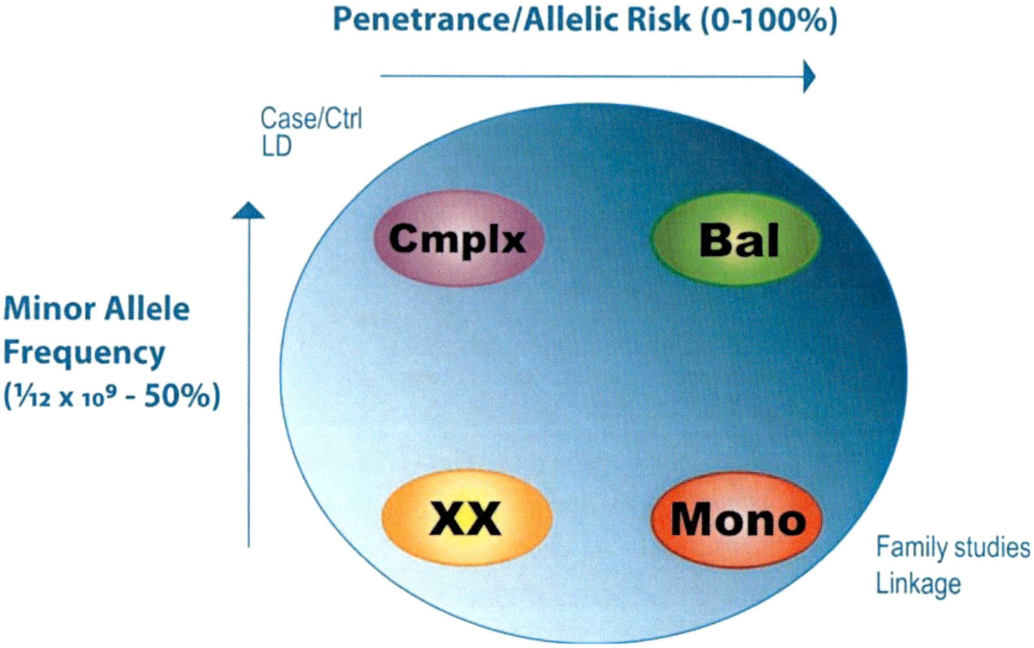


**Figure 1:** Monogenic *versus* complex phenotypes. Genetic variants may be anywhere between vanishingly rare (with a minor allele frequency at a minimum of one in the entire human population of approximately 6 billion, thus an allele frequency of $1/12x10^9$ chromosomes) up to 50% (after which the minor allele becomes the major allele). The physiological effect of a genetic variant may be individually very strong (high penetrance) or weak (low penetrance). Rare high penetrance alleles have historically been identified in families, and studied by genome-wide scanning with dense polymorphic anonymous markers, currently SNPs, followed by statistical linkage analysis and gene resequencing to identify causal variants in the family. Common low penetrance alleles are typically identified in large case/control cohorts, and studied by genome-wide SNP genotyping followed by simple statistical tests of differential allele frequency in the two classes, or by more sophisticated linkage disequilibrium (haplotype) mapping. Common high penetrance variants are unusual, since these would normally be either quickly fixed or eliminated from breeding populations. One circumstance under which such variants can be maintained is balanced selection, where there are opposing selections on heterozygotes *versus* homozygotes (malaria and sickle cell anemia being the best documented example). Rare variants of small functional effect are easily discovered through random resequencing but difficult to study mechanistically.

Even for nominally monogenic disorders the case of CF is far from typical. There are many medical examples in which a particular clinical diagnosis has been linked to high penetrance genetic variation in many different genes (*e.g.* ataxias

[5], sensorineural deafness [6-8], retinitis pigmentosa [9-11]). In Charcot-Marie-Tooth disease, causal genes have been found with a broad spectrum of cellular processes including specialized neuronal structures, protein turnover, vesicle fusion, microtubule transport, transcription factors, and even several aminoacyl tRNA-synthetases [12-19]. Clearly, some disease diagnoses represent a common end-point for different types of defects in gene function.

Somewhat less common in human genetics are examples of different allele-specific physiological outcomes. For obvious quantitative traits, such as plasma cholesterol, there are alleles in genes such as PCSK9 and APOB which may alternatively cause hyper- or hypocholesterolemia. These appear to correlate with gain or loss of function, such that loss of function alleles cause either high end or low end mean values in the measured trait, but not both, whereas gain of function alleles cause the opposite mean value. In a developmental example, Huntington disease is caused by triplet repeat expansions of a particular region of the encoding gene leading to some form of aberrant protein which causes neuronal toxicity through an incompletely understood mechanism. Formally such an allele is a clear gain of function (more specifically, a neomorph or new function *versus* a hypermorph or higher level of normal function). The alternative type of allele, a loss of function, is not known for the HD gene in humans, however the orthologous HD gene directed knockout is lethal in homozygous mice [20], and it is likely that a human loss of function allele would have a similar extreme phenotype clinically unrelated to Huntington disease.

There are also a few relatively well-documented examples of digenic inheritance in human disease, such that a clinical phenotype depends on simultaneous genetic variation in two different protein-coding genes. Retinitis pigmentosa resulting from simultaneous heterozygous mutations in the genes peripherin/RDS and ROM1 is perhaps the first documented example [21-28]. There are also likely digenic models for hemochromatosis [29, 30], Bardet-Biedl syndrome [31-36], and sensorineural deafness [37-46].

For many monogenic genetic disorders there is significant inter-patient variability. This may be due to allele specific effects of the primary causal gene, or to secondary genetic modifiers [47, 48]. In some cases, it can be attributed to well-understood environmental factors. For example, mutations of the gene encoding phenylalanine hydroxylase cause childhood onset phenylketonuria, but there is great variation in clinical severity among patients [49], and disease progression depends on dietary intake levels of phenylalanine, such that dietary restrictions have beneficial effects in clinical practice. Similarly, hereditary hemochromatosis is generally more severe clinically in males than females carrying the same risk alleles; this difference is generally interpreted as females being partially rescued from iron overload through regular blood loss during menstruation [50]. It is widely assumed that combinations of genetic and environmental factors play a major role in clinical severity for many monogenic disorders [51-53]. This has been difficult to address experimentally for most such disorders, as they are individually rare and affected patients are dispersed across the globe. Significant attempts have been made to identify genetic co-factors or modifiers for cystic fibrosis [54-57] and hemochromatosis [58, 59]. The distinction between digenic (or oligogenic) inheritance and genetic modifiers is somewhat semantic, depending on the ease of clinical ascertainment and severity of the physiological effect of the primary gene. In any case, the ability to combine spontaneous, mutagen-induced, and engineered genetic variants in model organisms such as mouse and fruit fly, make it clear that physiological states can be readily decomposed into the effects of multiple genes acting in concert in an individual. The commonly used term "synthetic phenotype" in *Drosophila* genetics refers to the same phenomenon as "digenic inheritance", or "major genetic modifier", terms used more often in human genetics.

Monogenic disorders have been operationally identified primarily by their observed transmission in families. In the case of weakly penetrant genetic variants, or those with major genetic modifiers, it is difficult or impossible to trace the inheritance of the physiological state in families (whether a disease, a quantitative measurement of a risk factor such as blood pressure or plasma cholesterol, or a benign trait which may also be either quantitative (such as height) or dichotomous), in a fashion consistent with a single chromosomal locus. In the latter cases, familial clusters of affected individuals fail to lead to a single linked chromosomal locus following genome-wide marker-assisted mapping, suggestive of multiple genetic inputs yielding the observed excess of phenotypes in the family. Furthermore, for every anecdotal case of traits "running in families", there are examples where this does not occur. Clearly, some physiological states require multiple genetic inputs of similar penetrance, and some allelic variants on their own have little physiological impact but can have high penetrance in combination with other variants or environmental cofactors. Assuming that all physiological traits can ultimately be defined by some quantitative variable or variables, although these may not be known, the effect of an individual variant allele can be visualized by its effect on the mean trait measurement. The further from the population mean a particular allelic variant is, the easier it is to detect and follow in individual pedigrees. The distinction between dominant and recessive acting alleles, while a practical complication in the detection of familial transmission, is not theoretically relevant to this point.

One example which highlights the distinction between these theoretical and practical issues is age-related macular degeneration (ARMD). ARMD has been strongly associated with genetic variation in the gene encoding complement factor H (CFH) [60-63]. The gene was identified independently by family-based linkage and by GWAS methods. The statistical signal in GWAS studies was stronger, as there is a very strong age-of-onset effect, which combined with the natural human life-span makes it impossible to observe familial transmission of the phenotype to younger generations. Factoring in the age-of-onset, the penetrance of CFH variants is extremely high, and this can be arguably considered a monogenic trait in the affected patients. If humans lived several hundred years, large dominant pedigrees for ARMD caused by CFH variants might well be ascertained.

Not all traits are diseases, although diseases receive the most attention from geneticists and are often the most readily ascertained through routine clinical surveillance. However, non-medical traits such as hair or eye color, height, and ability to smell or taste particular chemicals, have been studied genetically following large population-based sampling schemes without a medical phenotypic ascertainment [64-70].

## HOW MANY GENES ARE THERE?

Determining the number of actual genes encoded by human (or other) genomes has proved surprisingly challenging. Early estimates based on quantitative measurements of CpG islands led to estimates in excess of 80,000 protein coding genes in the human genome [71]. Since then, estimates have dropped repeatedly, to 28-34,000 based on comparison of exon comparison between the pufferfish *Tetraodon nigroviridis* and human genomes [72], to 21,037 based on large scale cDNA library sequencing [73, 74], and to 20,488 based on comparison of proposed human genes to sequences of other primate genomes [75]. As of mid-2010 (time of publication of the original version of this review), RefSeq, a widely used NCBI library of gene annotations, included 21,515 unique entries. These include hundreds of non-protein coding genes such as 363 small nucleolar RNA (SNO) genes, 28 SNAR genes and 637 micro-RNA (MIR) genes, so RefSeq probably includes approximately 20,000 protein coding genes. There does not appear to be a more recent set of easily accessible RefSeq statistics. In addition there are several ongoing manual gene curation efforts, specifically Havana, VEGA [76, 77] and CCDS [78], as well as the Mammalian Gene Collection (MGC). The CCDS currently has 18,472 curated unique protein coding genes, an increase of only 301 genes since mid-2010, out of 26,473 different coding sequences which presumably include multiple alternative isoforms for many genes. The MGC currently includes 17,592 genes for which full length cDNA clones are available; this number has not changed since mid-2010.

Although a consensus appears near that the total count of protein coding genes in the human genome is between 20-22,000, the advent of next-generation technologies has allowed further exploration of alternative splicing through deep resequencing of cDNA libraries. Several recent studies have documented novel alternatively spliced forms of known genes, as well as completely novel exons within previously studied genes [79, 80]. Thus the number of potential alternative protein isoforms is many times greater than the actual gene count. Moreover, proteomic approaches have documented a modest number of apparently *bona fide* translation products not otherwise accounted for in gene annotations from cDNA-based libraries [81, 82]. Proteomic methods also identify many different post-translational processed forms of some primary translation products, with modifications such as proteolytic fragmentation, phosphorylation, acetylation, glycosylation, *etc.* These methods are inherently sensitive to effective concentration of the analytes in biological samples, hence a truly comprehensive list of genome-encoded protein products seems an unrealistic outcome for the moment, although sensitivities especially in mass spectrometry are continually improving.

The human mitochondrial genome has been comparatively straightforward to annotate, as it resembles bacterial genomes in gene structure. It encodes 13 proteins, all involved in respiratory electron transport and oxidative phosphorylation, as well as 22 mitochondrion specific tRNAs, and the 12S and 16S ribosomal RNAs (rRNAs) required for the mitochondrial ribosome [83]. However, the human mitochondrion itself contains 900-1000 different proteins, based on direct proteomic analyses [84, 85]. Thus the large majority of mitochondrial proteins are encoded by nuclear chromosomal genes.

## STUDIES IN MODEL ORGANISMS

Saturation mutagenesis of model organisms has been widely used to identify mutable genes either for viability or for various interesting biological phenotypes [86]. In diploid organisms, dominant mutations are readily detected in heterozygotes after mutagenesis and breeding to rule out direct effects of the mutagen, although dominant lethality is obviously a difficult phenotype to recover. Recessive mutations can be easily studied in fungi by sporulation and reduction to haploidy, or in metazoans by establishing mutagenized heterozygous lines and inbreeding to create homozygotes, or particularly in the fly, *Drosophila melanogaster,* by complementation testing of mutagenized heterozygotes with chromosomal deletions of cytologically or molecularly defined extent. A longstanding question has been whether all biochemically defined genes (transcribed, spliced, protein coding/noncoding, *etc.*) have detectable mutational phenotypes. Even now, only a few historically saturated regions can be well-defined in terms of chromosomal extent, number of genetically mutable loci, and annotated protein-coding transcripts. With the advent of the complete genomic sequence plus extensive gene annotation, such studies can be revisited. This brief overview will focus on animals, leaving plant genetics aside.

The budding yeast *S. cerevisiae* is relatively straightforward to saturate since most genes lack introns. As a result at least for protein-coding genes, predictions can be made based on the occurrence of long contiguous open reading frames (ORFs). Although a lower ORF length cutoff is somewhat subjective, the yeast genome project in 2001 identified 6138 potential protein coding genes. Essentially all of these ORFs have been individually deleted, and the resulting strains analyzed under a few different sets of growth conditions [87-89]. According to the yeast deletion project, 18.2% of genes are essential for growth in rich medium (*i.e.* deletion strains fail to grow). A large-scale effort is currently under way to map genetic interactions by combining individually deleted genes in pairwise combination [90]. Ultimately over 36 million different pairwise tests must be performed; as of recently only 5.4 million tests have been reported but the results are intriguing. Approximately 170,000 interactions were detected, meaning that a strain carrying both deletions differed significantly in fitness from either single deletion strain. These interactions could be further studied by overlaying orthogonal annotations such as functional cellular modules, expression profiles, and protein-protein interaction profiles.

In *D. melanogaster*, historical saturation mutageneses have been performed for multiple genomic regions, while the essentially complete genome sequence of the organism together with a preliminary gene annotation was published in 2000 [91]. Overall the fly genome is believed to contain 13-14,000 protein coding genes. Saturation of the *zeste-white* region of the X chromosome has been performed repeatedly using various mutagens [92, 93]. Defining the region as that uncovered by a particular deletion (Df(1)$w^{rJ1}$, encompassing cytological bands 3A2-3C2), mutations were initially identified in 17 loci, 15 generating lethality and 2 generating visible phenotypes. Currently FlyBase annotates 31 genetic loci uncovered by this deletion, including genes with additional lethal, female sterile or visible phenotypes beyond the original 17. One caveat is that in rare cases, different mutations within the same gene may show genetic complementation, slightly overestimating the number of actual genes. Through transcriptional and comparative genomic analyses, Flybase annotates a total of 43 genes in this chromosomal region, including 19 identified only as candidate genes. In some cases genetically defined loci have not yet been assigned to structurally defined genes. Thus up to three fourths (31/43) of annotated genes may be mutable to an observable phenotype in this region.

In another chromosomal region also intensively studied in *D. melanogaster*, a strikingly different result was obtained. In a 2.9 million base region around the alcohol dehydrogenase (*Adh*) gene, Ashburner *et al.* documented 218 potential protein coding genes, but only 73 genetically definable loci in the region, of which only 49 have been assigned to specific structural genes so far [94]. Thus in the *Adh* region only a third of structurally defined genes may have individual mutational phenotypes, although it remains possible that some of these genes when mutated have a phenotype not readily observable in the laboratory.

Attempts at saturation mutagenesis using P-elements have been widely performed in *D. melanogaster* [95], however it is well documented that this transposon shows significant site specificity thus is not a good mutagen for genome-wide saturation. Efforts are currently under way to generate genome-wide saturation with transposable elements demonstrating less site specificity [96-98]. Although the larger number of genes in the fly compared to yeast makes comprehensive whole genome pairwise genetic mapping a daunting prospect, nonetheless *de novo* mutagenesis in genetically compromised fly strains (carrying either a chromosomal deletion of defined extent, or a heterozygous or homozygous mutation in a specific gene for which one is looking for potential modifier loci) is a routinely used tool to identify genetic interactions and members of shared physiological or regulatory pathways [99, 100]. Such modifier genes are often referred to as suppressors or enhancers of a phenotype; they may be either visible or lethal in nature, and as noted above correspond to digenic loci as defined usually in humans.

Although the small nematode worm *Caenorhabditis elegans* has a shorter experimental history than flies, ease of manipulation quickly made it a favorite among developmental geneticists. Moreover its small number of cells has permitted the unique opportunity to map all of embryonic and postembryonic development at the cellular level [101, 102]. The *C. elegans* genome has also been sequenced, and annotated to contain slightly fewer than 20,000 protein coding genes [103]. It remains puzzling why the worm, a much 'simpler' organism than the fly, has more genes, and in fact has about the same number as mammals. *C. elegans* has been the subject of a large number of genetic analyses, and pilot transposon-mediated mutagenesis has been performed [104]. However, the use of antisense RNA (RNAi), originally discovered in *C. elegans* [105], has permitted large scale gene knockdown experiments using libraries of gene-specific antisense targeting reagents, formally similar to germ-line loss-of-function mutagenesis [106, 107]. A study of RNAi targeting 16,757 genes from a predicted set of 19,427, detected mutant phenotypes for 1,722 probes, or 10% of targeted genes. Of these, slightly more than half caused embryonic lethality (RNAi was fed to hermaphrodite worms, allowing analysis of the effects in their progeny), with an estimated success rate of 70% compared to a set of genes with known germ-line embryonic lethal mutations. This suggests that 8% of genes can mutate to an embryonic lethal state in this organism. The overall low rate of observed phenotypes is difficult to interpret; it may simply be that most genes are functionally redundant in the worm. More likely, RNAi reagents may not be effective in completely reducing gene function for some genes.

Large scale random mutageneses have been performed in the mouse, *Mus musculus,* using both chemical and genetic mutagens [108-127]. Chemical mutageneses typically employ the DNA modifying agent ethylnitrosourea (ENU), but the prohibitive logistics of generating and maintaining the huge numbers of heterozygous F1 lines, required to create homozygous F2 animals to screen systematically for recessive mutations, have precluded a true genome-wide saturation. Genetic approaches primarily involve transposon-mediated insertional mutagenesis in mouse embryonic stem (ES) cells. There are now multiple libraries of embryonic stem cells carrying random transposon insertions, for which insertion sites have been mapped by direct sequencing. This potentially permits researchers to study the insertional phenotype for a gene of interest, if it happens to exist in a library. Although a large fraction of annotated protein-coding genes have targeted insertion cell lines, the vast majority of these transposon insertions have not been developed from ES cells into whole animals. Moreover, many of these insertions are not explicit gene knockouts as insertions are not targeted to coding exon for at least some of the insertion vectors.

As an alternative to random mutagenesis, many targeted gene knockout (KO) mouse strains have been constructed. Given the costs of development of mouse KOs, these are inevitably biased in favor of “interesting” genes, with the occurrence of known human genetic phenotypes being one component of interest.

However there are other reasons for genes to be defined as interesting, and it is unclear whether existing knockouts are biased toward genes certain to yield a detectable mutant phenotype in mice. In a review of mouse KOs as of 1995, it was reported that of 263 different genes targeted, 25% yielded an embryonic lethal phenotype, with another 10% leading to death within a few weeks postnatally [128-130]. Overall, 95% of knockouts were reported to show an observable phenotype, a very high proportion in comparison to the RNAi results with *C. elegans*. Subsequently it was reported that about 1/3 of direct knockouts yield embryonic lethality [131, 132]. In an analysis of mouse knockouts for 34 genes which are known targets of commercial therapeutics for important human diseases, all but one knockout strain showed some detectable phenotype in the mouse, most of which were directly relevant to the equivalent human disease [133]. In one case, that of insulin, the gene is known to be duplicated in the mouse genome, such that the double knockout is lethal whereas deletion of either gene alone does not yield a phenotype [134, 135]. In contrast, a knockout disrupting the insulin receptor gene, which exists in a single haploid copy in the mouse genome, is severely ill and dies shortly after birth, validating that the insulin pathway is intrinsically mutable in the mouse [136, 137].

The zebrafish, *Danio rerio,* has been the subject of many large scale mutageneses, since it was originally developed as a research organism by Streisinger [138-143]. In the absence of a gene-specific targeting system, most mutageneses have involved ENU treatment, followed by screens for interesting biological phenotypes. Identifying the actual mutated genes requires large scale recombinational mapping using anonymous polymorphic DNA markers, either microsatellites or SNPs, followed by positional cloning. Since the advent of high throughput sequencing, an intermediate approach called TILLING has been developed [144-146]. Instead of mapping and positional cloning, mutations in a given gene are found directly by sequencing pools of very large numbers of mutagenized animals (or plants, where the method was originally developed [145]) until a variant in the desired locus is detected. Subsets of the original positive pool are retested until eventually a single animal is identified carrying the mutation of interest. That animal is subsequently bred to test for a phenotypic effect of the mutation in either heterozygous or homozygous state. Using these approaches mutations have been found affecting many different physiological systems. Another effective approach in zebrafish is the use of antisense reagents. Modified antisense oligonucleotides, incorporating morpholino chemistry in the backbone to reduce enzymatic breakdown, can be injected in early fish embryos [147]. Oligos targeting genes are capable of producing very specific knockdowns, similar to the situation in worms. One limitation is that the morpholinos are eventually degraded or diluted, so only events of embryogenesis can be interrogated with this technology. Morpholinos have been used in other important experimental organisms such as the clawed frog *Xenopus laevis*, but genomic saturation screens have not been reported.

In sum, mutational studies in model organisms have demonstrated that a sizeable proportion of genes can be mutated to generate observable phenotypes. The exact number is difficult to determine from the available data sets, as there seem to be broad discrepancies between yeast and worm (small proportion of directed knockouts/knockdowns generating severe phenotypes) and flies (substantial portion of genes mutable) and mice (large proportion of directed knockouts generating phenotypes), nor is there a consensus as to the proportion of all genes that are essential for viability. It remains to be seen how these results will translate to human physiology, although it seems safe to assume that there must be substantial numbers of genes mutating to embryonic lethality, to viability with clinically observable phenotypes, or to complete viability without an easily detectable phenotype.

## HOW MANY MONOGENIC DISORDERS ARE KNOWN?

Keeping these points in mind, how many characterized monogenic phenotypes are there in humans? The entire human genetics literature is vast. There are at least three organizations attempting to curate the entire literature [148]. The best-known, Online Inheritance in Man (OMIM), functions as a mixed medical and genetic database, with entries for clinical disorders intermingled with entries for specific genes [149]. It has limited direct search functions. Querying with the universal term 0001 together with a limit for allelic variants, OMIM currently returns 2757 independent gene entries, while the statistics page reports 3444 phenotypes with a known molecular basis; the excess presumably resulting from some genes mutating to different phenotypes with independent OMIM entries [150]. Both of these numbers have increased substantially since mid-2010. The great majority of these represent monogenic disorders (as practically defined, high penetrance variants with reasonable evidence of genic pathogenicity based on the mutation itself). However, a fraction, possibly as many as 10%, may represent low penetrance risk alleles not clearly definable as part of monogenic disorders. OMIM does not curate all reported genetic variants in all genes, although entries are regularly updated when new interesting results are published.

The Human Gene Mutation Database (HGMD) curated at Cardiff University, attempts to consolidate all reported genetic variants in the literature [151]. It has an incomplete publically available version, and an up-to-date commercial version. The commercial HGMD currently includes 4411 independent gene entries and 120,004 mutation entries. It is difficult to assess how many of these correspond to monogenic conditions, *versus* low penetrance risk factors for complex phenotypes which HGMD has increasingly curated and included. As of mid-2010, HGMD indicated that 2491 genes contain at least one presumptive high penetrance allele (frameshift or premature stop codon, splice site, missense or indel) (*P. Stenson, pers comm*) [152]; that number is presumably also currently larger by several hundred new monogenic characterized genes. Differences in content between OMIM and HGMD are likely related to subtleties of literature curation of suggested *versus* confirmed genotype/phenotype causation, particularly in the case of low penetrance statistical outlier variants which are more heavily represented in HGMD than in OMIM.

The Human Genome Variation Society (HGVS) is developing a program of comprehensive curation which may eventually provide similar information as OMIM and HGMD combined [153, 154]. Currently the database HGVbaseG2P appears focused primarily on the results of GWAS studies for complex disorders.

The total number of molecularly characterized, true monogenic disorders remains slightly uncertain, but is around 2500-2700, representing 10-15% of annotated protein-coding genes. For the remaining 85-90% of genes, no high penetrance genetic variation has been causally linked to a physiological phenotype in humans.

## MUTATION RATES

Rates of spontaneous mutation in the human genome have been estimated in various ways. Combining early results of sequencing a small number of real or pseudogenes, Drake *et al.* suggested a rate of approximately $10^{-8}$ events per base pair per generation, equivalent to 64 new point mutations in each zygote [155]. By comparing partial human and chimpanzee sequences, Nachman and Crowell estimated a neutral rate of $1.3\text{-}3.4\text{x}10^{-8}$ events per bp per generation, equivalent to 91-238 new mutations per zygote [156]. More recently a direct test was performed by sequencing large segments of flow-sorted Y chromosomes from two males of a precisely known ancestral genealogy. Four new mutations were detected between the two samples, which correcting for chromosome size yielded a rate of $3\text{x}10^{-8}$ events per bp per generation, although the small number of individual events implies a reasonable sampling bias [157]. Finally, whole genomes were resequenced for a pedigree of two parents and two offspring, allowing the most direct possible measure. A total of 28 confirmed new mutations were detected, which after various corrections yielded a rate of $1.1\text{x}10^{-8}$ events per bp per generation, equivalent to 70 new events per zygote, with a substantially lower chance of sample bias [158]. Other studies are generally consistent with this rate, although providing some evidence for variable rates among families [159, 160].

These all represent essentially point mutation rates. It is now clear that larger structural rearrangements, generally termed copy number variants (CNVs), also arise at reasonably high frequencies in human populations, although it is difficult to assess true *de novo* mutation rates given the somewhat arbitrary definition of a CNV [161-163]. Many CNVs include genes and have potential functional consequences, although the proportion is not yet well established [164].

Not all mutations need have functional consequences. It is often considered that most of them are in fact functionally neutral, although this remains controversial. There is no easy way to assess this since most new mutations arise in non-coding regions of the genome, in either introns or intergenic regions. Our understanding of functional non-coding elements is still very incomplete, although large scale comparative genomics analyses hold promise to identify important functional elements, at least for those shared among the species compared [165]. The fraction of haploid sequence coding for protein is 1-2%, thus if each zygote receives 70 new genome-wide mutations, at most one of these is expected to be in a coding exon on average. Coding variants may be silent (synonymous) or change an amino acid (non-synonymous). Some studies have argued that a large proportion of non-synonymous changes have functional significance, although these may only be evident over evolution time frames and not be directly relevant to phenotypic interpretation in individuals [166-169]. It should be noted that synonymous changes may also have measurable effects on gene function, by influencing splicing *via* exon splice enhancer or repressor elements, or translation efficiency, or even codon bias given that different tRNAs charged with the same amino acid but with different anticodon sequences may not be present at equivalent cytoplasmic concentrations.

The rate of new highly deleterious mutations (usually termed U when summed over the entire genome), while certain to be substantially less than the rate of overall nucleotide change, cannot readily be inferred from the global mutation rate. A comparison of human and chimpanzee sequences, using some assumptions about evolutionary selection, led to an estimate for U of 1.6 per zygote [170]. A separate analysis, again using human/chimp comparison, obtained a value of U = 3 [156]. In a key study, Kondrashov combined and re-analyzed data for 20 different genes for which typical loss-of-function mutations lead to either autosomal dominant or X-linked high penetrance clinically ascertainable phenotypes [171]. Medical ascertainment rates per live birth include both new cases and familial cases, but by restricting to new cases alone, in populations where new cases are almost completely ascertained through routine health care surveillance, the median and mean total mutation rates for this set of genes can be calculated as 7 and 14 x $10^{-6}$ per live birth respectively. Assuming 20,000 protein coding genes, these are equivalent to U = 0.14-0.28 per genome per birth, substantially lower than the rates estimated by evolutionary comparison with other primates. A likely explanation for the discrepancy is again that evolutionary comparisons probably include functional effects too modest to yield a clinically ascertainable phenotype in individuals. Although a rate of U ~ 0.2 seems low, given a worldwide 2009 birth rate of approximately 139,000,000 [172], this represents almost 28 million new pathogenic mutations, or 1400 new mutations per gene worldwide. Even Canada, with a modest total population of about 34 million, has an annual birth rate of 378,000, and could have as many as 75,000 new pathogenic mutations per year nationwide. A significant caveat is that these rates may reflect a bias to genes with unusually high mutation rates, hence better studied by medical geneticists. A number of X-linked recessive mutating genes have lower mutation rates, such that a value of U = 0.02 may be a much more conservative genome-average estimate [173]. Even so, the entire human gene repertoire is clearly saturated annually worldwide with new pathogenic alleles. There is no shortage of high penetrance mutations in every gene segregating in human populations. The only reason why high penetrance monogenic disorders are relatively rare is that most mutations act recessively.

## DOMINANCE *VERSUS* RECESSIVITY, GAIN *VERSUS* LOSS OF FUNCTION

The proportion of monogenic disorders transmitted as dominant *versus* recessive traits in families is hard to determine from the curated databases. Similarly, it is difficult to assess the proportion of gain-of-function (gof) *versus* loss-of-function (lof) alleles, although this is not purely an issue of curation since for many mutant alleles the functional effect remains untested biochemically. Complete gene deletions represent the gold standard for a loss-of-function allele, however relatively few clean deletions of individual genes have been associated with monogenic conditions. Many contiguous deletion syndromes encompassing multiple genes are well documented, for most of which it is unclear whether the phenotype results from deletion of one particular gene in the deleted interval or from combined loss of several genes together. In one case, a recurrent deletion of about 1 million base pairs (Mbp) including about a dozen genes was found to be causal for Smith-Magenis syndrome [174]. Eventually rare cases were ascertained not carrying the deletion, in which point mutations were found in one particular gene, RAI1, which is now thought to be the main cause of the phenotype even in the deletion patients [175, 176].

Geneticists generally assume that major protein truncations arising through frameshift or premature stop codon mutations are pathogenic, but it would be naive to presume that these all represent complete loss-of-function since amino terminal protein fragments can encode partial or aberrant activity. Wilkie provides a useful compendium of molecular mechanisms that can lead to genetic dominance (haploinsufficiency, excessive gene expression, novel toxic protein function, *etc.*) with examples from model organisms and the human genetics literature [177]. For individual genes, the best way to distinguish between gain and loss of function phenotypes is to observe an array of alleles associated with a particular phenotype. For example, mutations of BRCA1 which increase the risk of early-onset breast cancer include a wide variety of missense, truncating and deletion mutations [178]. It is reasonable to presume that most of these represent loss of function alleles. Similarly, many independent alleles causing Huntington disease all involve expansion of the triplet repeat, while causal mutations elsewhere in the gene are not known for this phenotype, supportive of the interpretation that these are gain of function alleles. In contrast, Friedreich ataxia is typically caused by a triplet repeat expansion in the frataxin gene, however a small fraction of patients result from point alleles, both missense and truncating [179]. Conceivably, the ataxia could represent a loss of function phenotype, and the frequency of triplet repeat expansion alleles could result simply from an easily accessed mutational mechanism *versus* a specific novel function of the mutated protein. Alternatively, some specific point alleles could cause an aberrant novel protein function akin to that of the triplet repeat expansion. One glaring example is the gene encoding superoxide dismutase (SOD), mutations in which are causal for a fraction of cases of amyotrophic lateral sclerosis (ALS, or Lou Gehrig disease). There are many different causal mutations in SOD, almost all of which are missense variants [180]. However, a mouse gene knockout does not generate the ALS phenotype, whereas overexpressing equivalent missense variants recapitulates some aspects of the human disease [181]. The consensus in the field is that many or most of these missense variants, which arise in almost every residue of this small tightly folded protein, are somehow gain-of-function, although the exact biochemical mechanisms are still being elucidated. Thus it cannot be assumed that phenotypes arising from many different mutations in a gene invariably represent loss of function, although they probably do in most cases.

The clinical presentation of different mutant alleles yielding different phenotypes can sometimes clarify the interpretation. Hypercholesterolemia can be caused by one of a small number of missense mutations in the gene encoding protease

PCSK9. However, a much larger number of mutations including obligate protein truncations in the same gene are associated with hypocholesterolemia. The hyper alleles are typically heterozygous, whereas the hypo alleles are typically homozygous in the patients [182-186]. These observations strongly suggest that the hyper alleles are gain of function, and the hypo alleles are loss of function. The rarity of the gof alleles is consistent that, as anticipated, the map of genotype to the hyper phenotype is more highly constrained in this gene than the map from genotype to the hypo phenotype. Similarly, many presumptive lof alleles of the gene encoding apolipoprotein B exist, and lead to hypobetalipoproteinemia (with low LDL-cholesterol) due to lack of this major structural protein component of VLDL and LDL particles. In contrast, only a small number of specific amino acid missense variants are known in the ApoB gene causing hyperbetalipoproteinemia (with high LDL-cholesterol), with the mechanism believed to be failure of recognition of mutant ApoB protein by the LDL receptor, and thus inability of the body to remove LDL particles from the plasma [187-189].

Although most mutations are probably recessive, the ratio of dominant to recessive monogenic disorders is probably significantly skewed in human genetics in favor of dominants. High penetrance mutations segregating as heterozygotes have the potential to generate large multigenerational pedigrees transmitting the physiological trait. If the trait is a medical disorder, such as Huntington disease or Alagille Syndrome, such large families are relatively obvious to astute clinicians, particularly since the advent of effective molecular mapping in the 1980s. In contrast, recessive disorders often involve only a single affected in a nuclear family, since only ¼ of offspring segregate the trait (for an autosomal disorder). Ascertainment of recessive traits by clinicians typically involves unusually large sibships with recurrence of the phenotype in multiple children, or recurrence where parents are known to be closely related (such as first cousins). Generally, in small sibships with only one or a few affected children, environmental or stochastic etiologies are equally likely, and the hypothesis of recessive segregation is less easy to prove except by a genome-wide mapping experiment.

There is a strong practical reason to believe that most loss of function alleles behave as recessive genetic traits. As noted above, measured rates of mutation are consistent with fairly high rates of occurrence of novel pathogenic alleles in humans. If many new mutations are pathogenic, then severe effects on fitness of entire populations would ensue if most mutations acted dominantly with pathogenic effects in heterozygotes. Hence, most mutations are reasonably anticipated to have recessive effects on physiological traits (even if they have formally co-dominant effects on their immediate biochemical function through simple gene dosage effects). Indeed, protection from high mutation rates is likely one of the evolutionary forces driving diploidy and hence sex, although evolutionary biologists tend to focus on sex as a mechanism for generating rather than modulating genomic diversity.

The recessive nature of most mutations was appreciated early in the 20$^{th}$ century, and theoretical geneticists developed several alternative models to explain this [190, 191]. Wright suggested that two doses of genes provided near-saturating levels of activity for the ultimate physiological trait, such that loss of one dose had a non-linear and only weak effect on the trait (Fig. **2**) [192]. Fisher suggested instead that mutant alleles evolved in concert with genetic modifiers elsewhere in the genome. Wright's argument received strong theoretical support from the metabolic control rate theory of Kacser and Burns [193, 194]. These theoreticians sought to understand why rates of production of pathway end-products in engineered microorganisms were difficult to influence by changing amounts of individual enzymes in the reaction pathways, even when those enzymes were known to be 'rate-limiting' based on biochemical kinetic studies. They showed mathematically that control of flux rates through complex metabolic pathways *in vivo* was more likely to be distributed than localized to one gene product (Fig. **3**) [193]. The result would be that modest (even two-fold) changes in dosage of a gene product would be unlikely to affect overall flux through a metabolic pathway, although this point has been argued [195]. Assuming that some kind of biochemically quantitative pathways underlie all physiological traits, even those of embryonic development or neuropsychiatric function, metabolic control theory suggests at least in principle that gene dose response could readily be nonlinear, and that loss of even a full copy of gene function (leaving half of a full normal dosage) might have only partial effects in non-enzymatic systems [196].

Genetic studies in model organisms tend to support these arguments [177]. Although spontaneous or mutagen induced mouse mutants can span the full range of mutation types, engineered knockouts almost always involve deletions of large amounts of protein-coding potential, and are likely to function as lof alleles in most cases. Anecdotal evidence from mouse geneticists suggests that most such knockouts have no detectable phenotype in heterozygous state. A caveat to these observations of course is that some heterozygous knockout mice could have physiological phenotypes undetected by experimentalists. Similar results are seen in fly mutagenesis experiments, in which mutagenized flies are made heterozygous against a strain carrying a defined chromosomal deletion, with breakpoints defined either cytologically or now by DNA sequencing. In such experiments, many independent alleles can be obtained for most genes covered by the deletion, and their phenotypes compared against the deletion or when homozygosed individually or combined as compound heterozygotes against each other. The general result of such studies is that most alleles with a detectable phenotype in most genes act recessively, with no obvious heterozygous effect.

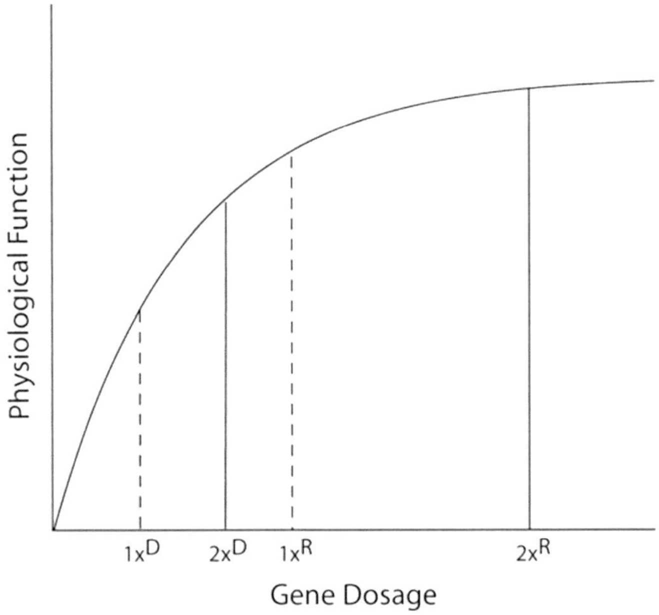


**Figure 2:** Saturation kinetics and genetic dominance. As suggested by Wright, if two doses of a wild type gene in a diploid organism provide saturating levels of physiological function ($2x^R$), then loss of one copy through mutation would still provide most functionality ($1x^R$). Such a situation would lead to recessive genetic behavior, whereby a heterozygote lof mutation would have little biological impact. Wright hypothesized that this would be the case for the majority of genes. In some cases, two wild type gene doses might provide function in the subsaturing range ($2x^D$), such that loss of one dose in heterozygous lof mutation carriers would have substantial biological effect ($1x^D$), leading to observed dominance. Dominance caused by unusual gain-of-function alleles is not an element of this model.

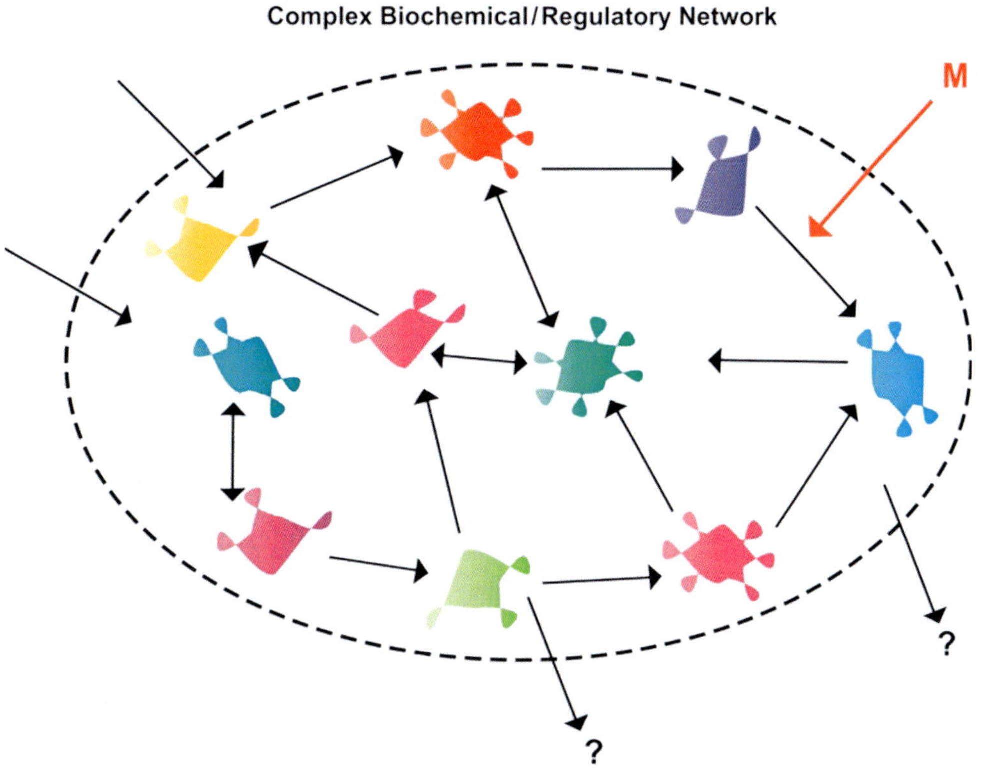


**Figure 3:** Metabolic control theory and gene networks. As suggested by Kacser and Burns [193], measures of enzymatic rates in purified *in vitro* systems do not necessarily reflect true *in vivo* situations. In reality, metabolic networks are complex, often with alternative routes from substrate inputs to catabolite outputs. In such cases, changing the rate of one specific step (indicated by arrow M), either *via* genetic mutation or through a targeted chemical antagonist or agonist (*i.e.* a drug) might lead to only modest, or unpredictable effects on total flux through the system and on the equilibrium levels of intermediate metabolites. This could explain both dominance as well as functional redundancy, whereby mutation of most genes even in homozygotes is not lethal, even though the genes are "required" in a mechanistic sense. Similar situations could be the case for developmental regulatory networks, where the flow of information rather than chemical metabolites is concerned.

In a unique analysis to test the Wright hypothesis directly, the mutational spectrum was considered for the microorganism *Chlamydomonas*, which normally grows in a haploid state. When semi-synthetic diploids are generated, mutations which originally arose in haploids prove nonetheless to be almost exclusively recessive [197]. This result strongly argues against recessivity arising from coordinate evolution of oligogenic modifiers in tandem with the primary mutation, and supports Wright's argument that recessivity of mutations is a consequence of the way in which metabolic pathways have ancestrally evolved. Efforts to extend the Kacser and Burns argument to other, nonmetabolic developmental pathways have shown that this is feasible [190]. Nonetheless, it remains conceivable that certain developmental programs might not be well buffered in higher eukaryotes. One example might be the human brain, whose size and complexity have increased tremendously in a very short evolutionary time of several million years, as compared to other organ systems which have changed little even among mammals. Copy number studies have documented a significantly higher incidence of CNVs, mostly deletions, in autistic or schizophrenic than in control patients, presumably mostly in the heterozygous state. If so, then loss of function point mutations might also cause frequent cases of schizophrenia or autism in the heterozygous dominant state; these both appear to have a high rate of sporadic etiology consistent with high mutation rates, heterozygous effects, and recent evolutionary history.

Given that most mutations in most genes act recessively, the rate of monogenic disorders ascertained clinically is substantially lower than the rate of appearance of new alleles based on the mutation rate calculations described above. Occurrence of patients with recessive disorders depends on a combination of new mutation rates and population demographics. In outbred large populations, individuals will rarely be born carrying two non-complementing mutations in *trans* in the same gene (either as homozygotes or compound heterozygotes). In contrast, in population isolates experiencing significant inbreeding due to either geographic or cultural factors (*i.e.* genetic drift), rates of recessive disorders may be substantially higher than otherwise expected. In addition, due to founder effects different genetic disorders should be predicted to occur at increased frequency in different populations. This is well-documented by medical geneticists, in diverse groups such as Finns, French Canadians, Ashkenazi Jews and Alberta Hutterites [173, 198-201]. In one town in Israel, geneticists identified 19 different recessive monogenic disorders arising in a population of less than 10,000 [202].

## DEFINING THE REST OF THE PHENOME

If only a small fraction of protein-coding genes have a molecularly characterized monogenic phenotype in humans, what does this mean for the complete phenome? If the complete phenome comprises all possible phenotypes of all genes in the genome, then the phenome is effectively infinite. However, if most mutations are loss-of-function alleles, and generally these cause similar phenotypes in a given gene, then the lof phenome is much smaller than the entire phenome, is probably finiteand only slightly larger than the number of genes. Further, if gof alleles are much rarer than lof alleles, then most of the theoretically possible gof phenome will never be observed given the total human population size. In contrast, the lof phenome probably exists near completion across the world now.

All these points together strongly suggest the following scenario: lof alleles of all genes in the genome arise regularly around the world; most of these represent recessive alleles; clinical ascertainment of homozygotes for these alleles will be individually rare except in founder or other population types with higher rates of consanguinity than expected for complete panmixis; compound heterozygotes will arise at similar and low rates essentially everywhere, with only slightly increased rate in founder populations. The actual rate of appearance of recessive cases that might be clinically ascertained for a given gene is very difficult to estimate, since it depends on current frequencies for deleterious alleles, which in turn depend on details of local population demographic history which for most populations are poorly known. However, even in presumptive outbred or non-founder populations there can be increased frequency of recessive deleterious alleles through unappreciated low levels of population mobility – the reduced number of observed *versus* potential haplotypes found in all non-African human populations provides an extreme example of this [203-205].

The inescapable conclusion is that recessive cases for lof alleles of most genes do occur regularly. Where then are these missing phenotypes? Three possibilities remain: many genes mutate to a neutral lof phenotype, or else they mutate to an embryonic lethal phenotype, or else they mutate to phenotypes not yet clinically defined or molecularly characterized as genetic conditions. Half of all spontaneous miscarriages show cytogenetically detectable chromosomal abnormalities [206]. With the advent of high density hybridization arrays, smaller rearrangements especially deletions can readily be detected, well below the resolution of light microscopy [207, 208]. It seems likely that an additional fraction of miscarriages, even earlier in fetal development, entail mutation of critical embryonic genes. For example, viable human mutation phenotypes are known for six of the 39 genes encoded by the four HOX gene clusters [209]. These are among the extreme rostral and caudal expressed genes HOXA1 [210], A2 [211], A11 [212], A13 [213, 214], D10 [215, 216] and D13 [217]. The majority of more centrally expressed HOX genes may mutate to recessive embryonic lethality, as may other early expressed pattern-forming genes. OMIM currently includes 3695 phenotype entries with suggestion of a genetic etiology, many of which probably do represent recessive cases arising in families too small to map a genetic locus.

The advent of dense SNP marker panels replacing microsatellite marker maps greatly facilitated the mapping of genetic loci, especially for traits in population isolates with not only increased recessivity but also homozygosity [218-221]. Nonetheless, minimal family size to provide adequate genome-wide statistical power for mapping based approaches, or number of affecteds in consanguineous sibships for homozygosity mapping, continued to constrain gene discovery for monogenic disorders. Recently however, the availability of new DNA sequencing technologies has drastically changed this paradigm. Beginning with the sequencing of genomes a few famous individuals, by now hundreds if not thousands of individual human genomes have been sequenced using next-generation technologies [222-229]. To reduce the cost of full genome sequencing, which remains onerous for most research lab budgets, hybridization capture methods have been developed by which selected subsets of the genome, such as protein-coding exons, can be sequenced in an approach called exome sequencing [230-233].

It is now clear that individual human genomes have 3-4 million differences from the human consensus sequence as assembled by the Human Genome Project and used universally as a basis for comparison to newly sequenced individual DNAs (current version hg19, also called GRCh37, viewable through the NCBI, Ensembl or UCSC genome browser web sites). Of these, 90-95% can be found in the dbSNP database, which now includes a large number of variants discovered through the 1000 Genomes Project [234-236]. In some cases, the human consensus sequence has a minor not a major allele at a variable site, so that most newly sequenced individuals actually carry the "variant". This was well appreciated prior to the advent of next-generation sequencing, and was inevitable given the small number of DNA sources used to generate sequencing libraries by the Human Genome Project. Excluding variants found with moderate (>1%) allele frequencies in the public SNP databases, individual human genomes still carry tens of thousands of nominally private mutations, distributed between unique and repetitive, coding and non-coding genomic sequences. Further restricting just to protein-coding exons and some adjacent intronic sequence critical for proper pre-mRNA splicing, individual genomes typically carry several hundred private non-synonymous variants. This number is still too large to define pathogenic causality generally, although a steadily increasing number of reports demonstrates the feasibility of array-capture exome resequencing to define pathogenic variants in affected individuals or families with rare clinical phenotypes under ideal conditions [237-245]. Operationally, assigning genotype/phenotype causality unambiguously requires either multiple patients mutated in the same gene with a rare monogenic phenotype, or alternatively strong validation of a candidate gene through animal model studies. Nonetheless, whole genome or exome sequencing has already revolutionized medical genetics, allowing the identification of causal genes for many more rare phenotypes, including some (such as prenatal lethality, or dominant disease with reduced viability or fertility) for which family studies are intrinsically impossible. It is now generally accepted that exome, and eventually whole genome sequencing, will become regular components of clinical medical diagnostics, although much work remains to be done to define the appropriate patient populations that will reasonably benefit from genomic diagnoses. Likewise there is active ongoing debate concerning the ethical and social issues surrounding genome-wide medical sequencing, including the interpretation (or not) of incidental findings (*i.e.* variants in genes apparently unrelated to the primary clinical ascertainment, some of which may be medically actionable) [246-252].

This review has focused primarily on protein coding genes. There are some examples of genes producing non-protein coding RNAs for which high penetrance mutational phenotypes occur (the H19 gene in mice [253], and the XIST gene in humans [254-256]). In general such assignments are challenging since we have little ability to define functional changes in non-coding RNAs except when they are extensively deleted. It remains to be seen whether naturally occurring variation in other such RNAs, such as microRNAs, can lead to monogenic phenotypes. One exception is in the mitochondrial genome. Well documented mutations in mitochondrial genes, including tRNA and rRNA coding genes, lead to clinical disorders of the mitochondrion, which although complex physiologically are generally definable by experienced clinicians [257, 258].

## PRACTICAL APPLICATIONS OF PHENOME SATURATION

Beyond the purely scientific value of determining high penetrance mutational phenotypes for as many genes as possible, there is also practical value in this program. The human genetics community has focused much of its effort for the past decade on defining risk variants for complex, often adult-onset degenerative diseases of great social and commercial importance. This work has mostly employed the analytical paradigm of high throughput SNP genotyping and case/control comparison using large patient cohorts (the GWAS approach). GWAS studies have succeeded in identifying many such risk factors, but in general the variants found individually add only small incremental risk of disease, and only a modest fraction of total genetic risk in populations is explainable summing over all such common alleles [259]. Moreover, it is unclear whether genes associated with complex disease through modest statistical effects will generally provide good therapeutic targets. There is currently lively debate whether the missing heritability lies in epistatic interactions among already identified lower penetrance common variants, or in aggregate large numbers of individually rare higher penetrance variants undetectable by GWAS. The emerging consensus is that either way, finding the missing heritability will require high throughput genomic resequencing in larger and larger cohorts. If one of the goals is to identify novel therapeutic targets for pharmaceutical intervention, this will be a long-term and monumentally expensive proposition.

In contrast, there are many good examples of rare monogenic traits of direct relevance to common disease. Elsewhere I have discussed the utility of rare "monogenic disorders" as a tool to define novel therapeutic targets applicable to large patient populations and hence large commercial markets [260]. Genes underlying such rare monogenic disorders can be found either by traditional family-based linkage studies [261-263], or for quantitative traits such as body mass, plasma glucose or cholesterol, by sequencing population outliers looking for excess variation in specific genes [264-268]. Both approaches, particularly the second, are increasingly feasible using whole genome or exome sequencing. The genes underlying such traits provide "low hanging fruit" for drug development, something desperately needed by the biotechnology and pharmaceutical industries, which are universally acknowledged to be suffering a dearth of good new products. For example, the human genome encodes 416 G-protein coupled receptors, not including olfactory receptors (S. Papillon, *pers. comm.)* Some of these are targets of drugs already on the market, and GPCRs have several major advantages for new drug development. As with the human genome in general, only about 10% of GPCRs have documented human monogenic phenotypes. Many of these rare human phenotypes are directly relevant to much more common disease models of considerable socioeconomic importance. Thus the GPCR phenome represents a fertile area for genetic analysis, requiring only the resequencing of this set of genes in clinically defined cases of medical interest [269].

Setting aside the current hype surrounding personalized medicine, there are several ways in which phenome saturation will have immediate diagnostic benefit. In pediatric hospital settings, a substantial fraction of admissions involve genetic disorders [270, 271]. Some of these, such as congenital heart malformations and neural tube defects, are seemingly genetically highly complex and causal genes have in general not been found. In others, such as sensorineural deafness [7, 8, 272] and mitochondrial disorders [237, 258, 273], there is significant genetic heterogeneity with many known genes, suggesting a clinical sequencing paradigm which is manageable by whole exome sequencing. Others involve genetically simple disorders. For example, pediatric hospitals often have dedicated cystic fibrosis units, as CF is one of the most commonly diagnosed pediatric genetic disorders with a known single gene basis. Given that there are already over 2500 molecular characterized genetic disorders, many of which are regularly ascertained in any large pediatric care setting, genetic analysis is critical in pediatrics even now. The definitive diagnosis of such conditions, either through molecular analyses or simply by phenotypic description, has high value to patients and families, and potentially to the health care system as well. Further definition of all remaining pediatric monogenic conditions will prove similarly useful, if only to distinguish rapidly between genetic and non-genetic causes of disease in individual patients, reducing the burden of otherwise time-consuming and expensive diagnostic odysseys.

## CONCLUSION

I have attempted to document that most of the human phenome remains to be ascertained and molecularly characterized. This is primarily due to challenges in clinical ascertainment, not to any theoretical issues with intrinsic deleterious mutation rates or the occurrence of previously uncharacterized phenotypes. Mutation rates in the human population are sufficiently high that pathogenic variants must exist in all functional genes at the present time. Most of these are likely to cause recessive phenotypes, thus ascertainment depends on identifying homozygous or compound heterozygous mutation carriers from among all medically identified patients. Historically, causal gene identification has depended on the identification of multiplex families, followed by mapping with panels of anonymous DNA polymorphisms, position cloning and resequencing. With the advent of next-generation sequencing technologies, it is now feasible to sequence whole genomes or exomes of medical patients without the need for multiplex affected families or statistical linkage mapping. However, the challenge remains of identifying truly causal variants against the total background of variation in the genomes of such patients. This will require collaboration among medical geneticists dispersed around the world, possibly with development of large genotype/phenotype databases, together with the use of model organisms to validate gene function and/or mutation pathogenicity.

With these ideas in mind, this is my own estimate of the lof human phenome. Time will tell how accurate these predictions are.

Known genes with phenotypes (~2500)

Embryonic lethality (5000)

Intellectual disability or other high penetrance neuropsychiatric condition (1000)

No clinical phenotype (3000)

Phenotypes to be defined or characterized (8-10000)

## WEB SITES

Main NCBI entry site to genome browser http://www.ncbi.nlm.nih.gov/

UCSC genome browser http://genome.ucsc.edu/

Ensembl genome browser http://www.ensembl.org/

HGMD curated mutation database http://www.hgmd.cf.ac.uk/ac/index.php

CCDS gene annotation database http://www.ncbi.nlm.nih.gov/projects/CCDS/ CcdsBrowse.cgi

Mammalian Gene Collection database http://mgc.nci.nih.gov/

Mitochondrial proteome database http://www.mitop.de:8080/mitop2/

Canadian rare genetic diseases http://raredisorders.ca/

Comparative genome annotations http://ecrbrowser.dcode.org/

GeneTests molecular diagnostic database http://www.ncbi.nlm.nih.gov/sites/Gene Tests/?db=GeneTests

BioGPS gene expression database http://biogps.gnf.org/#goto=welcome

Budding yeast mutation database http://www-sequence.stanford.edu/group/yeast_ deletion_project/deletions3.html

Fission yeast mutation database http://www.sanger.ac.uk/Projects/S_pombe/

Statistics Canada http://www.statcan.gc.ca

## CONFLICT OF INTEREST

The author(s) confirm that this chapter content has no conflict of interest.

## ACKNOWLEDGEMENTS

Many thanks to Mark Ludman, Christie Riddell, David Skidmore, Sarah Dyack, Andrew Orr, Grant Mitchell, Jacques Michaud, Gregor Andelfinger and Guy Rouleau for sharing insights gained from their wide clinical genetics experience. Thanks to Gerard Battail for discussions about genomic error correction mechanisms, and to Bill Gelbart, Pierre Drapeau, Louis Saint-Amant and Christian Neri for their insights into model organism developmental genetics. Thanks to D. Samuels for providing illustrations. MES is supported by Genome Canada, Genome Atlantic, Nova Scotia Health Research Foundation, Nova Scotia Research and Innovation Trust, IWK Health Centre Foundation, Dalhousie Medical Research Fund and Capital Health Research Fund, March of Dimes and the Centre de Recherche du CHU Ste-Justine.

## DISCLOSURE

This is an updated report of the earlier review article by Mark Samuels, "Saturation of the Human Phenome", Current Genomics, 11(7), 482-499 (2010).

## REFERENCES

[1] Freimer, N.; Sabatti, C. The human phenome project. *Nat Genet.* **2003**, *34*, 15-21.

[2] Oti, M.; Huynen, M.A.; Brunner, H.G. Phenome connections. *Trends Genet.* **2008**, *24*, 103-6.

[3] Oti, M.; Huynen, M.A.; Brunner, H.G. The biological coherence of human phenome databases. *Am J Hum Genet.* **2009**, *85*, 801-8.

[4] Scriver, C.R. After the genome--the phenome? *J Inherit Metab Dis.* **2004**, *27*, 305-17.

[5] Paulson, H.L. The spinocerebellar ataxias. *J Neuroophthalmol.* **2009**, *29*, 227-37.

[6] Dror, A.A.; Avraham, K.B. Hearing loss: mechanisms revealed by genetics and cell biology. *Annu Rev Genet.* **2009**, *43*, 411-37.

[7] Matsunaga, T. Value of genetic testing in the otological approach for sensorineural hearing loss. *Keio J Med.* **2009**, *58*, 216-22.

[8] Hilgert, N.; Smith, R.J.; Van Camp, G. Forty-six genes causing nonsyndromic hearing impairment: which ones should be analyzed in DNA diagnostics? *Mutat Res.* **2009**, *681*, 189-96.

[9] Berger, W.; Kloeckener-Gruissem, B.; Neidhardt, J. The molecular basis of human retinal and vitreoretinal diseases. *Prog Retin Eye Res.* **2010**, *epub*,

[10] Phelan, J.K.; Bok, D. A brief review of retinitis pigmentosa and the identified retinitis pigmentosa genes. *Mol Vis.* **2000**, *6*, 116-24.

[11] van Soest, S.; Westerveld, A.; de Jong, P.T.; Bleeker-Wagemakers, E.M.; Bergen, A.A. Retinitis pigmentosa: defined from a molecular point of view. *Surv Ophthalmol.* **1999**, *43*, 321-34.

[12] Banchs, I.; Casasnovas, C.; Alberti, A.; De Jorge, L.; Povedano, M.; Montero, J.; Martinez-Matos, J.A.; Volpini, V. Diagnosis of Charcot-Marie-Tooth disease. *J Biomed Biotechnol.* **2009**, *2009*, 985415.

[13] Barisic, N.; Claeys, K.G.; Sirotkovic-Skerlev, M.; Lofgren, A.; Nelis, E.; De Jonghe, P.; Timmerman, V. Charcot-Marie-Tooth disease: a clinico-genetic confrontation. *Ann Hum Genet.* **2008**, *72*, 416-41.

[14] Casasnovas, C.; Cano, L.M.; Alberti, A.; Cespedes, M.; Rigo, G. Charcot-Marie-tooth disease. *Foot Ankle Spec.* **2008**, *1*, 350-4.

[15] Jani-Acsadi, A.; Krajewski, K.; Shy, M.E. Charcot-Marie-Tooth neuropathies: diagnosis and management. *Semin Neurol.* **2008**, *28*, 185-94.

[16] Pareyson, D.; Marchesi, C.; Salsano, E. Hereditary predominantly motor neuropathies. *Curr Opin Neurol.* **2009**, *22*, 451-9.

[17] Parman, Y. Hereditary neuropathies. *Curr Opin Neurol.* **2007**, *20*, 542-7.

[18] Reilly, M.M. Sorting out the inherited neuropathies. *Pract Neurol.* **2007**, *7*, 93-105.

[19] Szigeti, K.; Nelis, E.; Lupski, J.R. Molecular diagnostics of Charcot-Marie-Tooth disease and related peripheral neuropathies. *Neuromolecular Med.* **2006**, *8*, 243-54.

[20] Dragatsis, I.; Efstratiadis, A.; Zeitlin, S. Mouse mutant embryos lacking huntingtin are rescued from lethality by wild-type extraembryonic tissues. *Development.* **1998**, *125*, 1529-39.

[21] Kajiwara, K.; Berson, E.L.; Dryja, T.P. Digenic retinitis pigmentosa due to mutations at the unlinked peripherin/RDS and ROM1 loci. *Science.* **1994**, *264*, 1604-8.

[22] Dryja, T.P.; Hahn, L.B.; Kajiwara, K.; Berson, E.L. Dominant and digenic mutations in the peripherin/RDS and ROM1 genes in retinitis pigmentosa. *Invest Ophthalmol Vis Sci.* **1997**, *38*, 1972-82.

[23] Goldberg, A.F.; Molday, R.S. Defective subunit assembly underlies a digenic form of retinitis pigmentosa linked to mutations in peripherin/rds and rom-1. *Proc Natl Acad Sci U S A.* **1996**, *93*, 13726-30.

[24] Jin, Z.B.; Mandai, M.; Yokota, T.; Higuchi, K.; Ohmori, K.; Ohtsuki, F.; Takakura, S.; Itabashi, T.; Wada, Y.; Akimoto, M.; Ooto, S.; Suzuki, T.; Hirami, Y.; Ikeda, H.; Kawagoe, N.; Oishi, A.; Ichiyama, S.; Takahashi, M.; Yoshimura, N.; Kosugi, S. Identifying pathogenic genetic background of simplex or multiplex retinitis pigmentosa patients: a large scale mutation screening study. *J Med Genet.* **2008**, *45*, 465-72.

[25] Kedzierski, W.; Nusinowitz, S.; Birch, D.; Clarke, G.; McInnes, R.R.; Bok, D.; Travis, G.H. Deficiency of rds/peripherin causes photoreceptor death in mouse models of digenic and dominant retinitis pigmentosa. *Proc Natl Acad Sci U S A.* **2001**, *98*, 7718-23.

[26] Keen, T.J.; Inglehearn, C.F. Mutations and polymorphisms in the human peripherin-RDS gene and their involvement in inherited retinal degeneration. *Hum Mutat.* **1996**, *8*, 297-303.

[27] Loewen, C.J.; Moritz, O.L.; Molday, R.S. Molecular characterization of peripherin-2 and rom-1 mutants responsible for digenic retinitis pigmentosa. *J Biol Chem.* **2001**, *276*, 22388-96.

[28] Sullivan, L.S.; Bowne, S.J.; Birch, D.G.; Hughbanks-Wheaton, D.; Heckenlively, J.R.; Lewis, R.A.; Garcia, C.A.; Ruiz, R.S.; Blanton, S.H.; Northrup, H.; Gire, A.I.; Seaman, R.; Duzkale, H.; Spellicy, C.J.; Zhu, J.; Shankar, S.P.; Daiger, S.P. Prevalence of disease-causing mutations in families with autosomal dominant retinitis pigmentosa: a screen of known genes in 200 families. *Invest Ophthalmol Vis Sci.* **2006**, *47*, 3052-64.

[29] Jacolot, S.; Le Gac, G.; Scotet, V.; Quere, I.; Mura, C.; Ferec, C. HAMP as a modifier gene that increases the phenotypic expression of the HFE pC282Y homozygous genotype. *Blood.* **2004**, *103*, 2835-40.

[30] Merryweather-Clarke, A.T.; Cadet, E.; Bomford, A.; Capron, D.; Viprakasit, V.; Miller, A.; McHugh, P.J.; Chapman, R.W.; Pointon, J.J.; Wimhurst, V.L.; Livesey, K.J.; Tanphaichitr, V.; Rochette, J.; Robson, K.J. Digenic inheritance of mutations in HAMP and HFE results in different types of haemochromatosis. *Hum Mol Genet.* **2003**, *12*, 2241-7.

[31] Badano, J.L.; Kim, J.C.; Hoskins, B.E.; Lewis, R.A.; Ansley, S.J.; Cutler, D.J.; Castellan, C.; Beales, P.L.; Leroux, M.R.; Katsanis, N. Heterozygous mutations in BBS1, BBS2 and BBS6 have a potential epistatic effect on Bardet-Biedl patients with two mutations at a second BBS locus. *Hum Mol Genet.* **2003**, *12*, 1651-9.

[32] Badano, J.L.; Leitch, C.C.; Ansley, S.J.; May-Simera, H.; Lawson, S.; Lewis, R.A.; Beales, P.L.; Dietz, H.C.; Fisher, S.; Katsanis, N. Dissection of epistasis in oligogenic Bardet-Biedl syndrome. *Nature.* **2006**, *439*, 326-30.

[33] Beales, P.L.; Badano, J.L.; Ross, A.J.; Ansley, S.J.; Hoskins, B.E.; Kirsten, B.; Mein, C.A.; Froguel, P.; Scambler, P.J.; Lewis, R.A.; Lupski, J.R.; Katsanis, N. Genetic interaction of BBS1 mutations with alleles at other BBS loci can result in non-Mendelian Bardet-Biedl syndrome. *Am J Hum Genet.* **2003**, *72*, 1187-99.

[34] Eichers, E.R.; Lewis, R.A.; Katsanis, N.; Lupski, J.R. Triallelic inheritance: a bridge between Mendelian and multifactorial traits. *Ann Med.* **2004**, *36*, 262-72.

[35] Fauser, S.; Munz, M.; Besch, D. Further support for digenic inheritance in Bardet-Biedl syndrome. *J Med Genet.* **2003**, *40*, e104.

[36] Katsanis, N. The oligogenic properties of Bardet-Biedl syndrome. *Hum Mol Genet.* **2004**, *13 Spec No 1*, R65-71.

[37] Adato, A.; Kalinski, H.; Weil, D.; Chaib, H.; Korostishevsky, M.; Bonne-Tamir, B. Possible interaction between USH1B and USH3 gene products as implied by apparent digenic deafness inheritance. *Am J Hum Genet.* **1999**, *65*, 261-5.

[38] Cama, E.; Melchionda, S.; Palladino, T.; Carella, M.; Santarelli, R.; Genovese, E.; Benettazzo, F.; Zelante, L.; Arslan, E. Hearing loss features in GJB2 biallelic mutations and GJB2/GJB6 digenic inheritance in a large Italian cohort. *Int J Audiol.* **2009**, *48*, 12-7.

[39] Chen, A.H.; Fukushima, K.; McGuirt, W.T.; Smith, R.J. DFNB15: autosomal recessive non-syndromic hearing loss gene-chromosome 3q, 19p or digenic recessive inheritance? *Adv Otorhinolaryngol.* **2000**, *56*, 171-5.

[40] Liu, X.Z.; Yuan, Y.; Yan, D.; Ding, E.H.; Ouyang, X.M.; Fei, Y.; Tang, W.; Yuan, H.; Chang, Q.; Du, L.L.; Zhang, X.; Wang, G.; Ahmad, S.; Kang, D.Y.; Lin, X.; Dai, P. Digenic inheritance of non-syndromic deafness caused by mutations at the gap junction proteins Cx26 and Cx31. *Hum Genet.* **2009**, *125*, 53-62.

[41] Morell, R.; Spritz, R.A.; Ho, L.; Pierpont, J.; Guo, W.; Friedman, T.B.; Asher, J.H., Jr. Apparent digenic inheritance of Waardenburg syndrome type 2 (WS2) and autosomal recessive ocular albinism (AROA). *Hum Mol Genet.* **1997**, *6*, 659-64.

[42] Nozu, K.; Inagaki, T.; Fu, X.J.; Nozu, Y.; Kaito, H.; Kanda, K.; Sekine, T.; Igarashi, T.; Nakanishi, K.; Yoshikawa, N.; Iijima, K.; Matsuo, M. Molecular analysis of digenic inheritance in Bartter syndrome with sensorineural deafness. *J Med Genet.* **2008**, *45*, 182-6.

[43] Pallares-Ruiz, N.; Blanchet, P.; Mondain, M.; Claustres, M.; Roux, A.F. A large deletion including most of GJB6 in recessive non syndromic deafness: a digenic effect? *Eur J Hum Genet.* **2002**, *10*, 72-6.

[44] Stevenson, V.A.; Ito, M.; Milunsky, J.M. Connexin-30 deletion analysis in connexin-26 heterozygotes. *Genet Test.* **2003**, *7*, 151-4.

[45] Yang, T.; Gurrola, J.G., 2nd; Wu, H.; Chiu, S.M.; Wangemann, P.; Snyder, P.M.; Smith, R.J. Mutations of KCNJ10 together with mutations of SLC26A4 cause digenic nonsyndromic hearing loss associated with enlarged vestibular aqueduct syndrome. *Am J Hum Genet.* **2009**, *84*, 651-7.

[46] Zheng, Q.Y.; Yan, D.; Ouyang, X.M.; Du, L.L.; Yu, H.; Chang, B.; Johnson, K.R.; Liu, X.Z. Digenic inheritance of deafness caused by mutations in genes encoding cadherin 23 and protocadherin 15 in mice and humans. *Hum Mol Genet.* **2005**, *14*, 103-11.

[47] Weatherall, D.J. Single gene disorders or complex traits: lessons from the thalassaemias and other monogenic diseases. *Bmj.* **2000**, *321*, 1117-20.

[48] Weatherall, D.J. Phenotype-genotype relationships in monogenic disease: lessons from the thalassaemias. *Nat Rev Genet.* **2001**, *2*, 245-55.

[49] Scriver, C.R.; Waters, P.J. Monogenic traits are not simple: lessons from phenylketonuria. *Trends Genet.* **1999**, *15*, 267-72.

[50] Charlton, R.; Bothwell, T. Current Clinical Practice: Hereditary Haemochromatosis: Recent Advances in Biology and Management. *Hematology.* **1999**, *4*, 185-194.

[51] Weatherall, M.W.; Higgs, D.R.; Weiss, H.; Weatherall, D.J.; Serjeant, G.R. Phenotype/genotype relationships in sickle cell disease: a pilot twin study. *Clin Lab Haematol.* **2005**, *27*, 384-90.

[52] Dipple, K.M.; McCabe, E.R. Modifier genes convert "simple" Mendelian disorders to complex traits. *Mol Genet Metab.* **2000**, *71*, 43-50.

[53] Dipple, K.M.; McCabe, E.R. Phenotypes of patients with "simple" Mendelian disorders are complex traits: thresholds, modifiers, and systems dynamics. *Am J Hum Genet.* **2000**, *66*, 1729-35.

[54] Boyle, M.P. Strategies for identifying modifier genes in cystic fibrosis. *Proc Am Thorac Soc.* **2007**, *4*, 52-7.

[55] Collaco, J.M.; Cutting, G.R. Update on gene modifiers in cystic fibrosis. *Curr Opin Pulm Med.* **2008**, *14*, 559-66.

[56] Merlo, C.A.; Boyle, M.P. Modifier genes in cystic fibrosis lung disease. *J Lab Clin Med.* **2003**, *141*, 237-41.

[57] Slieker, M.G.; Sanders, E.A.; Rijkers, G.T.; Ruven, H.J.; van der Ent, C.K. Disease modifying genes in cystic fibrosis. *J Cyst Fibros.* **2005**, *4 Suppl 2*, 7-13.

[58] Roy, C.N.; Andrews, N.C. Recent advances in disorders of iron metabolism: mutations, mechanisms and modifiers. *Hum Mol Genet.* **2001**, *10*, 2181-6.

[59] Wood, M.J.; Powell, L.W.; Ramm, G.A. Environmental and genetic modifiers of the progression to fibrosis and cirrhosis in hemochromatosis. *Blood.* **2008**, *111*, 4456-62.

[60] Edwards, A.O.; Ritter, R., 3rd; Abel, K.J.; Manning, A.; Panhuysen, C.; Farrer, L.A. Complement factor H polymorphism and age-related macular degeneration. *Science.* **2005**, *308*, 421-4.

[61] Hageman, G.S.; Anderson, D.H.; Johnson, L.V.; Hancox, L.S.; Taiber, A.J.; Hardisty, L.I.; Hageman, J.L.; Stockman, H.A.; Borchardt, J.D.; Gehrs, K.M.; Smith, R.J.; Silvestri, G.; Russell, S.R.; Klaver, C.C.; Barbazetto, I.; Chang, S.; Yannuzzi, L.A.; Barile, G.R.; Merriam, J.C.; Smith, R.T.; Olsh, A.K.; Bergeron, J.; Zernant, J.; Merriam, J.E.; Gold, B.; Dean, M.; Allikmets, R. A common haplotype in the complement regulatory gene factor H (HF1/CFH) predisposes individuals to age-related macular degeneration. *Proc Natl Acad Sci U S A.* **2005**, *102*, 7227-32.

[62] Haines, J.L.; Hauser, M.A.; Schmidt, S.; Scott, W.K.; Olson, L.M.; Gallins, P.; Spencer, K.L.; Kwan, S.Y.; Noureddine, M.; Gilbert, J.R.; Schnetz-Boutaud, N.; Agarwal, A.; Postel, E.A.; Pericak-Vance, M.A. Complement factor H variant increases the risk of age-related macular degeneration. *Science.* **2005**, *308*, 419-21.

[63] Klein, R.J.; Zeiss, C.; Chew, E.Y.; Tsai, J.Y.; Sackler, R.S.; Haynes, C.; Henning, A.K.; SanGiovanni, J.P.; Mane, S.M.; Mayne, S.T.; Bracken, M.B.; Ferris, F.L.; Ott, J.; Barnstable, C.; Hoh, J. Complement factor H polymorphism in age-related macular degeneration. *Science.* **2005**, *308*, 385-9.

[64] Drayna, D.; Coon, H.; Kim, U.K.; Elsner, T.; Cromer, K.; Otterud, B.; Baird, L.; Peiffer, A.P.; Leppert, M. Genetic analysis of a complex trait in the Utah Genetic Reference Project: a major locus for PTC taste ability on chromosome 7q and a secondary locus on chromosome 16p. *Hum Genet.* **2003**, *112*, 567-72.

[65] Wooding, S.; Kim, U.K.; Bamshad, M.J.; Larsen, J.; Jorde, L.B.; Drayna, D. Natural selection and molecular evolution in PTC, a bitter-taste receptor gene. *Am J Hum Genet.* **2004**, *74*, 637-46.

[66] Eriksson, N.; Macpherson, J.M.; Tung, J.Y.; Hon, L.S.; Naughton, B.; Saxonov, S.; Avey, L.; Wojcicki, A.; Pe'er, I.; Mountain, J. Web-based, participant-driven studies yield novel genetic associations for common traits. *PLoS Genet.* **2010**, *6*, e1000993.

[67] Weedon, M.N.; Frayling, T.M. Reaching new heights: insights into the genetics of human stature. *Trends Genet.* **2008**, *24*, 595-603.

[68] Sturm, R.A. Molecular genetics of human pigmentation diversity. *Hum Mol Genet.* **2009**, *18*, R9-17.

[69] Sturm, R.A.; Duffy, D.L.; Box, N.F.; Newton, R.A.; Shepherd, A.G.; Chen, W.; Marks, L.H.; Leonard, J.H.; Martin, N.G. Genetic association and cellular function of MC1R variant alleles in human pigmentation. *Ann N Y Acad Sci.* **2003**, *994*, 348-58.

[70] Sturm, R.A.; Larsson, M. Genetics of human iris colour and patterns. *Pigment Cell Melanoma Res.* **2009**, *22*, 544-62.

[71] Antequera, F.; Bird, A. Number of CpG islands and genes in human and mouse. *Proc Natl Acad Sci U S A.* **1993**, *90*, 11995-9.

[72] Roest Crollius, H.; Jaillon, O.; Bernot, A.; Dasilva, C.; Bouneau, L.; Fischer, C.; Fizames, C.; Wincker, P.; Brottier, P.; Quetier, F.; Saurin, W.; Weissenbach, J. Estimate of human gene number provided by genome-wide analysis using Tetraodon nigroviridis DNA sequence. *Nat Genet.* **2000**, *25*, 235-8.

[73] Imanishi, T.; Itoh, T.; Suzuki, Y.; O'Donovan, C.; Fukuchi, S.; Koyanagi, K.O.; Barrero, R.A.; Tamura, T.; Yamaguchi-Kabata, Y.; Tanino, M.; Yura, K.; Miyazaki, S.; Ikeo, K.; Homma, K.; Kasprzyk, A.; Nishikawa, T.; Hirakawa, M.; Thierry-Mieg, J.; Thierry-Mieg,

D.; Ashurst, J.; Jia, L.; Nakao, M.; Thomas, M.A.; Mulder, N.; Karavidopoulou, Y.; Jin, L.; Kim, S.; Yasuda, T.; Lenhard, B.; Eveno, E.; Suzuki, Y.; Yamasaki, C.; Takeda, J.; Gough, C.; Hilton, P.; Fujii, Y.; Sakai, H.; Tanaka, S.; Amid, C.; Bellgard, M.; Bonaldo Mde, F.; Bono, H.; Bromberg, S.K.; Brookes, A.J.; Bruford, E.; Carninci, P.; Chelala, C.; Couillault, C.; de Souza, S.J.; Debily, M.A.; Devignes, M.D.; Dubchak, I.; Endo, T.; Estreicher, A.; Eyras, E.; Fukami-Kobayashi, K.; Gopinath, G.R.; Graudens, E.; Hahn, Y.; Han, M.; Han, Z.G.; Hanada, K.; Hanaoka, H.; Harada, E.; Hashimoto, K.; Hinz, U.; Hirai, M.; Hishiki, T.; Hopkinson, I.; Imbeaud, S.; Inoko, H.; Kanapin, A.; Kaneko, Y.; Kasukawa, T.; Kelso, J.; Kersey, P.; Kikuno, R.; Kimura, K.; Korn, B.; Kuryshev, V.; Makalowska, I.; Makino, T.; Mano, S.; Mariage-Samson, R.; Mashima, J.; Matsuda, H.; Mewes, H.W.; Minoshima, S.; Nagai, K.; Nagasaki, H.; Nagata, N.; Nigam, R.; Ogasawara, O.; Ohara, O.; Ohtsubo, M.; Okada, N.; Okido, T.; Oota, S.; Ota, M.; Ota, T.; Otsuki, T.; Piatier-Tonneau, D.; Poustka, A.; Ren, S.X.; Saitou, N.; Sakai, K.; Sakamoto, S.; Sakate, R.; Schupp, I.; Servant, F.; Sherry, S.; Shiba, R.; Shimizu, N.; Shimoyama, M.; Simpson, A.J.; Soares, B.; Steward, C.; Suwa, M.; Suzuki, M.; Takahashi, A.; Tamiya, G.; Tanaka, H.; Taylor, T.; Terwilliger, J.D.; Unneberg, P.; Veeramachaneni, V.; Watanabe, S.; Wilming, L.; Yasuda, N.; Yoo, H.S.; Stodolsky, M.; Makalowski, W.; Go, M.; Nakai, K.; Takagi, T.; Kanehisa, M.; Sakaki, Y.; Quackenbush, J.; Okazaki, Y.; Hayashizaki, Y.; Hide, W.; Chakraborty, R.; Nishikawa, K.; Sugawara, H.; Tateno, Y.; Chen, Z.; Oishi, M.; Tonellato, P.; Apweiler, R.; Okubo, K.; Wagner, L.; Wiemann, S.; Strausberg, R.L.; Isogai, T.; Auffray, C.; Nomura, N.; Gojobori, T.; Sugano, S. Integrative annotation of 21,037 human genes validated by full-length cDNA clones. *PLoS Biol.* **2004**, *2*, e162.

[74] Yamasaki, C.; Koyanagi, K.O.; Fujii, Y.; Itoh, T.; Barrero, R.; Tamura, T.; Yamaguchi-Kabata, Y.; Tanino, M.; Takeda, J.; Fukuchi, S.; Miyazaki, S.; Nomura, N.; Sugano, S.; Imanishi, T.; Gojobori, T. Investigation of protein functions through data-mining on integrated human transcriptome database, H-Invitational database (H-InvDB). *Gene.* **2005**, *364*, 99-107.

[75] Clamp, M.; Fry, B.; Kamal, M.; Xie, X.; Cuff, J.; Lin, M.F.; Kellis, M.; Lindblad-Toh, K.; Lander, E.S. Distinguishing protein-coding and noncoding genes in the human genome. *Proc Natl Acad Sci U S A.* **2007**, *104*, 19428-33.

[76] Ashurst, J.L.; Chen, C.K.; Gilbert, J.G.; Jekosch, K.; Keenan, S.; Meidl, P.; Searle, S.M.; Stalker, J.; Storey, R.; Trevanion, S.; Wilming, L.; Hubbard, T. The Vertebrate Genome Annotation (Vega) database. *Nucleic Acids Res.* **2005**, *33*, D459-65.

[77] Wilming, L.G.; Gilbert, J.G.; Howe, K.; Trevanion, S.; Hubbard, T.; Harrow, J.L. The vertebrate genome annotation (Vega) database. *Nucleic Acids Res.* **2008**, *36*, D753-60.

[78] Pruitt, K.D.; Harrow, J.; Harte, R.A.; Wallin, C.; Diekhans, M.; Maglott, D.R.; Searle, S.; Farrell, C.M.; Loveland, J.E.; Ruef, B.J.; Hart, E.; Suner, M.M.; Landrum, M.J.; Aken, B.; Ayling, S.; Baertsch, R.; Fernandez-Banet, J.; Cherry, J.L.; Curwen, V.; Dicuccio, M.; Kellis, M.; Lee, J.; Lin, M.F.; Schuster, M.; Shkeda, A.; Amid, C.; Brown, G.; Dukhanina, O.; Frankish, A.; Hart, J.; Maidak, B.L.; Mudge, J.; Murphy, M.R.; Murphy, T.; Rajan, J.; Rajput, B.; Riddick, L.D.; Snow, C.; Steward, C.; Webb, D.; Weber, J.A.; Wilming, L.; Wu, W.; Birney, E.; Haussler, D.; Hubbard, T.; Ostell, J.; Durbin, R.; Lipman, D. The consensus coding sequence (CCDS) project: Identifying a common protein-coding gene set for the human and mouse genomes. *Genome Res.* **2009**, *19*, 1316-23.

[79] Pan, Q.; Shai, O.; Lee, L.J.; Frey, B.J.; Blencowe, B.J. Deep surveying of alternative splicing complexity in the human transcriptome by high-throughput sequencing. *Nat Genet.* **2008**, *40*, 1413-5.

[80] Sultan, M.; Schulz, M.H.; Richard, H.; Magen, A.; Klingenhoff, A.; Scherf, M.; Seifert, M.; Borodina, T.; Soldatov, A.; Parkhomchuk, D.; Schmidt, D.; O'Keeffe, S.; Haas, S.; Vingron, M.; Lehrach, H.; Yaspo, M.L. A global view of gene activity and alternative splicing by deep sequencing of the human transcriptome. *Science.* **2008**, *321*, 956-60.

[81] Knowles, D.G.; McLysaght, A. Recent *de novo* origin of human protein-coding genes. *Genome Res.* **2009**, *19*, 1752-9.

[82] Wu, D.D.; Irwin, D.M.; Zhang, Y.P. *De novo* origin of human protein-coding genes. *PLoS Genet.* **2011**, *7*, e1002379.

[83] Anderson, S.; Bankier, A.T.; Barrell, B.G.; de Bruijn, M.H.; Coulson, A.R.; Drouin, J.; Eperon, I.C.; Nierlich, D.P.; Roe, B.A.; Sanger, F.; Schreier, P.H.; Smith, A.J.; Staden, R.; Young, I.G. Sequence and organization of the human mitochondrial genome. *Nature.* **1981**, *290*, 457-65.

[84] Elstner, M.; Andreoli, C.; Ahting, U.; Tetko, I.; Klopstock, T.; Meitinger, T.; Prokisch, H. MitoP2: an integrative tool for the analysis of the mitochondrial proteome. *Mol Biotechnol.* **2008**, *40*, 306-15.

[85] Elstner, M.; Andreoli, C.; Klopstock, T.; Meitinger, T.; Prokisch, H. The mitochondrial proteome database: MitoP2. *Methods Enzymol.* **2009**, *457*, 3-20.

[86] Miklos, G.L.; Rubin, G.M. The role of the genome project in determining gene function: insights from model organisms. *Cell.* **1996**, *86*, 521-9.

[87] Kelly, D.E.; Lamb, D.C.; Kelly, S.L. Genome-wide generation of yeast gene deletion strains. *Comp Funct Genomics.* **2001**, *2*, 236-42.

[88] Que, Q.Q.; Winzeler, E.A. Large-scale mutagenesis and functional genomics in yeast. *Funct Integr Genomics.* **2002**, *2*, 193-8.

[89] Giaever, G.; Chu, A.M.; Ni, L.; Connelly, C.; Riles, L.; Veronneau, S.; Dow, S.; Lucau-Danila, A.; Anderson, K.; Andre, B.; Arkin, A.P.; Astromoff, A.; El-Bakkoury, M.; Bangham, R.; Benito, R.; Brachat, S.; Campanaro, S.; Curtiss, M.; Davis, K.; Deutschbauer, A.; Entian, K.D.; Flaherty, P.; Foury, F.; Garfinkel, D.J.; Gerstein, M.; Gotte, D.; Guldener, U.; Hegemann, J.H.; Hempel, S.; Herman, Z.; Jaramillo, D.F.; Kelly, D.E.; Kelly, S.L.; Kotter, P.; LaBonte, D.; Lamb, D.C.; Lan, N.; Liang, H.; Liao, H.; Liu, L.; Luo, C.; Lussier, M.; Mao, R.; Menard, P.; Ooi, S.L.; Revuelta, J.L.; Roberts, C.J.; Rose, M.; Ross-Macdonald, P.; Scherens, B.; Schimmack, G.; Shafer, B.; Shoemaker, D.D.; Sookhai-Mahadeo, S.; Storms, R.K.; Strathern, J.N.; Valle, G.; Voet, M.; Volckaert, G.; Wang, C.Y.; Ward, T.R.; Wilhelmy, J.; Winzeler, E.A.; Yang, Y.; Yen, G.; Youngman, E.; Yu, K.; Bussey, H.; Boeke, J.D.; Snyder, M.; Philippsen, P.; Davis, R.W.; Johnston, M. Functional profiling of the Saccharomyces cerevisiae genome. *Nature.* **2002**, *418*, 387-91.

[90] Costanzo, M.; Baryshnikova, A.; Bellay, J.; Kim, Y.; Spear, E.D.; Sevier, C.S.; Ding, H.; Koh, J.L.; Toufighi, K.; Mostafavi, S.; Prinz, J.; St Onge, R.P.; VanderSluis, B.; Makhnevych, T.; Vizeacoumar, F.J.; Alizadeh, S.; Bahr, S.; Brost, R.L.; Chen, Y.; Cokol, M.; Deshpande, R.; Li, Z.; Lin, Z.Y.; Liang, W.; Marback, M.; Paw, J.; San Luis, B.J.; Shuteriqi, E.; Tong, A.H.; van Dyk, N.; Wallace, I.M.; Whitney, J.A.; Weirauch, M.T.; Zhong, G.; Zhu, H.; Houry, W.A.; Brudno, M.; Ragibizadeh, S.; Papp, B.; Pal, C.; Roth, F.P.; Giaever, G.; Nislow, C.; Troyanskaya, O.G.; Bussey, H.; Bader, G.D.; Gingras, A.C.;

Morris, Q.D.; Kim, P.M.; Kaiser, C.A.; Myers, C.L.; Andrews, B.J.; Boone, C. The genetic landscape of a cell. *Science.* **2010**, *327*, 425-31.

[91] Adams, M.D.; Celniker, S.E.; Holt, R.A.; Evans, C.A.; Gocayne, J.D.; Amanatides, P.G.; Scherer, S.E.; Li, P.W.; Hoskins, R.A.; Galle, R.F.; George, R.A.; Lewis, S.E.; Richards, S.; Ashburner, M.; Henderson, S.N.; Sutton, G.G.; Wortman, J.R.; Yandell, M.D.; Zhang, Q.; Chen, L.X.; Brandon, R.C.; Rogers, Y.H.; Blazej, R.G.; Champe, M.; Pfeiffer, B.D.; Wan, K.H.; Doyle, C.; Baxter, E.G.; Helt, G.; Nelson, C.R.; Gabor, G.L.; Abril, J.F.; Agbayani, A.; An, H.J.; Andrews-Pfannkoch, C.; Baldwin, D.; Ballew, R.M.; Basu, A.; Baxendale, J.; Bayraktaroglu, L.; Beasley, E.M.; Beeson, K.Y.; Benos, P.V.; Berman, B.P.; Bhandari, D.; Bolshakov, S.; Borkova, D.; Botchan, M.R.; Bouck, J.; Brokstein, P.; Brottier, P.; Burtis, K.C.; Busam, D.A.; Butler, H.; Cadieu, E.; Center, A.; Chandra, I.; Cherry, J.M.; Cawley, S.; Dahlke, C.; Davenport, L.B.; Davies, P.; de Pablos, B.; Delcher, A.; Deng, Z.; Mays, A.D.; Dew, I.; Dietz, S.M.; Dodson, K.; Doup, L.E.; Downes, M.; Dugan-Rocha, S.; Dunkov, B.C.; Dunn, P.; Durbin, K.J.; Evangelista, C.C.; Ferraz, C.; Ferriera, S.; Fleischmann, W.; Fosler, C.; Gabrielian, A.E.; Garg, N.S.; Gelbart, W.M.; Glasser, K.; Glodek, A.; Gong, F.; Gorrell, J.H.; Gu, Z.; Guan, P.; Harris, M.; Harris, N.L.; Harvey, D.; Heiman, T.J.; Hernandez, J.R.; Houck, J.; Hostin, D.; Houston, K.A.; Howland, T.J.; Wei, M.H.; Ibegwam, C.; Jalali, M.; Kalush, F.; Karpen, G.H.; Ke, Z.; Kennison, J.A.; Ketchum, K.A.; Kimmel, B.E.; Kodira, C.D.; Kraft, C.; Kravitz, S.; Kulp, D.; Lai, Z.; Lasko, P.; Lei, Y.; Levitsky, A.A.; Li, J.; Li, Z.; Liang, Y.; Lin, X.; Liu, X.; Mattei, B.; McIntosh, T.C.; McLeod, M.P.; McPherson, D.; Merkulov, G.; Milshina, N.V.; Mobarry, C.; Morris, J.; Moshrefi, A.; Mount, S.M.; Moy, M.; Murphy, B.; Murphy, L.; Muzny, D.M.; Nelson, D.L.; Nelson, D.R.; Nelson, K.A.; Nixon, K.; Nusskern, D.R.; Pacleb, J.M.; Palazzolo, M.; Pittman, G.S.; Pan, S.; Pollard, J.; Puri, V.; Reese, M.G.; Reinert, K.; Remington, K.; Saunders, R.D.; Scheeler, F.; Shen, H.; Shue, B.C.; Siden-Kiamos, I.; Simpson, M.; Skupski, M.P.; Smith, T.; Spier, E.; Spradling, A.C.; Stapleton, M.; Strong, R.; Sun, E.; Svirskas, R.; Tector, C.; Turner, R.; Venter, E.; Wang, A.H.; Wang, X.; Wang, Z.Y.; Wassarman, D.A.; Weinstock, G.M.; Weissenbach, J.; Williams, S.M.; WoodageT; Worley, K.C.; Wu, D.; Yang, S.; Yao, Q.A.; Ye, J.; Yeh, R.F.; Zaveri, J.S.; Zhan, M.; Zhang, G.; Zhao, Q.; Zheng, L.; Zheng, X.H.; Zhong, F.N.; Zhong, W.; Zhou, X.; Zhu, S.; Zhu, X.; Smith, H.O.; Gibbs, R.A.; Myers, E.W.; Rubin, G.M.; Venter, J.C. The genome sequence of Drosophila melanogaster. *Science.* **2000**, *287*, 2185-95.

[92] Liu, C.P.; Lim, J.K. Complementation analysis of methyl methane-sulfonate-induced recessive lethal mutations in the zeste-white region of the X chromosome of Drosophila melanogaster. *Genetics.* **1975**, *79*, 601-11.

[93] Lim, J.K.; Snyder, L.A. Cytogenetic and complementation analyses of recessive lethal mutations induced in the X chromosome of Drosophila by three alkylating agents. *Genet Res.* **1974**, *24*, 1-10.

[94] Ashburner, M.; Misra, S.; Roote, J.; Lewis, S.E.; Blazej, R.; Davis, T.; Doyle, C.; Galle, R.; George, R.; Harris, N.; Hartzell, G.; Harvey, D.; Hong, L.; Houston, K.; Hoskins, R.; Johnson, G.; Martin, C.; Moshrefi, A.; Palazzolo, M.; Reese, M.G.; Spradling, A.; Tsang, G.; Wan, K.; Whitelaw, K.; Celniker, S.; *et al.* An exploration of the sequence of a 2.9-Mb region of the genome of Drosophila melanogaster: the Adh region. *Genetics.* **1999**, *153*, 179-219.

[95] Cooley, L.; Kelley, R.; Spradling, A. Insertional mutagenesis of the Drosophila genome with single P elements. *Science.* **1988**, *239*, 1121-8.

[96] Myrick, K.V.; Huet, F.; Mohr, S.E.; Alvarez-Garcia, I.; Lu, J.T.; Smith, M.A.; Crosby, M.A.; Gelbart, W.M. Large-scale functional annotation and expanded implementations of the P{wHy} hybrid transposon in the Drosophila melanogaster genome. *Genetics.* **2009**, *182*, 653-60.

[97] Hacker, U.; Nystedt, S.; Barmchi, M.P.; Horn, C.; Wimmer, E.A. piggyBac-based insertional mutagenesis in the presence of stably integrated P elements in Drosophila. *Proc Natl Acad Sci U S A.* **2003**, *100*, 7720-5.

[98] Thibault, S.T.; Singer, M.A.; Miyazaki, W.Y.; Milash, B.; Dompe, N.A.; Singh, C.M.; Buchholz, R.; Demsky, M.; Fawcett, R.; Francis-Lang, H.L.; Ryner, L.; Cheung, L.M.; Chong, A.; Erickson, C.; Fisher, W.W.; Greer, K.; Hartouni, S.R.; Howie, E.; Jakkula, L.; Joo, D.; Killpack, K.; Laufer, A.; Mazzotta, J.; Smith, R.D.; Stevens, L.M.; Stuber, C.; Tan, L.R.; Ventura, R.; Woo, A.; Zakrajsek, I.; Zhao, L.; Chen, F.; Swimmer, C.; Kopczynski, C.; Duyk, G.; Winberg, M.L.; Margolis, J. A complementary transposon tool kit for Drosophila melanogaster using P and piggyBac. *Nat Genet.* **2004**, *36*, 283-7.

[99] St Johnston, D. The art and design of genetic screens: Drosophila melanogaster. *Nat Rev Genet.* **2002**, *3*, 176-88.

[100] Ryder, E.; Russell, S. Transposable elements as tools for genomics and genetics in Drosophila. *Brief Funct Genomic Proteomic.* **2003**, *2*, 57-71.

[101] Sulston, J.E.; Albertson, D.G.; Thomson, J.N. The Caenorhabditis elegans male: postembryonic development of nongonadal structures. *Dev Biol.* **1980**, *78*, 542-76.

[102] Sulston, J.E.; Schierenberg, E.; White, J.G.; Thomson, J.N. The embryonic cell lineage of the nematode Caenorhabditis elegans. *Dev Biol.* **1983**, *100*, 64-119.

[103] Hillier, L.W.; Coulson, A.; Murray, J.I.; Bao, Z.; Sulston, J.E.; Waterston, R.H. Genomics in C. elegans: so many genes, such a little worm. *Genome Res.* **2005**, *15*, 1651-60.

[104] Granger, L.; Martin, E.; Segalat, L. Mos as a tool for genome-wide insertional mutagenesis in Caenorhabditis elegans: results of a pilot study. *Nucleic Acids Res.* **2004**, *32*, e117.

[105] Montgomery, M.K.; Fire, A. Double-stranded RNA as a mediator in sequence-specific genetic silencing and co-suppression. *Trends Genet.* **1998**, *14*, 255-8.

[106] Fraser, A.G.; Kamath, R.S.; Zipperlen, P.; Martinez-Campos, M.; Sohrmann, M.; Ahringer, J. Functional genomic analysis of C. elegans chromosome I by systematic RNA interference. *Nature.* **2000**, *408*, 325-30.

[107] Kamath, R.S.; Fraser, A.G.; Dong, Y.; Poulin, G.; Durbin, R.; Gotta, M.; Kanapin, A.; Le Bot, N.; Moreno, S.; Sohrmann, M.; Welchman, D.P.; Zipperlen, P.; Ahringer, J. Systematic functional analysis of the Caenorhabditis elegans genome using RNAi. *Nature.* **2003**, *421*, 231-7.

[108] Alessandrini, F.; Jakob, T.; Wolf, A.; Wolf, E.; Balling, R.; Hrabe de Angelis, M.; Ring, J.; Behrendt, H. Enu mouse mutagenesis: generation of mouse mutants with aberrant plasma IgE levels. *Int Arch Allergy Immunol.* **2001**, *124*, 25-8.

[109] Barbaric, I.; Wells, S.; Russ, A.; Dear, T.N. Spectrum of ENU-induced mutations in phenotype-driven and gene-driven screens in the mouse. *Environ Mol Mutagen.* **2007**, *48*, 124-42.

[110] Boles, M.K.; Wilkinson, B.M.; Wilming, L.G.; Liu, B.; Probst, F.J.; Harrow, J.; Grafham, D.; Hentges, K.E.; Woodward, L.P.; Maxwell, A.; Mitchell, K.; Risley, M.D.; Johnson, R.; Hirschi, K.; Lupski, J.R.; Funato, Y.; Miki, H.; Marin-Garcia, P.; Matthews, L.; Coffey, A.J.; Parker, A.; Hubbard, T.J.; Rogers, J.; Bradley, A.; Adams, D.J.; Justice, M.J. Discovery of candidate disease genes in ENU-induced mouse mutants by large-scale

sequencing, including a splice-site mutation in nucleoredoxin. *PLoS Genet.* **2009**, *5*, e1000759.

[111] Brown, S.D.; Balling, R. Systematic approaches to mouse mutagenesis. *Curr Opin Genet Dev.* **2001**, *11*, 268-73.

[112] Clark, A.T.; Goldowitz, D.; Takahashi, J.S.; Vitaterna, M.H.; Siepka, S.M.; Peters, L.L.; Frankel, W.N.; Carlson, G.A.; Rossant, J.; Nadeau, J.H.; Justice, M.J. Implementing large-scale ENU mutagenesis screens in North America. *Genetica.* **2004**, *122*, 51-64.

[113] Cordes, S.P. N-ethyl-N-nitrosourea mutagenesis: boarding the mouse mutant express. *Microbiol Mol Biol Rev.* **2005**, *69*, 426-39.

[114] Hrabe de Angelis, M.; Balling, R. Large scale ENU screens in the mouse: genetics meets genomics. *Mutat Res.* **1998**, *400*, 25-32.

[115] Hrabe de Angelis, M.; Strivens, M. Large-scale production of mouse phenotypes: the search for animal models for inherited diseases in humans. *Brief Bioinform.* **2001**, *2*, 170-80.

[116] Hrabe de Angelis, M.H.; Flaswinkel, H.; Fuchs, H.; Rathkolb, B.; Soewarto, D.; Marschall, S.; Heffner, S.; Pargent, W.; Wuensch, K.; Jung, M.; Reis, A.; Richter, T.; Alessandrini, F.; Jakob, T.; Fuchs, E.; Kolb, H.; Kremmer, E.; Schaeble, K.; Rollinski, B.; Roscher, A.; Peters, C.; Meitinger, T.; Strom, T.; Steckler, T.; Holsboer, F.; Klopstock, T.; Gekeler, F.; Schindewolf, C.; Jung, T.; Avraham, K.; Behrendt, H.; Ring, J.; Zimmer, A.; Schughart, K.; Pfeffer, K.; Wolf, E.; Balling, R. Genome-wide, large-scale production of mutant mice by ENU mutagenesis. *Nat Genet.* **2000**, *25*, 444-7.

[117] Nishimura, I.; Drake, T.A.; Lusis, A.J.; Lyons, K.M.; Nadeau, J.H.; Zernik, J. ENU large-scale mutagenesis and quantitative trait linkage (QTL) analysis in mice: novel technologies for searching polygenetic determinants of craniofacial abnormalities. *Crit Rev Oral Biol Med.* **2003**, *14*, 320-30.

[118] Nolan, P.M.; Peters, J.; Vizor, L.; Strivens, M.; Washbourne, R.; Hough, T.; Wells, C.; Glenister, P.; Thornton, C.; Martin, J.; Fisher, E.; Rogers, D.; Hagan, J.; Reavill, C.; Gray, I.; Wood, J.; Spurr, N.; Browne, M.; Rastan, S.; Hunter, J.; Brown, S.D. Implementation of a large-scale ENU mutagenesis program: towards increasing the mouse mutant resource. *Mamm Genome.* **2000**, *11*, 500-6.

[119] Rathkolb, B.; Fuchs, E.; Kolb, H.J.; Renner-Muller, I.; Krebs, O.; Balling, R.; Hrabe de Angelis, M.; Wolf, E. Large-scale N-ethyl-N-nitrosourea mutagenesis of mice--from phenotypes to genes. *Exp Physiol.* **2000**, *85*, 635-44.

[120] Sakuraba, Y.; Sezutsu, H.; Takahasi, K.R.; Tsuchihashi, K.; Ichikawa, R.; Fujimoto, N.; Kaneko, S.; Nakai, Y.; Uchiyama, M.; Goda, N.; Motoi, R.; Ikeda, A.; Karashima, Y.; Inoue, M.; Kaneda, H.; Masuya, H.; Minowa, O.; Noguchi, H.; Toyoda, A.; Sakaki, Y.; Wakana, S.; Noda, T.; Shiroishi, T.; Gondo, Y. Molecular characterization of ENU mouse mutagenesis and archives. *Biochem Biophys Res Commun.* **2005**, *336*, 609-16.

[121] Soewarto, D.; Fella, C.; Teubner, A.; Rathkolb, B.; Pargent, W.; Heffner, S.; Marschall, S.; Wolf, E.; Balling, R.; Hrabe de Angelis, M. The large-scale Munich ENU-mouse-mutagenesis screen. *Mamm Genome.* **2000**, *11*, 507-10.

[122] Stanford, W.L.; Cohn, J.B.; Cordes, S.P. Gene-trap mutagenesis: past, present and beyond. *Nat Rev Genet.* **2001**, *2*, 756-68.

[123] Araki, M.; Araki, K.; Yamamura, K. International Gene Trap Project: towards gene-driven saturation mutagenesis in mice. *Curr Pharm Biotechnol.* **2009**, *10*, 221-9.

[124] Guan, C.; Ye, C.; Yang, X.; Gao, J. A review of current large-scale mouse knockout efforts. *Genesis. 48*, 73-85.

[125] Nord, A.S.; Chang, P.J.; Conklin, B.R.; Cox, A.V.; Harper, C.A.; Hicks, G.G.; Huang, C.C.; Johns, S.J.; Kawamoto, M.; Liu, S.; Meng, E.C.; Morris, J.H.; Rossant, J.; Ruiz, P.; Skarnes, W.C.; Soriano, P.; Stanford, W.L.; Stryke, D.; von Melchner, H.; Wurst, W.; Yamamura, K.; Young, S.G.; Babbitt, P.C.; Ferrin, T.E. The International Gene Trap Consortium Website: a portal to all publicly available gene trap cell lines in mouse. *Nucleic Acids Res.* **2006**, *34*, D642-8.

[126] Skarnes, W.C.; von Melchner, H.; Wurst, W.; Hicks, G.; Nord, A.S.; Cox, T.; Young, S.G.; Ruiz, P.; Soriano, P.; Tessier-Lavigne, M.; Conklin, B.R.; Stanford, W.L.; Rossant, J. A public gene trap resource for mouse functional genomics. *Nat Genet.* **2004**, *36*, 543-4.

[127] Stanford, W.L.; Epp, T.; Reid, T.; Rossant, J. Gene trapping in embryonic stem cells. *Methods Enzymol.* **2006**, *420*, 136-62.

[128] Brandon, E.P.; Idzerda, R.L.; McKnight, G.S. Targeting the mouse genome: a compendium of knockouts (Part III). *Curr Biol.* **1995**, *5*, 873-81.

[129] Brandon, E.P.; Idzerda, R.L.; McKnight, G.S. Targeting the mouse genome: a compendium of knockouts (Part II). *Curr Biol.* **1995**, *5*, 758-65.

[130] Brandon, E.P.; Idzerda, R.L.; McKnight, G.S. Knockouts. Targeting the mouse genome: a compendium of knockouts (Part I). *Curr Biol.* **1995**, *5*, 625-34.

[131] Kuhn, R.; Wurst, W. Overview on mouse mutagenesis. *Methods Mol Biol.* **2009**, *530*, 1-12.

[132] Hurst, L.D.; Smith, N.G. Do essential genes evolve slowly? *Curr Biol.* **1999**, *9*, 747-50.

[133] Zambrowicz, B.P.; Sands, A.T. Knockouts model the 100 best-selling drugs--will they model the next 100? *Nat Rev Drug Discov.* **2003**, *2*, 38-51.

[134] Leroux, L.; Desbois, P.; Lamotte, L.; Duvillie, B.; Cordonnier, N.; Jackerott, M.; Jami, J.; Bucchini, D.; Joshi, R.L. Compensatory responses in mice carrying a null mutation for Ins1 or Ins2. *Diabetes.* **2001**, *50 Suppl 1*, S150-3.

[135] Duvillie, B.; Cordonnier, N.; Deltour, L.; Dandoy-Dron, F.; Itier, J.M.; Monthioux, E.; Jami, J.; Joshi, R.L.; Bucchini, D. Phenotypic alterations in insulin-deficient mutant mice. *Proc Natl Acad Sci U S A.* **1997**, *94*, 5137-40.

[136] Joshi, R.L.; Lamothe, B.; Cordonnier, N.; Mesbah, K.; Monthioux, E.; Jami, J.; Bucchini, D. Targeted disruption of the insulin receptor gene in the mouse results in neonatal lethality. *Embo J.* **1996**, *15*, 1542-7.

[137] Accili, D.; Drago, J.; Lee, E.J.; Johnson, M.D.; Cool, M.H.; Salvatore, P.; Asico, L.D.; Jose, P.A.; Taylor, S.I.; Westphal, H. Early neonatal death in mice homozygous for a null allele of the insulin receptor gene. *Nat Genet.* **1996**, *12*, 106-9.

[138] Mullins, M.C.; Hammerschmidt, M.; Haffter, P.; Nusslein-Volhard, C. Large-scale mutagenesis in the zebrafish: in search of genes controlling development in a vertebrate. *Curr Biol.* **1994**, *4*, 189-202.

[139] Haffter, P.; Nusslein-Volhard, C. Large scale genetics in a small vertebrate, the zebrafish. *Int J Dev Biol.* **1996**, *40*, 221-7.

[140] Geisler, R.; Rauch, G.J.; Geiger-Rudolph, S.; Albrecht, A.; van Bebber, F.; Berger, A.; Busch-Nentwich, E.; Dahm, R.; Dekens, M.P.; Dooley, C.; Elli, A.F.; Gehring, I.; Geiger, H.; Geisler, M.; Glaser, S.; Holley, S.; Huber, M.; Kerr, A.; Kirn, A.; Knirsch, M.; Konantz, M.; Kuchler, A.M.; Maderspacher, F.; Neuhauss, S.C.; Nicolson, T.; Ober, E.A.; Praeg, E.; Ray, R.; Rentzsch, B.; Rick, J.M.; Rief, E.; Schauerte, H.E.; Schepp, C.P.; Schonberger, U.; Schonthaler, H.B.; Seiler, C.; Sidi, S.; Sollner, C.; Wehner, A.; Weiler,

C.; Nusslein-Volhard, C. Large-scale mapping of mutations affecting zebrafish development. *BMC Genomics.* **2007**, *8*, 11.

[141] de Bruijn, E.; Cuppen, E.; Feitsma, H. Highly Efficient ENU Mutagenesis in Zebrafish. *Methods Mol Biol.* **2009**, *546*, 3-12.

[142] Draper, B.W.; McCallum, C.M.; Stout, J.L.; Slade, A.J.; Moens, C.B. A high-throughput method for identifying N-ethyl-N-nitrosourea (ENU)-induced point mutations in zebrafish. *Methods Cell Biol.* **2004**, *77*, 91-112.

[143] Grunwald, D.J.; Streisinger, G. Induction of recessive lethal and specific locus mutations in the zebrafish with ethyl nitrosourea. *Genet Res.* **1992**, *59*, 103-16.

[144] Sood, R.; English, M.A.; Jones, M.; Mullikin, J.; Wang, D.M.; Anderson, M.; Wu, D.; Chandrasekharappa, S.C.; Yu, J.; Zhang, J.; Paul Liu, P. Methods for reverse genetic screening in zebrafish by resequencing and TILLING. *Methods.* **2006**, *39*, 220-7.

[145] Henikoff, S.; Till, B.J.; Comai, L. TILLING. Traditional mutagenesis meets functional genomics. *Plant Physiol.* **2004**, *135*, 630-6.

[146] Moens, C.B.; Donn, T.M.; Wolf-Saxon, E.R.; Ma, T.P. Reverse genetics in zebrafish by TILLING. *Brief Funct Genomic Proteomic.* **2008**, *7*, 454-9.

[147] Nasevicius, A.; Ekker, S.C. Effective targeted gene 'knockdown' in zebrafish. *Nat Genet.* **2000**, *26*, 216-20.

[148] George, R.A.; Smith, T.D.; Callaghan, S.; Hardman, L.; Pierides, C.; Horaitis, O.; Wouters, M.A.; Cotton, R.G. General mutation databases: analysis and review. *J Med Genet.* **2008**, *45*, 65-70.

[149] Hamosh, A.; Scott, A.F.; Amberger, J.; Valle, D.; McKusick, V.A. Online Mendelian Inheritance in Man (OMIM). *Hum Mutat.* **2000**, *15*, 57-61.

[150] Amberger, J.; Bocchini, C.A.; Scott, A.F.; Hamosh, A. McKusick's Online Mendelian Inheritance in Man (OMIM). *Nucleic Acids Res.* **2009**, *37*, D793-6.

[151] Stenson, P.D.; Mort, M.; Ball, E.V.; Howells, K.; Phillips, A.D.; Thomas, N.S.; Cooper, D.N. The Human Gene Mutation Database: 2008 update. *Genome Med.* **2009**, *1*, 13.

[152] Cooper, D.N.; Chen, J.M.; Ball, E.V.; Howells, K.; Mort, M.; Phillips, A.D.; Chuzhanova, N.; Krawczak, M.; Kehrer-Sawatzki, H.; Stenson, P.D. Genes, mutations, and human inherited disease at the dawn of the age of personalized genomics. *Hum Mutat.* **2010**, *31*, 631-55.

[153] Ring, H.Z.; Kwok, P.Y.; Cotton, R.G. Human Variome Project: an international collaboration to catalogue human genetic variation. *Pharmacogenomics.* **2006**, *7*, 969-72.

[154] Fredman, D.; Munns, G.; Rios, D.; Sjoholm, F.; Siegfried, M.; Lenhard, B.; Lehvaslaiho, H.; Brookes, A.J. HGVbase: a curated resource describing human DNA variation and phenotype relationships. *Nucleic Acids Res.* **2004**, *32*, D516-9.

[155] Drake, J.W.; Charlesworth, B.; Charlesworth, D.; Crow, J.F. Rates of spontaneous mutation. *Genetics.* **1998**, *148*, 1667-86.

[156] Nachman, M.W.; Crowell, S.L. Estimate of the mutation rate per nucleotide in humans. *Genetics.* **2000**, *156*, 297-304.

[157] Xue, Y.; Wang, Q.; Long, Q.; Ng, B.L.; Swerdlow, H.; Burton, J.; Skuce, C.; Taylor, R.; Abdellah, Z.; Zhao, Y.; MacArthur, D.G.; Quail, M.A.; Carter, N.P.; Yang, H.; Tyler-Smith, C. Human Y chromosome base-substitution mutation rate measured by direct sequencing in a deep-rooting pedigree. *Curr Biol.* **2009**, *19*, 1453-7.

[158] Roach, J.C.; Glusman, G.; Smit, A.F.; Huff, C.D.; Hubley, R.; Shannon, P.T.; Rowen, L.; Pant, K.P.; Goodman, N.; Bamshad, M.; Shendure, J.; Drmanac, R.; Jorde, L.B.; Hood, L.;

Galas, D.J. Analysis of genetic inheritance in a family quartet by whole-genome sequencing. *Science.* **2010**, *328*, 636-9.

[159] Awadalla, P.; Gauthier, J.; Myers, R.A.; Casals, F.; Hamdan, F.F.; Griffing, A.R.; Cote, M.; Henrion, E.; Spiegelman, D.; Tarabeux, J.; Piton, A.; Yang, Y.; Boyko, A.; Bustamante, C.; Xiong, L.; Rapoport, J.L.; Addington, A.M.; DeLisi, J.L.; Krebs, M.O.; Joober, R.; Millet, B.; Fombonne, E.; Mottron, L.; Zilversmit, M.; Keebler, J.; Daoud, H.; Marineau, C.; Roy-Gagnon, M.H.; Dube, M.P.; Eyre-Walker, A.; Drapeau, P.; Stone, E.A.; Lafreniere, R.G.; Rouleau, G.A. Direct measure of the *de novo* mutation rate in autism and schizophrenia cohorts. *Am J Hum Genet.* **2010**, *87*, 316-24.

[160] Conrad, D.F.; Keebler, J.E.; DePristo, M.A.; Lindsay, S.J.; Zhang, Y.; Casals, F.; Idaghdour, Y.; Hartl, C.L.; Torroja, C.; Garimella, K.V.; Zilversmit, M.; Cartwright, R.; Rouleau, G.A.; Daly, M.; Stone, E.A.; Hurles, M.E.; Awadalla, P. Variation in genome-wide mutation rates within and between human families. *Nat Genet.* **2011**, *43*, 712-4.

[161] Carson, A.R.; Feuk, L.; Mohammed, M.; Scherer, S.W. Strategies for the detection of copy number and other structural variants in the human genome. *Hum Genomics.* **2006**, *2*, 403-14.

[162] Feuk, L.; Carson, A.R.; Scherer, S.W. Structural variation in the human genome. *Nat Rev Genet.* **2006**, *7*, 85-97.

[163] Pinto, D.; Marshall, C.; Feuk, L.; Scherer, S.W. Copy-number variation in control population cohorts. *Hum Mol Genet.* **2007**, *16 Spec No. 2*, R168-73.

[164] Kumar, D. Disorders of the genome architecture: a review. *Genomic Med.* **2008**, *2*, 69-76.

[165] Lindblad-Toh, K.; Garber, M.; Zuk, O.; Lin, M.F.; Parker, B.J.; Washietl, S.; Kheradpour, P.; Ernst, J.; Jordan, G.; Mauceli, E.; Ward, L.D.; Lowe, C.B.; Holloway, A.K.; Clamp, M.; Gnerre, S.; Alfoldi, J.; Beal, K.; Chang, J.; Clawson, H.; Cuff, J.; Di Palma, F.; Fitzgerald, S.; Flicek, P.; Guttman, M.; Hubisz, M.J.; Jaffe, D.B.; Jungreis, I.; Kent, W.J.; Kostka, D.; Lara, M.; Martins, A.L.; Massingham, T.; Moltke, I.; Raney, B.J.; Rasmussen, M.D.; Robinson, J.; Stark, A.; Vilella, A.J.; Wen, J.; Xie, X.; Zody, M.C.; Baldwin, J.; Bloom, T.; Chin, C.W.; Heiman, D.; Nicol, R.; Nusbaum, C.; Young, S.; Wilkinson, J.; Worley, K.C.; Kovar, C.L.; Muzny, D.M.; Gibbs, R.A.; Cree, A.; Dihn, H.H.; Fowler, G.; Jhangiani, S.; Joshi, V.; Lee, S.; Lewis, L.R.; Nazareth, L.V.; Okwuonu, G.; Santibanez, J.; Warren, W.C.; Mardis, E.R.; Weinstock, G.M.; Wilson, R.K.; Delehaunty, K.; Dooling, D.; Fronik, C.; Fulton, L.; Fulton, B.; Graves, T.; Minx, P.; Sodergren, E.; Birney, E.; Margulies, E.H.; Herrero, J.; Green, E.D.; Haussler, D.; Siepel, A.; Goldman, N.; Pollard, K.S.; Pedersen, J.S.; Lander, E.S.; Kellis, M. A high-resolution map of human evolutionary constraint using 29 mammals. *Nature.* **2011**, *478*, 476-82.

[166] Kryukov, G.V.; Pennacchio, L.A.; Sunyaev, S.R. Most rare missense alleles are deleterious in humans: implications for complex disease and association studies. *Am J Hum Genet.* **2007**, *80*, 727-39.

[167] Sunyaev, S.; Ramensky, V.; Koch, I.; Lathe, W., 3rd; Kondrashov, A.S.; Bork, P. Prediction of deleterious human alleles. *Hum Mol Genet.* **2001**, *10*, 591-7.

[168] Yampolsky, L.Y.; Kondrashov, F.A.; Kondrashov, A.S. Distribution of the strength of selection against amino acid replacements in human proteins. *Hum Mol Genet.* **2005**, *14*, 3191-201.

[169] Boyko, A.R.; Williamson, S.H.; Indap, A.R.; Degenhardt, J.D.; Hernandez, R.D.; Lohmueller, K.E.; Adams, M.D.; Schmidt, S.; Sninsky, J.J.; Sunyaev, S.R.; White, T.J.;

Nielsen, R.; Clark, A.G.; Bustamante, C.D. Assessing the evolutionary impact of amino acid mutations in the human genome. *PLoS Genet.* **2008**, *4*, e1000083.
[170] Eyre-Walker, A.; Keightley, P.D. High genomic deleterious mutation rates in hominids. *Nature.* **1999**, *397*, 344-7.
[171] Kondrashov, A.S. Direct estimates of human per nucleotide mutation rates at 20 loci causing Mendelian diseases. *Hum Mutat.* **2003**, *21*, 12-27.
[172] World Population Data sheet. Population Reference Bureau **2009**, 1-19.
[173] Sankaranarayanan, K. Ionizing radiation and genetic risks IX. Estimates of the frequencies of mendelian diseases and spontaneous mutation rates in human populations: a 1998 perspective. *Mutat Res.* **1998**, *411*, 129-78.
[174] Fischer, H.; Oswald, H.P.; Duba, H.C.; Doczy, L.; Simma, B.; Utermann, G.; Haas, O.A. Constitutional interstitial deletion of 17(p11.2) (Smith-Magenis syndrome): a clinically recognizable microdeletion syndrome. Report of two cases and review of the literature. *Klin Padiatr.* **1993**, *205*, 162-6.
[175] Elsea, S.H.; Williams, S.R. Smith-Magenis syndrome: haploinsufficiency of RAI1 results in altered gene regulation in neurological and metabolic pathways. *Expert Rev Mol Med.* **2011**, *13*, e14.
[176] Slager, R.E.; Newton, T.L.; Vlangos, C.N.; Finucane, B.; Elsea, S.H. Mutations in RAI1 associated with Smith-Magenis syndrome. *Nat Genet.* **2003**, *33*, 466-8.
[177] Wilkie, A.O. The molecular basis of genetic dominance. *J Med Genet.* **1994**, *31*, 89-98.
[178] Frank, T.S.; Deffenbaugh, A.M.; Reid, J.E.; Hulick, M.; Ward, B.E.; Lingenfelter, B.; Gumpper, K.L.; Scholl, T.; Tavtigian, S.V.; Pruss, D.R.; Critchfield, G.C. Clinical characteristics of individuals with germline mutations in BRCA1 and BRCA2: analysis of 10,000 individuals. *J Clin Oncol.* **2002**, *20*, 1480-90.
[179] Delatycki, M.B.; Williamson, R.; Forrest, S.M. Friedreich ataxia: an overview. *J Med Genet.* **2000**, *37*, 1-8.
[180] Vucic, S.; Kiernan, M.C. Pathophysiology of neurodegeneration in familial amyotrophic lateral sclerosis. *Curr Mol Med.* **2009**, *9*, 255-72.
[181] Julien, J.P.; Kriz, J. Transgenic mouse models of amyotrophic lateral sclerosis. *Biochim Biophys Acta.* **2006**, *1762*, 1013-24.
[182] Abifadel, M.; Rabes, J.P.; Devillers, M.; Munnich, A.; Erlich, D.; Junien, C.; Varret, M.; Boileau, C. Mutations and polymorphisms in the proprotein convertase subtilisin kexin 9 (PCSK9) gene in cholesterol metabolism and disease. *Hum Mutat.* **2009**, *30*, 520-9.
[183] Soutar, A.K.; Naoumova, R.P. Mechanisms of disease: genetic causes of familial hypercholesterolemia. *Nat Clin Pract Cardiovasc Med.* **2007**, *4*, 214-25.
[184] Horton, J.D.; Cohen, J.C.; Hobbs, H.H. Molecular biology of PCSK9: its role in LDL metabolism. *Trends Biochem Sci.* **2007**, *32*, 71-7.
[185] Rahalkar, A.R.; Hegele, R.A. Monogenic pediatric dyslipidemias: classification, genetics and clinical spectrum. *Mol Genet Metab.* **2008**, *93*, 282-94.
[186] Tarugi, P.; Averna, M.; Di Leo, E.; Cefalu, A.B.; Noto, D.; Magnolo, L.; Cattin, L.; Bertolini, S.; Calandra, S. Molecular diagnosis of hypobetalipoproteinemia: an ENID review. *Atherosclerosis.* **2007**, *195*, e19-27.
[187] Whitfield, A.J.; Barrett, P.H.; van Bockxmeer, F.M.; Burnett, J.R. Lipid disorders and mutations in the APOB gene. *Clin Chem.* **2004**, *50*, 1725-32.
[188] Farese, R.V., Jr.; Linton, M.F.; Young, S.G. Apolipoprotein B gene mutations affecting cholesterol levels. *J Intern Med.* **1992**, *231*, 643-52.

[189] Varret, M.; Abifadel, M.; Rabes, J.P.; Boileau, C. Genetic heterogeneity of autosomal dominant hypercholesterolemia. *Clin Genet.* **2008**, *73*, 1-13.

[190] Phadnis, N.; Fry, J.D. Widespread correlations between dominance and homozygous effects of mutations: implications for theories of dominance. *Genetics.* **2005**, *171*, 385-92.

[191] Porteous, J.W. Dominance--one hundred and fifteen years after Mendel's paper. *J Theor Biol.* **1996**, *182*, 223-32.

[192] Wright, S. Physiological and evolutionary theories of dominance. *American Naturalist.* **1934**, *68*, 24-53.

[193] Kacser, H.; Burns, J.A. The control of flux. *Symp Soc Exp Biol.* **1973**, *27*, 65-104.

[194] Kacser, H.; Burns, J.A. The molecular basis of dominance. *Genetics.* **1981**, *97*, 639-66.

[195] Grossniklaus, U.; Madhusudhan, M.S.; Nanjundiah, V. Nonlinear enzyme kinetics can lead to high metabolic flux control coefficients: implications for the evolution of dominance. *J Theor Biol.* **1996**, *182*, 299-302.

[196] Bagheri, H.C.; Wagner, G.P. Evolution of dominance in metabolic pathways. *Genetics.* **2004**, *168*, 1713-35.

[197] Orr, H.A. A test of Fisher's theory of dominance. *Proc Natl Acad Sci U S A.* **1991**, *88*, 11413-5.

[198] Charrow, J. Ashkenazi Jewish genetic disorders. *Fam Cancer.* **2004**, *3*, 201-6.

[199] Boycott, K.M.; Parboosingh, J.S.; Chodirker, B.N.; Lowry, R.B.; McLeod, D.R.; Morris, J.; Greenberg, C.R.; Chudley, A.E.; Bernier, F.P.; Midgley, J.; Moller, L.B.; Innes, A.M. Clinical genetics and the Hutterite population: a review of Mendelian disorders. *Am J Med Genet A.* **2008**, *146A*, 1088-98.

[200] Laberge, A.M.; Michaud, J.; Richter, A.; Lemyre, E.; Lambert, M.; Brais, B.; Mitchell, G.A. Population history and its impact on medical genetics in Quebec. *Clin Genet.* **2005**, *68*, 287-301.

[201] Peltonen, L.; Jalanko, A.; Varilo, T. Molecular genetics of the Finnish disease heritage. *Hum Mol Genet.* **1999**, *8*, 1913-23.

[202] Zlotogora, J.; Shalev, S.; Habiballah, H.; Barjes, S. Genetic disorders among Palestinian Arabs: 3. Autosomal recessive disorders in a single village. *Am J Med Genet.* **2000**, *92*, 343-5.

[203] Conrad, D.F.; Jakobsson, M.; Coop, G.; Wen, X.; Wall, J.D.; Rosenberg, N.A.; Pritchard, J.K. A worldwide survey of haplotype variation and linkage disequilibrium in the human genome. *Nat Genet.* **2006**, *38*, 1251-60.

[204] Costas, J.; Salas, A.; Phillips, C.; Carracedo, A. Human genome-wide screen of haplotype-like blocks of reduced diversity. *Gene.* **2005**, *349*, 219-25.

[205] Altshuler, D.; Brooks, L.D.; Chakravarti, A.; Collins, F.S.; Daly, M.J.; Donnelly, P. A haplotype map of the human genome. *Nature.* **2005**, *437*, 1299-320.

[206] Hassold, T.; Chen, N.; Funkhouser, J.; Jooss, T.; Manuel, B.; Matsuura, J.; Matsuyama, A.; Wilson, C.; Yamane, J.A.; Jacobs, P.A. A cytogenetic study of 1000 spontaneous abortions. *Ann Hum Genet.* **1980**, *44*, 151-78.

[207] Philipp, T.; Philipp, K.; Reiner, A.; Beer, F.; Kalousek, D.K. Embryoscopic and cytogenetic analysis of 233 missed abortions: factors involved in the pathogenesis of developmental defects of early failed pregnancies. *Hum Reprod.* **2003**, *18*, 1724-32.

[208] Schaeffer, A.J.; Chung, J.; Heretis, K.; Wong, A.; Ledbetter, D.H.; Lese Martin, C. Comparative genomic hybridization-array analysis enhances the detection of aneuploidies

and submicroscopic imbalances in spontaneous miscarriages. *Am J Hum Genet.* **2004**, *74*, 1168-74.

[209] Kosaki, K.; Kosaki, R.; Suzuki, T.; Yoshihashi, H.; Takahashi, T.; Sasaki, K.; Tomita, M.; McGinnis, W.; Matsuo, N. Complete mutation analysis panel of the 39 human HOX genes. *Teratology.* **2002**, *65*, 50-62.

[210] Bosley, T.M.; Alorainy, I.A.; Salih, M.A.; Aldhalaan, H.M.; Abu-Amero, K.K.; Oystreck, D.T.; Tischfield, M.A.; Engle, E.C.; Erickson, R.P. The clinical spectrum of homozygous HOXA1 mutations. *Am J Med Genet A.* **2008**, *146A*, 1235-40.

[211] Alasti, F.; Sadeghi, A.; Sanati, M.H.; Farhadi, M.; Stollar, E.; Somers, T.; Van Camp, G. A mutation in HOXA2 is responsible for autosomal-recessive microtia in an Iranian family. *Am J Hum Genet.* **2008**, *82*, 982-91.

[212] Horvat-Switzer, R.D.; Thompson, A.A. HOXA11 mutation in amegakaryocytic thrombocytopenia with radio-ulnar synostosis syndrome inhibits megakaryocytic differentiation *in vitro*. *Blood Cells Mol Dis.* **2006**, *37*, 55-63.

[213] Goodman, F.R.; Bacchelli, C.; Brady, A.F.; Brueton, L.A.; Fryns, J.P.; Mortlock, D.P.; Innis, J.W.; Holmes, L.B.; Donnenfeld, A.E.; Feingold, M.; Beemer, F.A.; Hennekam, R.C.; Scambler, P.J. Novel HOXA13 mutations and the phenotypic spectrum of hand-foot-genital syndrome. *Am J Hum Genet.* **2000**, *67*, 197-202.

[214] Mortlock, D.P.; Innis, J.W. Mutation of HOXA13 in hand-foot-genital syndrome. *Nat Genet.* **1997**, *15*, 179-80.

[215] Dobbs, M.B.; Gurnett, C.A.; Pierce, B.; Exner, G.U.; Robarge, J.; Morcuende, J.A.; Cole, W.G.; Templeton, P.A.; Foster, B.; Bowcock, A.M. HOXD10 M319K mutation in a family with isolated congenital vertical talus. *J Orthop Res.* **2006**, *24*, 448-53.

[216] Shrimpton, A.E.; Levinsohn, E.M.; Yozawitz, J.M.; Packard, D.S., Jr.; Cady, R.B.; Middleton, F.A.; Persico, A.M.; Hootnick, D.R. A HOX gene mutation in a family with isolated congenital vertical talus and Charcot-Marie-Tooth disease. *Am J Hum Genet.* **2004**, *75*, 92-6.

[217] Muragaki, Y.; Mundlos, S.; Upton, J.; Olsen, B.R. Altered growth and branching patterns in synpolydactyly caused by mutations in HOXD13. *Science.* **1996**, *272*, 548-51.

[218] Gunderson, K.L.; Kuhn, K.M.; Steemers, F.J.; Ng, P.; Murray, S.S.; Shen, R. Whole-genome genotyping of haplotype tag single nucleotide polymorphisms. *Pharmacogenomics.* **2006**, *7*, 641-8.

[219] Sellick, G.S.; Longman, C.; Tolmie, J.; Newbury-Ecob, R.; Geenhalgh, L.; Hughes, S.; Whiteford, M.; Garrett, C.; Houlston, R.S. Genomewide linkage searches for Mendelian disease loci can be efficiently conducted using high-density SNP genotyping arrays. *Nucleic Acids Res.* **2004**, *32*, e164.

[220] Miyazawa, H.; Kato, M.; Awata, T.; Kohda, M.; Iwasa, H.; Koyama, N.; Tanaka, T.; Huqun; Kyo, S.; Okazaki, Y.; Hagiwara, K. Homozygosity haplotype allows a genomewide search for the autosomal segments shared among patients. *Am J Hum Genet.* **2007**, *80*, 1090-102.

[221] Lander, E.S.; Botstein, D. Homozygosity mapping: a way to map human recessive traits with the DNA of inbred children. *Science.* **1987**, *236*, 1567-70.

[222] Bentley, D.R. Whole-genome re-sequencing. *Curr Opin Genet Dev.* **2006**, *16*, 545-52.

[223] Levy, S.; Sutton, G.; Ng, P.C.; Feuk, L.; Halpern, A.L.; Walenz, B.P.; Axelrod, N.; Huang, J.; Kirkness, E.F.; Denisov, G.; Lin, Y.; MacDonald, J.R.; Pang, A.W.; Shago, M.; Stockwell, T.B.; Tsiamouri, A.; Bafna, V.; Bansal, V.; Kravitz, S.A.; Busam, D.A.; Beeson,

K.Y.; McIntosh, T.C.; Remington, K.A.; Abril, J.F.; Gill, J.; Borman, J.; Rogers, Y.H.; Frazier, M.E.; Scherer, S.W.; Strausberg, R.L.; Venter, J.C. The diploid genome sequence of an individual human. *PLoS Biol.* **2007**, *5*, e254.

[224] Kim, J.I.; Ju, Y.S.; Park, H.; Kim, S.; Lee, S.; Yi, J.H.; Mudge, J.; Miller, N.A.; Hong, D.; Bell, C.J.; Kim, H.S.; Chung, I.S.; Lee, W.C.; Lee, J.S.; Seo, S.H.; Yun, J.Y.; Woo, H.N.; Lee, H.; Suh, D.; Lee, S.; Kim, H.J.; Yavartanoo, M.; Kwak, M.; Zheng, Y.; Lee, M.K.; Park, H.; Kim, J.Y.; Gokcumen, O.; Mills, R.E.; Zaranek, A.W.; Thakuria, J.; Wu, X.; Kim, R.W.; Huntley, J.J.; Luo, S.; Schroth, G.P.; Wu, T.D.; Kim, H.; Yang, K.S.; Park, W.Y.; Kim, H.; Church, G.M.; Lee, C.; Kingsmore, S.F.; Seo, J.S. A highly annotated whole-genome sequence of a Korean individual. *Nature.* **2009**, *460*, 1011-5.

[225] McKernan, K.J.; Peckham, H.E.; Costa, G.L.; McLaughlin, S.F.; Fu, Y.; Tsung, E.F.; Clouser, C.R.; Duncan, C.; Ichikawa, J.K.; Lee, C.C.; Zhang, Z.; Ranade, S.S.; Dimalanta, E.T.; Hyland, F.C.; Sokolsky, T.D.; Zhang, L.; Sheridan, A.; Fu, H.; Hendrickson, C.L.; Li, B.; Kotler, L.; Stuart, J.R.; Malek, J.A.; Manning, J.M.; Antipova, A.A.; Perez, D.S.; Moore, M.P.; Hayashibara, K.C.; Lyons, M.R.; Beaudoin, R.E.; Coleman, B.E.; Laptewicz, M.W.; Sannicandro, A.E.; Rhodes, M.D.; Gottimukkala, R.K.; Yang, S.; Bafna, V.; Bashir, A.; MacBride, A.; Alkan, C.; Kidd, J.M.; Eichler, E.E.; Reese, M.G.; De La Vega, F.M.; Blanchard, A.P. Sequence and structural variation in a human genome uncovered by short-read, massively parallel ligation sequencing using two-base encoding. *Genome Res.* **2009**, *19*, 1527-41.

[226] Wang, J.; Wang, W.; Li, R.; Li, Y.; Tian, G.; Goodman, L.; Fan, W.; Zhang, J.; Li, J.; Zhang, J.; Guo, Y.; Feng, B.; Li, H.; Lu, Y.; Fang, X.; Liang, H.; Du, Z.; Li, D.; Zhao, Y.; Hu, Y.; Yang, Z.; Zheng, H.; Hellmann, I.; Inouye, M.; Pool, J.; Yi, X.; Zhao, J.; Duan, J.; Zhou, Y.; Qin, J.; Ma, L.; Li, G.; Yang, Z.; Zhang, G.; Yang, B.; Yu, C.; Liang, F.; Li, W.; Li, S.; Li, D.; Ni, P.; Ruan, J.; Li, Q.; Zhu, H.; Liu, D.; Lu, Z.; Li, N.; Guo, G.; Zhang, J.; Ye, J.; Fang, L.; Hao, Q.; Chen, Q.; Liang, Y.; Su, Y.; San, A.; Ping, C.; Yang, S.; Chen, F.; Li, L.; Zhou, K.; Zheng, H.; Ren, Y.; Yang, L.; Gao, Y.; Yang, G.; Li, Z.; Feng, X.; Kristiansen, K.; Wong, G.K.; Nielsen, R.; Durbin, R.; Bolund, L.; Zhang, X.; Li, S.; Yang, H.; Wang, J. The diploid genome sequence of an Asian individual. *Nature.* **2008**, *456*, 60-5.

[227] Wheeler, D.A.; Srinivasan, M.; Egholm, M.; Shen, Y.; Chen, L.; McGuire, A.; He, W.; Chen, Y.J.; Makhijani, V.; Roth, G.T.; Gomes, X.; Tartaro, K.; Niazi, F.; Turcotte, C.L.; Irzyk, G.P.; Lupski, J.R.; Chinault, C.; Song, X.Z.; Liu, Y.; Yuan, Y.; Nazareth, L.; Qin, X.; Muzny, D.M.; Margulies, M.; Weinstock, G.M.; Gibbs, R.A.; Rothberg, J.M. The complete genome of an individual by massively parallel DNA sequencing. *Nature.* **2008**, *452*, 872-6.

[228] Bentley, D.R.; Balasubramanian, S.; Swerdlow, H.P.; Smith, G.P.; Milton, J.; Brown, C.G.; Hall, K.P.; Evers, D.J.; Barnes, C.L.; Bignell, H.R.; Boutell, J.M.; Bryant, J.; Carter, R.J.; Keira Cheetham, R.; Cox, A.J.; Ellis, D.J.; Flatbush, M.R.; Gormley, N.A.; Humphray, S.J.; Irving, L.J.; Karbelashvili, M.S.; Kirk, S.M.; Li, H.; Liu, X.; Maisinger, K.S.; Murray, L.J.; Obradovic, B.; Ost, T.; Parkinson, M.L.; Pratt, M.R.; Rasolonjatovo, I.M.; Reed, M.T.; Rigatti, R.; Rodighiero, C.; Ross, M.T.; Sabot, A.; Sankar, S.V.; Scally, A.; Schroth, G.P.; Smith, M.E.; Smith, V.P.; Spiridou, A.; Torrance, P.E.; Tzonev, S.S.; Vermaas, E.H.; Walter, K.; Wu, X.; Zhang, L.; Alam, M.D.; Anastasi, C.; Aniebo, I.C.; Bailey, D.M.; Bancarz, I.R.; Banerjee, S.; Barbour, S.G.; Baybayan, P.A.; Benoit, V.A.; Benson, K.F.; Bevis, C.; Black, P.J.; Boodhun, A.; Brennan, J.S.; Bridgham, J.A.; Brown, R.C.; Brown, A.A.; Buermann, D.H.; Bundu, A.A.; Burrows, J.C.; Carter, N.P.; Castillo, N.; Chiara,

E.C.M.; Chang, S.; Neil Cooley, R.; Crake, N.R.; Dada, O.O.; Diakoumakos, K.D.; Dominguez-Fernandez, B.; Earnshaw, D.J.; Egbujor, U.C.; Elmore, D.W.; Etchin, S.S.; Ewan, M.R.; Fedurco, M.; Fraser, L.J.; Fuentes Fajardo, K.V.; Scott Furey, W.; George, D.; Gietzen, K.J.; Goddard, C.P.; Golda, G.S.; Granieri, P.A.; Green, D.E.; Gustafson, D.L.; Hansen, N.F.; Harnish, K.; Haudenschild, C.D.; Heyer, N.I.; Hims, M.M.; Ho, J.T.; Horgan, A.M.; Hoschler, K.; Hurwitz, S.; Ivanov, D.V.; Johnson, M.Q.; James, T.; Huw Jones, T.A.; Kang, G.D.; Kerelska, T.H.; Kersey, A.D.; Khrebtukova, I.; Kindwall, A.P.; Kingsbury, Z.; Kokko-Gonzales, P.I.; Kumar, A.; Laurent, M.A.; Lawley, C.T.; Lee, S.E.; Lee, X.; Liao, A.K.; Loch, J.A.; Lok, M.; Luo, S.; Mammen, R.M.; Martin, J.W.; McCauley, P.G.; McNitt, P.; Mehta, P.; Moon, K.W.; Mullens, J.W.; Newington, T.; Ning, Z.; Ling Ng, B.; Novo, S.M.; O'Neill, M.J.; Osborne, M.A.; Osnowski, A.; Ostadan, O.; Paraschos, L.L.; Pickering, L.; Pike, A.C.; Pike, A.C.; Chris Pinkard, D.; Pliskin, D.P.; Podhasky, J.; Quijano, V.J.; Raczy, C.; Rae, V.H.; Rawlings, S.R.; Chiva Rodriguez, A.; Roe, P.M.; Rogers, J.; Rogert Bacigalupo, M.C.; Romanov, N.; Romieu, A.; Roth, R.K.; Rourke, N.J.; Ruediger, S.T.; Rusman, E.; Sanches-Kuiper, R.M.; Schenker, M.R.; Seoane, J.M.; Shaw, R.J.; Shiver, M.K.; Short, S.W.; Sizto, N.L.; Sluis, J.P.; Smith, M.A.; Ernest Sohna Sohna, J.; Spence, E.J.; Stevens, K.; Sutton, N.; Szajkowski, L.; Tregidgo, C.L.; Turcatti, G.; Vandevondele, S.; Verhovsky, Y.; Virk, S.M.; Wakelin, S.; Walcott, G.C.; Wang, J.; Worsley, G.J.; Yan, J.; Yau, L.; Zuerlein, M.; Rogers, J.; Mullikin, J.C.; Hurles, M.E.; McCooke, N.J.; West, J.S.; Oaks, F.L.; Lundberg, P.L.; Klenerman, D.; Durbin, R.; Smith, A.J. Accurate whole human genome sequencing using reversible terminator chemistry. *Nature.* **2008**, *456*, 53-9.

[229] Ahn, S.M.; Kim, T.H.; Lee, S.; Kim, D.; Ghang, H.; Kim, D.S.; Kim, B.C.; Kim, S.Y.; Kim, W.Y.; Kim, C.; Park, D.; Lee, Y.S.; Kim, S.; Reja, R.; Jho, S.; Kim, C.G.; Cha, J.Y.; Kim, K.H.; Lee, B.; Bhak, J.; Kim, S.J. The first Korean genome sequence and analysis: full genome sequencing for a socio-ethnic group. *Genome Res.* **2009**, *19*, 1622-9.

[230] Bau, S.; Schracke, N.; Kranzle, M.; Wu, H.; Stahler, P.F.; Hoheisel, J.D.; Beier, M.; Summerer, D. Targeted next-generation sequencing by specific capture of multiple genomic loci using low-volume microfluidic DNA arrays. *Anal Bioanal Chem.* **2009**, *393*, 171-5.

[231] Porreca, G.J.; Zhang, K.; Li, J.B.; Xie, B.; Austin, D.; Vassallo, S.L.; LeProust, E.M.; Peck, B.J.; Emig, C.J.; Dahl, F.; Gao, Y.; Church, G.M.; Shendure, J. Multiplex amplification of large sets of human exons. *Nat Methods.* **2007**, *4*, 931-6.

[232] Hodges, E.; Rooks, M.; Xuan, Z.; Bhattacharjee, A.; Benjamin Gordon, D.; Brizuela, L.; Richard McCombie, W.; Hannon, G.J. Hybrid selection of discrete genomic intervals on custom-designed microarrays for massively parallel sequencing. *Nat Protoc.* **2009**, *4*, 960-74.

[233] Hodges, E.; Xuan, Z.; Balija, V.; Kramer, M.; Molla, M.N.; Smith, S.W.; Middle, C.M.; Rodesch, M.J.; Albert, T.J.; Hannon, G.J.; McCombie, W.R. Genome-wide *in situ* exon capture for selective resequencing. *Nat Genet.* **2007**, *39*, 1522-7.

[234] Via, M.; Gignoux, C.; Burchard, E.G. The 1000 Genomes Project: new opportunities for research and social challenges. *Genome Med.* **2010**, *2*, 3.

[235] Siva, N. 1000 Genomes project. *Nat Biotechnol.* **2008**, *26*, 256.

[236] Kuehn, B.M. 1000 Genomes Project promises closer look at variation in human genome. *Jama.* **2008**, *300*, 2715.

[237] Vasta, V.; Ng, S.B.; Turner, E.H.; Shendure, J.; Hahn, S.H. Next generation sequence analysis for mitochondrial disorders. *Genome Med.* **2009**, *1*, 100.

[238] Hedges, D.J.; Burges, D.; Powell, E.; Almonte, C.; Huang, J.; Young, S.; Boese, B.; Schmidt, M.; Pericak-Vance, M.A.; Martin, E.; Zhang, X.; Harkins, T.T.; Zuchner, S. Exome sequencing of a multigenerational human pedigree. *PLoS One.* **2009**, *4*, e8232.

[239] Ng, S.B.; Buckingham, K.J.; Lee, C.; Bigham, A.W.; Tabor, H.K.; Dent, K.M.; Huff, C.D.; Shannon, P.T.; Jabs, E.W.; Nickerson, D.A.; Shendure, J.; Bamshad, M.J. Exome sequencing identifies the cause of a mendelian disorder. *Nat Genet.* **2009**,

[240] Ng, S.B.; Turner, E.H.; Robertson, P.D.; Flygare, S.D.; Bigham, A.W.; Lee, C.; Shaffer, T.; Wong, M.; Bhattacharjee, A.; Eichler, E.E.; Bamshad, M.; Nickerson, D.A.; Shendure, J. Targeted capture and massively parallel sequencing of 12 human exomes. *Nature.* **2009**, *461*, 272-6.

[241] Choi, M.; Scholl, U.I.; Ji, W.; Liu, T.; Tikhonova, I.R.; Zumbo, P.; Nayir, A.; Bakkaloglu, A.; Ozen, S.; Sanjad, S.; Nelson-Williams, C.; Farhi, A.; Mane, S.; Lifton, R.P. Genetic diagnosis by whole exome capture and massively parallel DNA sequencing. *Proc Natl Acad Sci U S A.* **2009**, *106*, 19096-101.

[242] Hoischen, A.; Gilissen, C.; Arts, P.; Wieskamp, N.; van der Vliet, W.; Vermeer, S.; Steehouwer, M.; de Vries, P.; Meijer, R.; Seiqueros, J.; Knoers, N.V.; Buckley, M.F.; Scheffer, H.; Veltman, J.A. Massively parallel sequencing of ataxia genes after array-based enrichment. *Hum Mutat.* **2010**, *31*, 494-9.

[243] Raca, G.; Jackson, C.; Warman, B.; Bair, T.; Schimmenti, L.A. Next generation sequencing in research and diagnostics of ocular birth defects. *Mol Genet Metab.* **2010**,

[244] Biesecker, L.G. Exome sequencing makes medical genomics a reality. *Nat Genet.* **2010**, *42*, 13-4.

[245] Lalonde, E.; Albrecht, S.; Ha, K.C.; Jacob, K.; Bolduc, N.; Polychronakos, C.; Dechelotte, P.; Majewski, J.; Jabado, N. Unexpected allelic heterogeneity and spectrum of mutations in Fowler syndrome revealed by next-generation exome sequencing. *Hum Mutat.* **2010**,

[246] Thorogood, A.; Knoppers, B.M.; Dondorp, W.J.; de Wert, G.M. Whole Genome Sequencing and the Physician. *Clin Genet.* **2012**,

[247] Knoppers, B.M. Pediatric Research and the Communication of Not-so Incidental Findings. *Pediatrics and Child Health.* **2012**, *in press*,

[248] Avard, D.; Black, L.; Samuel, J.; Griener, G.; Knoppers, B. Best Practices for Health Research Involving Children and Adolescents. *Centre of Genomics and Policy, McGill University.* **2012**, 1-160.

[249] Kuehn, B.M. NIH shifts focus from sequencing genes to fostering clinical applications. *JAMA.* **2012**, *307*, 132.

[250] Hayden, E.C. Sequencing set to alter clinical landscape. *Nature.* **2012**, *482*, 288.

[251] Gonzaga-Jauregui, C.; Lupski, J.R.; Gibbs, R.A. Human genome sequencing in health and disease. *Annu Rev Med.* **2012**, *63*, 35-61.

[252] Bick, D.; Dimmock, D. Whole exome and whole genome sequencing. *Curr Opin Pediatr.* **2011**, *23*, 594-600.

[253] Gabory, A.; Ripoche, M.A.; Yoshimizu, T.; Dandolo, L. The H19 gene: regulation and function of a non-coding RNA. *Cytogenet Genome Res.* **2006**, *113*, 188-93.

[254] Brown, C.J.; Ballabio, A.; Rupert, J.L.; Lafreniere, R.G.; Grompe, M.; Tonlorenzi, R.; Willard, H.F. A gene from the region of the human X inactivation centre is expressed exclusively from the inactive X chromosome. *Nature.* **1991**, *349*, 38-44.

[255] Leeb, M.; Steffen, P.A.; Wutz, A. X chromosome inactivation sparked by non-coding RNAs. *RNA Biol.* **2009**, *6*, 94-9.

[256] Senner, C.E.; Brockdorff, N. Xist gene regulation at the onset of X inactivation. *Curr Opin Genet Dev.* **2009**, *19*, 122-6.

[257] DiMauro, S. Lessons from mitochondrial DNA mutations. *Semin Cell Dev Biol.* **2001**, *12*, 397-405.

[258] Tuppen, H.A.; Blakely, E.L.; Turnbull, D.M.; Taylor, R.W. Mitochondrial DNA mutations and human disease. *Biochim Biophys Acta.* **2010**, *1797*, 113-28.

[259] Manolio, T.A.; Collins, F.S.; Cox, N.J.; Goldstein, D.B.; Hindorff, L.A.; Hunter, D.J.; McCarthy, M.I.; Ramos, E.M.; Cardon, L.R.; Chakravarti, A.; Cho, J.H.; Guttmacher, A.E.; Kong, A.; Kruglyak, L.; Mardis, E.; Rotimi, C.N.; Slatkin, M.; Valle, D.; Whittemore, A.S.; Boehnke, M.; Clark, A.G.; Eichler, E.E.; Gibson, G.; Haines, J.L.; Mackay, T.F.; McCarroll, S.A.; Visscher, P.M. Finding the missing heritability of complex diseases. *Nature.* **2009**, *461*, 747-53.

[260] Brinkman, R.R.; Dube, M.P.; Rouleau, G.A.; Orr, A.C.; Samuels, M.E. Human monogenic disorders - a source of novel drug targets. *Nat Rev Genet.* **2006**, *7*, 249-60.

[261] Shimkets, R.A.; Lifton, R.P. Recent advances in the molecular genetics of hypertension. *Curr Opin Nephrol Hypertens.* **1996**, *5*, 162-5.

[262] Porter, J.R.; Barrett, T.G. Monogenic syndromes of abnormal glucose homeostasis: clinical review and relevance to the understanding of the pathology of insulin resistance and beta cell failure. *J Med Genet.* **2005**, *42*, 893-902.

[263] Lifton, R.P. Genetic dissection of human blood pressure variation: common pathways from rare phenotypes. *Harvey Lect.* **2004**, *100*, 71-101.

[264] Ahituv, N.; Kavaslar, N.; Schackwitz, W.; Ustaszewska, A.; Martin, J.; Hebert, S.; Doelle, H.; Ersoy, B.; Kryukov, G.; Schmidt, S.; Yosef, N.; Ruppin, E.; Sharan, R.; Vaisse, C.; Sunyaev, S.; Dent, R.; Cohen, J.; McPherson, R.; Pennacchio, L.A. Medical sequencing at the extremes of human body mass. *Am J Hum Genet.* **2007**, *80*, 779-91.

[265] Cohen, J.C.; Kiss, R.S.; Pertsemlidis, A.; Marcel, Y.L.; McPherson, R.; Hobbs, H.H. Multiple rare alleles contribute to low plasma levels of HDL cholesterol. *Science.* **2004**, *305*, 869-72.

[266] Cohen, J.C.; Pertsemlidis, A.; Fahmi, S.; Esmail, S.; Vega, G.L.; Grundy, S.M.; Hobbs, H.H. Multiple rare variants in NPC1L1 associated with reduced sterol absorption and plasma low-density lipoprotein levels. *Proc Natl Acad Sci U S A.* **2006**, *103*, 1810-5.

[267] Romeo, S.; Pennacchio, L.A.; Fu, Y.; Boerwinkle, E.; Tybjaerg-Hansen, A.; Hobbs, H.H.; Cohen, J.C. Population-based resequencing of ANGPTL4 uncovers variations that reduce triglycerides and increase HDL. *Nat Genet.* **2007**, *39*, 513-6.

[268] Ji, W.; Foo, J.N.; O'Roak, B.J.; Zhao, H.; Larson, M.G.; Simon, D.B.; Newton-Cheh, C.; State, M.W.; Levy, D.; Lifton, R.P. Rare independent mutations in renal salt handling genes contribute to blood pressure variation. *Nat Genet.* **2008**, *40*, 592-9.

[269] Prickett, T.D.; Wei, X.; Cardenas-Navia, I.; Teer, J.K.; Lin, J.C.; Walia, V.; Gartner, J.; Jiang, J.; Cherukuri, P.F.; Molinolo, A.; Davies, M.A.; Gershenwald, J.E.; Stemke-Hale, K.; Rosenberg, S.A.; Margulies, E.H.; Samuels, Y. Exon capture analysis of G protein-coupled receptors identifies activating mutations in GRM3 in melanoma. *Nat Genet.* **2011**, *43*, 1119-26.

[270] McCandless, S.E.; Brunger, J.W.; Cassidy, S.B. The burden of genetic disease on inpatient care in a children's hospital. *Am J Hum Genet.* **2004**, *74*, 121-7.

[271] Scriver, C.R.; Neal, J.L.; Saginur, R.; Clow, A. The frequency of genetic disease and congenital malformation among patients in a pediatric hospital. *Can Med Assoc J.* **1973**, *108*, 1111-5.

[272] Lin, X.; Tang, W.; Ahmad, S.; Lu, J.; Colby, C.C.; Zhu, J.; Yu, Q. Applications of targeted gene capture and next-generation sequencing technologies in studies of human deafness and other genetic disabilities. *Hear Res.* **2012**,

[273] Scaglia, F. Nuclear gene defects in mitochondrial disorders. *Methods Mol Biol.* **2012**, *837*, 17-34.

CHAPTER 11

# Malaria Functional Genomics and Beyond

**Jianbing Mu[1,*], Karl B. Seydel[2], Adam Bates[1], Junhui Duan[3], Lubin Jiang[4] and Xin-zhuan Su[1]**

*[1]Laboratory of Malaria and Vector Research, National Institute of Allergy and Infectious Diseases, National Institutes of Health, Bethesda, MD, USA; [2]Michigan State University, East Lansing, MI, USA; [3]Laboratory of Malaria Immunology and Vaccinology, National Institute of Allergy and Infectious Diseases, National Institutes of Health, Bethesda, MD, USA; and [4]Institut Pasteur of Shanghai, Chinese Academy of Sciences, Shanghai, China*

**Abstract:** With the completion and near completion of many malaria parasite genome-sequencing projects, efforts are now being directed toward a better understanding of gene functions and discovery of vaccine and drug targets. Novel biologic pathways present in the parasite have been identified through homology searches and inter- and intra-species comparison of the parasite genome. Genetic mapping approaches, including linkage and population association analyses, have provided valuable insights into parasite evolution, virulence, drug resistance, and immune invasion. Genome-wide searches for loci under various selection pressures have lead to discovery of parasite genetic loci or genes playing a role in drug resistance or encoding for protective antigens. In addition, the *Plasmodium falciparum* genome sequence provides the basis for the development of various microarrays to monitor gene expression and to detect nucleotide substitution and deletion/amplification, for genome-wide profiling of the parasite proteome, and for investigation of epigenetic regulation of gene expression such as chromatin modification, and nucleosome position. In this brief review, we will highlight some recent advances and studies in characterizing gene function and related phenotype in *P. falciparum* that were made possible by the genome sequence, particularly the development of a genome-wide diversity map and various high-throughput genotyping methods for genome-wide association studies.

**Keywords:** Malaria, microarray, genome diversity, SNP, recombination, comparative genomics.

## MALARIA PARASITES AND GENOMES

Malaria is one of the most important tropical parasitic diseases in humans, causing great morbidity and mortality in many developing countries. Approximately 300–

***Address correspondence to Jianbing Mu:** LMVR, NIAID, NIH, 12735 Twinbrook Parkway, Room 3E-32C, Rockville, MD 20852, USA; Tel: (301) 451-5996; Fax: (301) 402-2201; E-mail: jmu@niaid.nih.gov

**Christian Neri (Ed)**

500 million clinical cases and nearly million deaths are reported each year [1]. Human malaria is caused by five species of the *Plasmodium* parasites, namely *Plasmodium falciparum, Plasmodium vivax, Plasmodium ovale, Plasmodium malariae*, and *Plasmodium knowlesi* [2]. Among these, *P. falciparum* causes the most severe forms of the disease. The life cycle of the malaria parasite involves multiple tissues and different stages (sexual and asexual) in two distinct hosts: mosquitoes and humans. Within its complicated life cycle, the parasite has a diploid genome for a short period of time during its development in the mosquito vector and has a haploid genome throughout the majority of its life cycle. The haploid parasite is amenable to the application of many genetic and genomic tools that play an important role in functional genomics research.

*P. falciparum* has 14 chromosomes containing ~23 million base-pair nucleotides with high AT content (~82%) and is predicted to have approximately 5,500 genes [3-5]. The number of protein-coding genes in *P. falciparum* is comparable to those in free-living yeasts, but the latter organism has a considerably smaller genome in comparison to *P. falciparum*. In addition to differences in coding capacity, the *P. falciparum* genome also has a greater number of hypothetical proteins (~60%) with limited homology to genes with known functions; the functions of these proteins are therefore unknown. Additionally, a large number of the current gene models in the *P. falciparum* genome database may contain errors [6,7], although recent efforts in cDNA sequencing have greatly improved the accuracy of genome annotation [8-11].

Various genomic approaches have been applied to define possible gene functions since completion of the *P. falciparum* genome-sequencing project in 2002 [5] (Fig. **1**), and significant progress has been made. Here we briefly review some of these developments, focusing on progress in genetic mapping using high-throughput genotyping.

## GENOME DIVERSITY AND GENETIC MAPPING

Genetic diversity is considered to contribute to the majority of phenotypic differences; therefore the function of a gene can be inferred either from linkage or association of genetic polymorphisms to differences in phenotypes [12,13].

Different genetic markers, such as restriction fragment-length polymorphism (RFLP), microsatellite (MS), and single nucleotide polymorphism (SNP), have been developed to measure the genetic diversity of the *P. falciparum* genome [14-16]. Three *P. falciparum* genetic crosses (3D7×HB3, HB3×Dd2 and 7G8×GB4) have been carried out to date. These genetic crosses have been successfully applied to identify genes in *P. falciparum* involved in drug resistance, such as *pfcrt* in chloroquine (CQ) resistance [17-19], *pfdhfr* in sulfadoxine-pyrimethamine (SP) resistance [20], and *pfrh5* in determination of the species-specific pathway of *P. falciparum* invasion [21]; More recently, the HB3×Dd2 genetic crosses was successfully employed to map and identify a gene family encoding CLAG3 proteins that may play a role in transport of nutrients into the *P. falciparum* parasites [22]. Analyses of progeny from HB3×Dd2 and/or 7G8×GB4 crosses also mapped parasite responses to a large number of compounds to multiple genetic loci and genes [23,24]. These and other analyses have redefined the roles of the *pfcrt* and *pfmdr1* genes in drug response of parasites [25,26]. An attempt has also been made to locate the genetic loci that contribute to reduced artemisinin susceptibility in *P. falciparum* using the progeny of the HB3 × Dd2 cross [27]. Additionally, many genetic crosses have been performed using the rodent malaria parasites, including *Plasmodium yoelii* [28-30] and *Plasmodium chabaudi* [31-33]. A number of excellent reviews and protocols for these genetic crosses can be found elsewhere [34-36]. Although dramatic success in studying parasite gene functions has been made using genetic crosses, the cost and intensive labor required in performing a genetic cross and cloning parasite progeny have limited its application for larger-scale functional analysis in human malaria parasites.

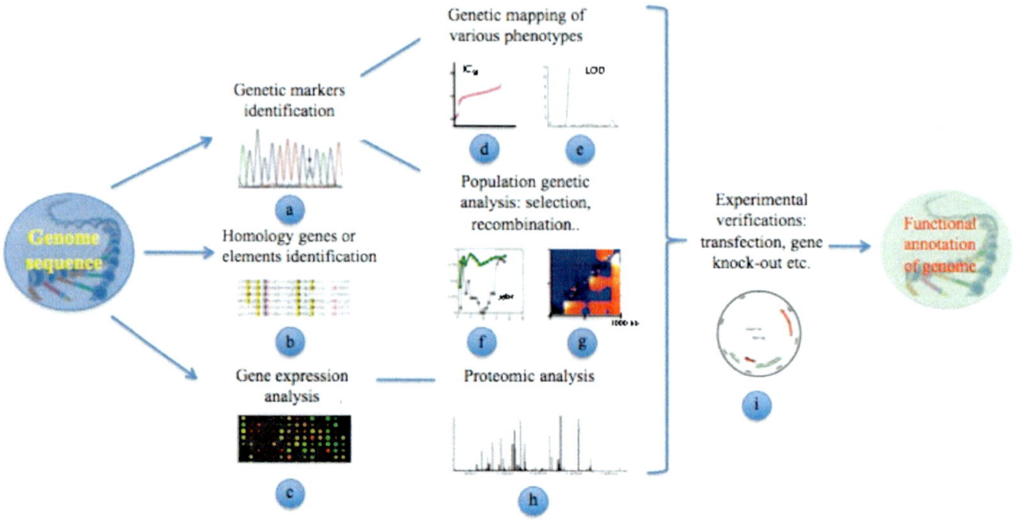


**Fig. (1).** Genetic mapping, comparative genomic analysis, and combination of transcriptomic, epigenetic and proteomic approaches can play important roles in understanding gene functions in *Plasmodium falciparum*. **a**. Sequence analysis to identify genetic polymorphisms. **b**. Sequence comparison to search homolog genes or elements. **c**, Microarray chips to evaluate gene expression at mRNA level. **d**, **e**. Analysis of different phenotype and genotype data to locate the candidate genes/loci associated with drug resistance and other traits. **f**, **g**. Population genetic analysis to detect genetic loci under selection or with elevated recombination frequency. **h**. Protein expression analysis and association with developmental stages. **i**. Predicated functions of candidate genes can be studied using genetic knock-out and other methods.

With multiple technologic advances, particularly development of high-throughput genotyping, since the publication of the *P. falciparum* genome, genetic markers used for mapping purposes in *P. falciparum* have shifted from RFLP to MS and then to SNPs. A project to systematically identify SNPs in the *P. falciparum* genome was initiated approximately a decade ago [37,38]. By resequencing approximately 20% of the genome from four parasites (HB3, Dd2, 7G8, and D10) that originated from different geographic locations, ~4,000 SNPs were identified after alignment of the sequences with that of 3D7 that was available in the public databases. With the joining of two groups from major sequencing centers at the Broad Institute and the Wellcome Trust Sanger Institute, more parasite genomes have been re-sequenced, and a much larger data set is now available [39,40]. Currently, approximately 180,000 SNPs have been identified from 18 full or partially sequenced *P. falciparum* strains (http://www.ncbi.nlm.nih.gov/projects/SNP/), although some of the SNPs could be errors from sequence alignments that required further verification. Additional strains from global populations are being sequenced using parallel sequencing, which will provide information for better understanding parasite genome diversity, population structure, and gene functions [41].

In addition to the re-sequencing approach, high-density tiling arrays have also been developed to study gene expression and genome diversity in malaria parasites [42-46]. Nucleotide change—including nucleotide substitution, deletion, and insertion—results in a reproducible loss or reduction of hybridization signal and therefore allow identification of the region with genetic changes. Several studies have used microarrays to detect nucleotide substitution and copy number variation (CNV) [42,43,47,48]. Approximately 20,000 single–feature polymorphisms (SFPs) from 14 field isolates and laboratory lines were identified using an Affymetrix array containing 298,782 oligonucleotide probes [42]. A similar study using a higher-density array (PFSANGER array) containing 2.5 million probes also detected more than 40,000 SFPs from five *P. falciparum* isolates [43]. Moreover, high-throughput SNP typing arrays have been developed for studying parasite population and genetic mapping. Several different platforms are currently available: one utilizes a standard Affymetrix hybridization array [42], another is based on molecular inversion probe (MIP) technology [49], and a third used NembleGen array platform [50]. Both the standard Affymetrix hybridization array and MIP chips can interrogate ~3000 SNPs initially. The MIP array has been upgraded to contain ~8000 SNPs and is currently available for public uses at MR4 (http://www.mr4.org/). The Affymetrix standard hybridization chip has been applied to evaluate linkage disequilibrium (LD) and natural selection near the *pfcrt* loci on chromosome 7 in parasite populations from different continents [47], and the MIP array has been used to genotype parasite isolates that have different phenotypic variation. The genotypic data obtained from the MIP array was applied to scan the parasite genome for population recombination events, recent positive selection signatures and association of genetic loci with multiple drug-resistant phenotypes in *P. falciparum* [49]. Undoubtedly, microarrays will play an important role in studying parasite genomes and in genetic mapping in the near future.

Significant insights have been gained in the population structure of malaria parasites as well as mapping candidate genes using currently available SNPs. Population structure or admixture can potentially lead to both false-positive and false-negative results in association studies; thus, structure analysis should be investigated prior to an association study. Analysis of SNPs from chromosome 3 among 99 globally collected parasites showed that malaria parasites could be clustered into different major groups according to their geographic origins independent of time of collection [51]. Similar results were also obtained in a study using the Affymetrix hybridization SNP array, which showed that continental boundaries between parasite populations gave rise to most population structure [52]; however, caution should be exercised in interpreting population structure results when using markers that are likely under selection. For example, SNPs from 49 transporter genes in *P. falciparum,* many of which were likely under drug selection, clustered African parasites into two groups according to parasite response to CQ [51], whereas SNPs in the gene encoding apical membrane antigen 1 (*pfama-1*), a target of host immunity, grouped parasites into six populations that were independent of geographic origin [53]. Therefore, results of population structure analysis will be influenced by the type and nature of genetic marker used.

Recombination can play an important role in shaping the parasite genome. For example, recombination changes the size of the LD region (or haplotype block) in the genome and generates new parasite variants that may evade host immunity. Locating recombination hotspots or coldspots in the genome could therefore provide insight into genome evolution and parasite transmission dynamics. Studies of genetic recombination in single chromosome [51] as well as whole genome [49] among different *P. falciparum* populations showed that recombination hotspots were located largely at the ends of the chromosome that contains many multifamily genes such as *var, rifin,* and *stevor*. Many of these recombination hot or cold spots appeared to be conserved among parasite populations, although the population recombination rate varied greatly, ranging from ~400/Mb in American parasites to over $10^5$/Mb in African parasites. Variation in recombination rate in different populations has been shown to affect the size of LD and haplotype blocks in the genome and therefore should be considered in the design of any association studies [51]. Recent studies on genetic recombination in the progeny of genetic cross using either high-throughput 454 re-sequence [54] or a high-density tilling array [48], revealed high recombination rates and hotspots. Motifs enriched in recombination hotspots and the potential motif binding proteins were also identified [48], although further investigation is required to validate the discovery.

Another important source of information on parasite-host interaction and parasite genome evolution is determining genetic loci that are under recent positive selection. As shown in the human genome, positively selected genes can be classified into groups according to the broad biologic processes of gene functions, such as gametogenesis, spermatogenesis, fertilization, metabolism of carbohydrates, lipids, and phosphates, and vitamin transport [55-57]. For malaria parasites, systematic evaluations of selection have been performed at some candidate genetic loci associated with drug response, such as *pfcrt* [58] and *pfdhfr* [59], as well as at the whole genome level [49]. As expected, genes that confer drug resistance are under strong positive selection, although the strength of selection may vary among different geographic samples. For example, a ~200-kb region containing *pfcrt* was shown to be under positive selection in a population isolated from Africa and Southeast Asia [58], whereas only ~70 kb was under such selection in a Laotian parasite population [59]. Interestingly, several parasite molecules, including ABC transporters, iron transporter, a member of SURFIN and some conserved *Plasmodium* proteins, were found to be under significant positive selection by genome-wide scan [49]. In addition to drug selection, host immunity is also a strong force in shaping the parasite genome, such as generation of polymorphisms in antigenic gene families (diversifying selection). Therefore, screening of highly polymorphic genes in the *P. falciparum* genome may lead to discovery of novel vaccine candidate genes [60]. Studies have showed that a large number of genes are under balancing selection in the *P. falciparum* genome [61], although polymorphism in these candidates may also pose some challenges for vaccine development.

With the availability of the high-density SNP markers in *P. falciparum*, genome-wide association studies (GWAS) have been carried out for identifying genetic variants associated with parasite phenotypes of interest. Using the 3K MIP array,

189 cloned parasites collected from different geographic location were genotyped, and parasites responses to seven different antimalarial drugs or large-scale chemical compounds *in vitro* were measured. GWAS showed that many novel candidate genes were associated with parasite response to many antimalarial drugs [49] or chemical compounds [23]. In a similar study, a gene with unknown function was associated to *P. falciparum* response to halofantrine, mefloquine, and lumefantrine, which was validated using genetic transformation of the parasite [61]. More recently, a locus on chromosome 13 was associated with delayed clearance of parasites in artemisinin treated patients [62]. The major challenge for these studies is how to validate and determine the biological functions of the candidate genes.

In addition to SNP markers, CNV can also be informative in characterizing gene functions. Copy number changes have been linked to various diseases and biologic processes and can contribute to phenotypic variation in many organisms [63,64]. In *P. falciparum*, recent studies have shown that amplification of *pfmdr1*, *pfgch1*, and *pfdxr* may be important for parasite resistance to mefloquine, antifolate drug, and fosmidomycin, respectively [42,47,65,66].

Genome-wide analyses of genetic diversity in *P. falciparum* have also led to identification of several candidate genes or loci for novel vaccine and drug targets [49,61]. It is expected that many more such loci will be discovered in the near future with the increasing availability of phenotypic and genomic data. In particular, differences in parasite responses to larger number of chemical compounds can be used as phenotypes to identify the parasite targets of the chemical compounds and to infer gene functions [23,24]. Indeed, combination of high throughput chemical screening and GWAS or other genome-wide approaches will expose many opportunities for developing new antimalarial drugs and for studying gene functions [67-71].

## COMPARATIVE GENOMICS AND HOMOLOGOUS GENE SEARCHES

Along with the dramatic efforts being put forth to search for genomic diversity, comparative analysis of the parasite genome sequences to discover homologous genes plays an important role in elucidating the functions of many predicted proteins. Searches based on biased G/C content and RNA folding potential have led to identification of a large number of noncoding RNA (ncRNA), including splicing RNA, small nucleolar RNA (snoRNA), and telomerase RNA in *P. falciparum* [72,73]. In addition to ncRNA with known functions, several candidate genes appear to be specific to *Plasmodium* spp. and lie adjacent to members of the *var* gene family, possibly contributing to control of allelic expression of this multigene family [74,75]. Alternatively, these genes could be acting as recombination hotspots, generating diversity that can contribute to immune evasion. Although no parasite-encoded microRNA (miRNA) genes have been found in *P. falciparum* to date and the function of miRNA-mediated control on gene expression in malaria parasites remains controversial, analysis of potential RNA folding using RNAmicro [76] revealed five novel ncRNA that might act as precursors for miRNA [73]. Further studies are needed to illustrate the role of these ncRNA in gene regulation.

Another example of the use of homology search to characterize gene functions is that of the erythrocyte binding-like (EBL) and reticulocyte-binding-like protein (RBL) gene families that are involved in parasite invasion of erythrocytes. Both of the gene families consist of multiple members located on different chromosomes of *P. falciparum* or other *Plasmodium* spp. Characterization of these families has been greatly enhanced by use of homology searches among different *Plasmodium* spp. A detailed summary of the functional characteristics of these genes can be found in excellent reviews elsewhere [77-79]. More recently, a gene family of intramembrane serine proteases encoded by eight different orthologous genes was discovered in the *P. falciparum* genome [80]. This gene family, termed rhomboid-like proteins (ROMs), is likely involved in host-parasite interaction and is present in the genomes of all Apicomplexan parasites whose genomes have been sequenced. Other families of essential proteases, including those implicated in parasite egress from the erythrocyte such as falcipain-2, plasmepsin II, and a family of putative papain-like proteases termed SERA, have been identified using homology search [81,82]. These proteases may provide key targets for development of new chemotherapeutic treatment strategies.

Comparative genetics have also been important in locating genetic regulatory elements, particularly the apicomplexan AP2 (ApiAP2), in the *P. falciparum* genome [83,84]. These discoveries have drastically changed the landscape of transcriptional regulation in *P. falciparum*. More than 20 ApiAP2 genes have been identified on different chromosomes, all of which were previously annotated as hypothetical proteins [4,85]. Detailed investigation of different members of this gene family suggests an essential role in regulating parasite development [84,86-88]. These data—combined with large catalogs of potential *cis*-acting sequences obtained from *in silico* discovery of transcription regulatory elements [89] —have enhanced our understanding of the role of transcriptional regulation in *P. falciparum*.

Comprehensive comparative genomics have opened the door for other newly emerging fields in *P. falciparum*, such as secretome and epigenome research. Through close examination of known exported proteins in *P. falciparum*, conserved motifs termed *Plasmodium* export element (PEXEL) and vacuolar transport signal (VTS) have been identified from the parasite genome [90,91]. These elements are necessary for export of hundreds of proteins from the parasites that serve to remodel the host erythrocyte [90-94]. The roles of these unique modifications on the infected erythrocyte have already been examined using genetic knockout and functional screens [92]. Molecules with secretion signals can be evaluated for antivirulence targets, as some of these genes are important in knob formation and/or involved in the increased rigidity of the infected erythrocytes [92]. A central portal through which most or all of these exported proteins are transported though the erythrocyte has been recently characterized [95]. Further investigations into the functions of these exported proteins are needed to gain additional insight into the *P. falciparum* secretome.

## VARIATION IN GENE EXPRESSION AT mRNA AND PROTEIN LEVELS

Gene functions can also be predicted by monitoring the dynamic of mRNA or protein expression in combination with related phenotypes or developmental stages. Prior to availability of the whole genome sequence, methods for measuring gene expression level in *P. falciparum* were mostly based on gene-by-gene methods such as Northern blot, reverse transcriptase-PCR, and complementary DNA (cDNA) libraries. Although these methods have contributed significantly to our understanding of gene expression and function, DNA microarray—which utilizes the genomic sequence for designing oligonucleotide probes—becomes the favored platform for studying gene expression in malaria parasites due to its higher resolution, reproducibility, and coverage. In *P. falciparum*, microarray transcriptome analysis identified clusters of genes with similar expression patterns that were differentially regulated across the life cycle [96]. Further analysis using an improved clustering approach called ontology-based pattern identification (OPI) in combination with evidence-based annotation revealed 320 gene clusters representing various biologic processes, leading to functional predictions for hundreds of previously uncharacterized malaria genes [97-99]. Comprehensive comparison of *in vitro* [89,96,100,101] and *in vivo* [102] gene expression patterns among different parasite isolates as well as expression level polymorphisms (ELPs) in a genetic cross [103,104] have allowed identification of hundreds of transcription regulatory elements and regulatory hotspots. Interestingly, although both *in vitro* and *in vivo* experiments demonstrated that gene transcription in *P. falciparum* parasites is rigidly programmed throughout the erythrocytic cycle, expression profiles showed dramatic differences for parasites grown in these two different environments [102]. At least three distinct physiologic states, which related to glycolytic growth, starvation response, and a general stress response, were found in *P. falciparum* parasites isolated directly from patients; only one state could match the *in vitro* parasite life stage [102]. A recent study suggested, however, that the "hidden" state of expression might be, in fact, transcripts from gametocytes [105].

The determinants of these transcriptional regulations remain elusive, although increasing identification of genetic regulatory elements and expression quantitative loci (eQTLs) have narrowed down the genetic regions for further investigations. Comparison of the gene expression profile of genetically modified parasites, such as drug-selected [106-108] or gene knock-out parasites [109], with their parental wild-type parasites will allow identification of genes that interact with those functionally modified genes.

The mechanism of gene expression variation has been linked not only to DNA sequence alterations but also to epigenetic modifications and other mechanism in *P. falciparum* [110-114]. Identification of the genes involved in epigenetic control in the *P. falciparum* genome, such as histone acetylation, methylation, and phosphorylation [11,88,115-121], has brought a new level to our understanding of parasite biology and the discovery of new drug targets. The most extensively studied gene family is the *var* gene family, which encodes hypervariable surface antigens and displays mutually exclusive expression in infected red blood cells [122,123]. Switching of *var* gene expression states from active to silent or *vice versa* may be associated with chromatin modifications [111,124], locations of active genes in the nucleus [125,126], and presence of regulatory introns [127,128]. Histone acetylation has been associated with *var* gene activation [125,126], whereas trimethylation of lysine 9 of histone H3 (H3K9me3) was found to silent the genes in *P. falciparum* parasites [124,129]. Recent studies of genes involved in *P. falciparum* invasion pathway have also indicated the epigenetic mechanism on the regulation of the gene expression. Differential expression of the invasion genes, such as *rh4, eba-165, clag 3.1, clag 3.2,* and *eba-140*, correlated with differential acetylation of H3K9, 14 and H4K5, 8, 12, 16 in their 5' UTRs [113,130,131]. Further global analysis of transcriptional variation in isogenic parasite lines showed that more genes involved in host-parasite interactions are under clonally variant expression and are controlled at the epigenetic level [114]. Genome-wide analysis of histone 3 modification using a chromatin immunoprecipitation (ChIP) assay revealed the cycle-regulated H3K4me3 and H3K9ac at asexual developmental stages in *P. falciparum* [132,133]. Disruption of one of the key genes in chromatin modification (*pfSir2*) that encodes a histone deactylase caused changes in the H3K9me3 profile, and inhibitors of this enzyme showed high potency against cultured *P. falciparum* parasites *in vitro* [134]. Given the observed differences in the epigenetic code compared with all other organisms studied, *Plasmodium*-specific epigenetic enzyme inhibitors could be explored for new therapeutic agents against *P. falciparum* [121,134].

Gene functions have also been predicted by large-scale comparative analysis of protein expression level in *P. falciparum*. Although it can be difficult to obtain sufficient material and to prevent contamination from host cells, two large-scale studies using high-throughput proteomics have detected many stage-specific predicted gene products consistent with results from transcript profiling studies

[135,136]. This genomics-based approach has also been widely applied in studies of drug targets [137-139], organelle composition [140], stage- and sex-specific gene functions [45,141,142], validation of data from genomic annotation, post-translational modifications [143-145]. Interestingly, a panel of novel *P. falciparum* antigens associated with host sterile protective immunity to malaria has been identified recently using this approach, although further functional validation is required [146,147]. With the completion of its human and insect host genome project, genomic, metabolomic and proteomic analyses of host-pathogen interactions have shed light on many malaria genes' functions [148-151]. This important research topic has been reviewed elsewhere recently [152-154]. Combined transcriptomic, epigenetic, and proteomic data also allowed uncovering regulatory mechanisms of gene expression in *P. falciparum* [45,144]. With the availability of newer methodologies, analysis of expression variation at the protein level may permit investigation of protein interaction and discovery of targets for new drugs and vaccines.

## FUTURE PROSPECTS

Functional genomic research in *P. falciparum* will undoubtedly continue to contribute greatly to our battle against this deadly parasite. As more phenotypic data become available, the ability to identify gene function will be greatly enhanced by high-throughput, genome-wide approaches. High-throughput assays for parasite phenotypes such as drug response, variation in invasion efficiency, population expression profiling, and variation in parasite metabolites can lead to gene function assignment with the use of genomic data. Moreover, next-generation sequencing methods are emerging as the dominant genomic technologies and can be applied in a variety of contexts for functional genomics research, including whole-genome sequencing, targeted resequencing, deep transcriptome analysis to complement microarray analysis, genome-wide epigenetic modifications, chromosome interaction, *etc*. In addition, application of novel genetic manipulation tools such as transposon mutagenesis (piggyBac) [155] and improved transfection methods [156] will be extremely valuable for generating functional mutations and for verifying gene functions. Genomics and related technologies have greatly facilitated the research on malaria and will continue to power new discoveries in the near future.

## CONFLICT OF INTEREST

The author(s) confirm that this chapter content has no conflict of interest.

## ACKNOWLEDGEMENTS

This work was supported by the Intramural Research Program of the Division of Intramural Research, National Institute of Allergy and Infectious Diseases, National Institutes of Health. We thank NIAID intramural editor Brenda Rae Marshall for assistance.



## DISCLOSURE

This is an updated report of the earlier review article by Mu J, Seydel KB, Bates A, Su XZ. Recent Progress in Functional Genomic Research in *Plasmodium falciparum*. Curr Genomics. 2010 Jun; 11(4): 279-86.

## ABBREVIATIONS

| | | |
|---|---|---|
| ApiAP2 | = | Apicomplexan AP2 |
| cDNA | = | Complementary DNA |
| ChIP | = | Chromatin immunoprecipitation |
| CNV | = | Copy number variation |
| EBL | = | Erythrocyte binding-like |
| ELP | = | Expression level polymorphisms |

| | | |
|---|---|---|
| eQTL | = | Expression quantitative trait loci |
| GWAS | = | Genome-wide association study |
| H3K9me3 | = | Trimethylation of lysine 9 of histone H3 |
| LD | = | Linkage disequilibrium |
| MIP | = | Molecular inversion probe |
| miRNA | = | MicroRNA |
| MS | = | Microsatellite |
| ncRNA | = | Noncoding RNA |
| OPI | = | Ontology-based pattern identification |
| PEXEL | = | *Plasmodium* export element |
| *PfAMA-1* | = | *P. falciparum* apical membrane antigen 1 |
| RFLP | = | Restriction fragment-length polymorphism |
| RBL | = | Reticulocyte-binding like protein |
| ROM | = | Rhomboid-like protein |
| SFP | = | Single-feature polymorphism |
| snoRNA | = | Small nucleolar RNA |
| SNP | = | Single nucleotide polymorphism |
| SP | = | Sulfadoxine-pyrimethamine |
| VTS | = | Vacuolar transport signal |

## REFERENCES

[1] Murray, C. J.; Rosenfeld, L. C.; Lim, S. S.; Andrews, K. G.; Foreman, K. J.; Haring, D.; Fullman, N.; Naghavi, M.; Lozano, R.; Lopez, A. D., Global malaria mortality between 1980 and 2010: a systematic analysis. *Lancet* **2012**, *379* (9814), 413-31.

[2] White, N. J., Plasmodium knowlesi: the fifth human malaria parasite. *Clin Infect Dis* **2008**, *46* (2), 172-173.

[3] Pain, A.; Hertz-Fowler, C., Plasmodium genomics: latest milestone. *Nat Rev Microbiol* **2009**, *7* (3), 180-181.

[4] Aravind, L.; Iyer, L. M.; Wellems, T. E.; Miller, L. H., Plasmodium biology: genomic gleanings. *Cell* **2003**, *115* (7), 771-785.

[5] Gardner, M. J.; Hall, N.; Fung, E.; White, O.; Berriman, M.; Hyman, R. W.; Carlton, J. M.; Pain, A.; Nelson, K. E.; Bowman, S.; Paulsen, I. T.; James, K.; Eisen, J. A.; Rutherford, K.; Salzberg, S. L.; Craig, A.; Kyes, S.; Chan, M. S.; Nene, V.; Shallom, S. J.; Suh, B.; Peterson, J.; Angiuoli, S.; Pertea, M.; Allen, J.; Selengut, J.; Haft, D.; Mather, M. W.; Vaidya, A. B.; Martin, D. M.; Fairlamb, A. H.; Fraunholz, M. J.; Roos, D. S.; Ralph, S. A.; McFadden, G. I.; Cummings, L. M.; Subramanian, G. M.; Mungall, C.; Venter, J. C.; Carucci, D. J.; Hoffman, S. L.; Newbold, C.; Davis, R. W.; Fraser, C. M.; Barrell, B., Genome sequence of the human malaria parasite Plasmodium falciparum. *Nature* **2002**, *419* (6906), 498-511.

[6] Lu, F.; Jiang, H.; Ding, J.; Mu, J.; Valenzuela, J. G.; Ribeiro, J. M.; Su, X. Z., cDNA sequences reveal considerable gene prediction inaccuracy in the Plasmodium falciparum genome. *BMC Genomics* **2007**, *8*, 255.

[7] Wakaguri, H.; Suzuki, Y.; Katayama, T.; Kawashima, S.; Kibukawa, E.; Hiranuka, K.; Sasaki, M.; Sugano, S.; Watanabe, J., Full-Malaria/Parasites and Full-Arthropods: databases of full-length cDNAs of parasites and arthropods, update 2009. *Nucleic Acids Res*, 2009; Vol. 37, pp D520-520..

[8] Lopez-Barragan, M. J.; Lemieux, J.; Quinones, M.; Williamson, K. C.; Molina-Cruz, A.; Cui, K.; Barillas-Mury, C.; Zhao, K.; Su, X. Z., Directional gene expression and antisense transcripts in sexual and asexual stages of Plasmodium falciparum. *BMC Genomics* **2011**, *12*, 587.

[9] Otto, T. D.; Wilinski, D.; Assefa, S.; Keane, T. M.; Sarry, L. R.; Bohme, U.; Lemieux, J.; Barrell, B.; Pain, A.; Berriman, M.; Newbold, C.; Llinas, M., New insights into the blood-stage transcriptome of Plasmodium falciparum using RNA-Seq. *Mol Microbiol* **2010**, *76* (1), 12-24.

[10] Sorber, K.; Dimon, M. T.; DeRisi, J. L., RNA-Seq analysis of splicing in Plasmodium falciparum uncovers new splice junctions, alternative splicing and splicing of antisense transcripts. *Nucleic Acids Res* **2011**, *39* (9), 3820-35.

[11] Bartfai, R.; Hoeijmakers, W. A.; Salcedo-Amaya, A. M.; Smits, A. H.; Janssen-Megens, E.; Kaan, A.; Treeck, M.; Gilberger, T. W.; Francoijs, K. J.; Stunnenberg, H. G., H2A.Z demarcates intergenic regions of the plasmodium falciparum epigenome that are dynamically marked by H3K9ac and H3K4me3. *PLoS Pathog* **2010**, *6* (12), e1001223.

[12] Anderson, T. J., Mapping drug resistance genes in Plasmodium falciparum by genome-wide association. *Curr Drug Targets Infect Disord* **2004**, *4* (1), 65-78.

[13] Su, X.; Hayton, K.; Wellems, T. E., Genetic linkage and association analyses for trait mapping in Plasmodium falciparum. *Nat Rev Genet* **2007**, *8* (7), 497-506.

[14] Su, X.; Ferdig, M. T.; Huang, Y.; Huynh, C. Q.; Liu, A.; You, J.; Wootton, J. C.; Wellems, T. E., A genetic map and recombination parameters of the human malaria parasite Plasmodium falciparum. *Science* **1999**, *286* (5443), 1351-1353.

[15] Wellems, T. E.; Su, X. Z.; Ferdig, M.; Fidock, D. A., Genome projects, genetic analysis, and the changing landscape of malaria research. *Curr Opin Microbiol* **1999**, *2* (4), 415-419.

[16] Su, X. Z.; Wootton, J. C., Genetic mapping in the human malaria parasite Plasmodium falciparum. *Mol Microbiol* **2004**, *53* (6), 1573-1582.

[17] Su, X.; Kirkman, L. A.; Fujioka, H.; Wellems, T. E., Complex polymorphisms in an approximately 330 kDa protein are linked to chloroquine-resistant P. falciparum in Southeast Asia and Africa. *Cell* **1997**, *91* (5), 593-603.

[18] Fidock, D. A.; Nomura, T.; Talley, A. K.; Cooper, R. A.; Dzekunov, S. M.; Ferdig, M. T.; Ursos, L. M.; Sidhu, A. B.; Naude, B.; Deitsch, K. W.; Su, X. Z.; Wootton, J. C.; Roepe, P. D.; Wellems, T. E., Mutations in the P. falciparum digestive vacuole transmembrane protein PfCRT and evidence for their role in chloroquine resistance. *Mol Cell* **2000**, *6* (4), 861-871.

[19] Wellems, T. E.; Panton, L. J.; Gluzman, I. Y.; do Rosario, V. E.; Gwadz, R. W.; Walker-Jonah, A.; Krogstad, D. J., Chloroquine resistance not linked to mdr-like genes in a Plasmodium falciparum cross. *Nature* **1990**, *345* (6272), 253-255.

[20] Peterson, D. S.; Walliker, D.; Wellems, T. E., Evidence that a point mutation in dihydrofolate reductase-thymidylate synthase confers resistance to pyrimethamine in falciparum malaria. *Proc Natl Acad Sci U S A* **1988**, *85* (23), 9114-9118.

[21] Hayton, K.; Gaur, D.; Liu, A.; Takahashi, J.; Henschen, B.; Singh, S.; Lambert, L.; Furuya, T.; Bouttenot, R.; Doll, M.; Nawaz, F.; Mu, J.; Jiang, L.; Miller, L. H.; Wellems, T. E., Erythrocyte binding protein PfRH5 polymorphisms determine species-specific pathways of Plasmodium falciparum invasion. *Cell Host Microbe* **2008**, *4* (1), 40-51.

[22] Nguitragool, W.; Bokhari, A. A.; Pillai, A. D.; Rayavara, K.; Sharma, P.; Turpin, B.; Aravind, L.; Desai, S. A., Malaria parasite clag3 genes determine channel-mediated nutrient uptake by infected red blood cells. *Cell* **2011**, *145* (5), 665-677.

[23] Yuan, J.; Cheng, K. C.; Johnson, R. L.; Huang, R.; Pattaradilokrat, S.; Liu, A.; Guha, R.; Fidock, D. A.; Inglese, J.; Wellems, T. E.; Austin, C. P.; Su, X. Z., Chemical genomic profiling for antimalarial therapies, response signatures, and molecular targets. *Science* **2011**, *333* (6043), 724-729.

[24] Yuan, J.; Johnson, R. L.; Huang, R.; Wichterman, J.; Jiang, H.; Hayton, K.; Fidock, D. A.; Wellems, T. E.; Inglese, J.; Austin, C. P.; Su, X. Z., Genetic mapping of targets mediating differential chemical phenotypes in Plasmodium falciparum. *Nat Chem Biol* **2009**, *5* (10), 765-771.

[25] Sanchez, C. P.; Mayer, S.; Nurhasanah, A.; Stein, W. D.; Lanzer, M., Genetic linkage analyses redefine the roles of PfCRT and PfMDR1 in drug accumulation and susceptibility in Plasmodium falciparum. *Mol Microbiol* **2011**, *82* (4), 865-878.

[26] Sa, J. M.; Twu, O.; Hayton, K.; Reyes, S.; Fay, M. P.; Ringwald, P.; Wellems, T. E., Geographic patterns of Plasmodium falciparum drug resistance distinguished by differential responses to amodiaquine and chloroquine. *Proc Natl Acad Sci U S A* **2009**, *106* (45), 18883-18889.

[27] Beez, D.; Sanchez, C. P.; Stein, W. D.; Lanzer, M., Genetic predisposition favors the acquisition of stable artemisinin resistance in malaria parasites. *Antimicrob Agents Chemother* **2011**, *55* (1), 50-55.

[28] Li, J.; Pattaradilokrat, S.; Zhu, F.; Jiang, H.; Liu, S.; Hong, L.; Fu, Y.; Koo, L.; Xu, W.; Pan, W.; Carlton, J. M.; Kaneko, O.; Carter, R.; Wootton, J. C.; Su, X. Z., Linkage maps from multiple genetic crosses and loci linked to growth-related virulent phenotype in Plasmodium yoelii. *Proc Natl Acad Sci U S A* **2011**, *108* (31), E374-382.

[29] Walliker, D.; Carter, R.; Morgan, S., Genetic recombination in malaria parasites. *Nature* **1971**, *232* (5312), 561-562.

[30] Pattaradilokrat, S.; Culleton, R. L.; Cheesman, S. J.; Carter, R., Gene encoding erythrocyte binding ligand linked to blood stage multiplication rate phenotype in Plasmodium yoelii yoelii. *Proc Natl Acad Sci U S A* **2009**, *106* (17), 7161-7166.

[31] Martinelli, A.; Cheesman, S.; Hunt, P.; Culleton, R.; Raza, A.; Mackinnon, M.; Carter, R., A genetic approach to the *de novo* identification of targets of strain-specific immunity in malaria parasites. *Proc Natl Acad Sci U S A* **2005**, *102* (3), 814-819.

[32] Culleton, R.; Martinelli, A.; Hunt, P.; Carter, R., Linkage group selection: rapid gene discovery in malaria parasites. *Genome Res* **2005**, *15* (1), 92-97.

[33] Modrzynska, K. K.; Creasey, A.; Loewe, L.; Cezard, T.; Borges, S. T.; Martinelli, A.; Rodrigues, L.; Cravo, P.; Blaxter, M.; Carter, R.; Hunt, P., Quantitative genome re-sequencing defines multiple mutations conferring chloroquine resistance in rodent malaria. *BMC Genomics* **2012**, *13* (1), 106.

[34] Pattaradilokrat, S.; Li, J.; Su, X. Z., Protocol for production of a genetic cross of the rodent malaria parasites. *J Vis Exp* **2011**.

[35] de Koning-Ward, T. F.; Janse, C. J.; Waters, A. P., The development of genetic tools for dissecting the biology of malaria parasites. *Annu Rev Microbiol* **2000**, *54*, 157-185.

[36] Carlton, J. M.; Hayton, K.; Cravo, P. V.; Walliker, D., Of mice and malaria mutants: unravelling the genetics of drug resistance using rodent malaria models. *Trends Parasitol* **2001**, *17* (5), 236-242.

[37] Mu, J.; Duan, J.; Makova, K. D.; Joy, D. A.; Huynh, C. Q.; Branch, O. H.; Li, W. H.; Su, X. Z., Chromosome-wide SNPs reveal an ancient origin for Plasmodium falciparum. *Nature* **2002**, *418* (6895), 323-326.

[38] Mu, J.; Ferdig, M. T.; Feng, X.; Joy, D. A.; Duan, J.; Furuya, T.; Subramanian, G.; Aravind, L.; Cooper, R. A.; Wootton, J. C.; Xiong, M.; Su, X. Z., Multiple transporters associated with malaria parasite responses to chloroquine and quinine. *Mol Microbiol* **2003**, *49* (4), 977-989.

[39] Jeffares, D. C.; Pain, A.; Berry, A.; Cox, A. V.; Stalker, J.; Ingle, C. E.; Thomas, A.; Quail, M. A.; Siebenthall, K.; Uhlemann, A. C.; Kyes, S.; Krishna, S.; Newbold, C.; Dermitzakis, E. T.; Berriman, M., Genome variation and evolution of the malaria parasite Plasmodium falciparum. *Nat Genet* **2007**, *39* (1), 120-125.

[40] Volkman, S. K.; Sabeti, P. C.; DeCaprio, D.; Neafsey, D. E.; Schaffner, S. F.; Milner, D. A., Jr.; Daily, J. P.; Sarr, O.; Ndiaye, D.; Ndir, O.; Mboup, S.; Duraisingh, M. T.; Lukens, A.; Derr, A.; Stange-Thomann, N.; Waggoner, S.; Onofrio, R.; Ziaugra, L.; Mauceli, E.; Gnerre, S.; Jaffe, D. B.; Zainoun, J.; Wiegand, R. C.; Birren, B. W.; Hartl, D. L.; Galagan, J. E.; Lander, E. S.; Wirth, D. F., A genome-wide map of diversity in Plasmodium falciparum. *Nat Genet* **2007**, *39* (1), 113-119.

[41] Carlton, J. M.; Escalante, A. A.; Neafsey, D.; Volkman, S. K., Comparative evolutionary genomics of human malaria parasites. *Trends Parasitol* **2008**, *24* (12), 545-550.

[42] Kidgell, C.; Volkman, S. K.; Daily, J.; Borevitz, J. O.; Plouffe, D.; Zhou, Y.; Johnson, J. R.; Le Roch, K.; Sarr, O.; Ndir, O.; Mboup, S.; Batalov, S.; Wirth, D. F.; Winzeler, E. A., A systematic map of genetic variation in Plasmodium falciparum. *PLoS Pathog* **2006**, *2* (6), e57.

[43] Jiang, H.; Yi, M.; Mu, J.; Zhang, L.; Ivens, A.; Klimczak, L. J.; Huyen, Y.; Stephens, R. M.; Su, X. Z., Detection of genome-wide polymorphisms in the AT-rich Plasmodium falciparum genome using a high-density microarray. *BMC Genomics* **2008**, *9*, 398.

[44] Bozdech, Z.; Mok, S.; Hu, G.; Imwong, M.; Jaidee, A.; Russell, B.; Ginsburg, H.; Nosten, F.; Day, N. P.; White, N. J.; Carlton, J. M.; Preiser, P. R., The transcriptome of Plasmodium vivax reveals divergence and diversity of transcriptional regulation in malaria parasites. *Proc Natl Acad Sci U S A* **2008**, *105* (42), 16290-16295.

[45] Hall, N.; Karras, M.; Raine, J. D.; Carlton, J. M.; Kooij, T. W.; Berriman, M.; Florens, L.; Janssen, C. S.; Pain, A.; Christophides, G. K.; James, K.; Rutherford, K.; Harris, B.; Harris, D.; Churcher, C.; Quail, M. A.; Ormond, D.; Doggett, J.; Trueman, H. E.; Mendoza, J.; Bidwell, S. L.; Rajandream, M. A.; Carucci, D. J.; Yates, J. R., 3rd; Kafatos, F. C.; Janse, C. J.; Barrell, B.; Turner, C. M.; Waters, A. P.; Sinden, R. E., A comprehensive survey of the Plasmodium life cycle by genomic, transcriptomic, and proteomic analyses. *Science* **2005**, *307* (5706), 82-86.

[46] Carret, C. K.; Horrocks, P.; Konfortov, B.; Winzeler, E.; Qureshi, M.; Newbold, C.; Ivens, A., Microarray-based comparative genomic analyses of the human malaria parasite Plasmodium falciparum using Affymetrix arrays. *Mol Biochem Parasitol* **2005**, *144* (2), 177-186.

[47] Dharia, N. V.; Sidhu, A. B.; Cassera, M. B.; Westenberger, S. J.; Bopp, S. E.; Eastman, R. T.; Plouffe, D.; Batalov, S.; Park, D. J.; Volkman, S. K.; Wirth, D. F.; Zhou, Y.; Fidock, D. A.; Winzeler, E. A., Use of high-density tiling microarrays to identify mutations globally and elucidate mechanisms of drug resistance in Plasmodium falciparum. *Genome Biol* **2009**, *10* (2), R21.

[48] Jiang, H.; Li, N.; Gopalan, V.; Zilversmit, M. M.; Varma, S.; Nagarajan, V.; Li, J.; Mu, J.; Hayton, K.; Henschen, B.; Yi, M.; Stephens, R.; McVean, G.; Awadalla, P.; Wellems, T. E.; Su, X. Z., High recombination rates and hotspots in a Plasmodium falciparum genetic cross. *Genome Biol* **2011**, *12* (4), R33.

[49] Mu, J.; Myers, R. A.; Jiang, H.; Liu, S.; Ricklefs, S.; Waisberg, M.; Chotivanich, K.; Wilairatana, P.; Krudsood, S.; White, N. J.; Udomsangpetch, R.; Cui, L.; Ho, M.; Ou, F.; Li, H.; Song, J.; Li, G.; Wang, X.; Seila, S.; Sokunthea, S.; Socheat, D.; Sturdevant, D. E.; Porcella, S. F.; Fairhurst, R. M.; Wellems, T. E.; Awadalla, P.; Su, X. Z., Plasmodium falciparum genome-wide scans for positive selection, recombination hot spots and resistance to antimalarial drugs. *Nat Genet* **2010**, *42* (3), 268-271.

[50] Tan, J. C.; Miller, B. A.; Tan, A.; Patel, J. J.; Cheeseman, I. H.; Anderson, T. J.; Manske, M.; Maslen, G.; Kwiatkowski, D. P.; Ferdig, M. T., An optimized microarray platform for assaying genomic variation in Plasmodium falciparum field populations. *Genome Biol* **2011**, *12* (4), R35.

[51] Mu, J.; Awadalla, P.; Duan, J.; McGee, K. M.; Joy, D. A.; McVean, G. A.; Su, X. Z., Recombination hotspots and population structure in Plasmodium falciparum. *PLoS Biol* **2005**, *3* (10), e335.

[52] Neafsey, D. E.; Schaffner, S. F.; Volkman, S. K.; Park, D.; Montgomery, P.; Milner, D. A., Jr.; Lukens, A.; Rosen, D.; Daniels, R.; Houde, N.; Cortese, J. F.; Tyndall, E.; Gates, C.; Stange-Thomann, N.; Sarr, O.; Ndiaye, D.; Ndir, O.; Mboup, S.; Ferreira, M. U.; Moraes Sdo, L.; Dash, A. P.; Chitnis, C. E.; Wiegand, R. C.; Hartl, D. L.; Birren, B. W.; Lander, E. S.; Sabeti, P. C.; Wirth, D. F., Genome-wide SNP genotyping highlights the role of natural selection in Plasmodium falciparum population divergence. *Genome Biol* **2008**, *9* (12), R171.

[53] Duan, J.; Mu, J.; Thera, M. A.; Joy, D.; Kosakovsky Pond, S. L.; Diemert, D.; Long, C.; Zhou, H.; Miura, K.; Ouattara, A.; Dolo, A.; Doumbo, O.; Su, X. Z.; Miller, L., Population structure of the genes encoding the polymorphic Plasmodium falciparum apical membrane antigen 1: Implications for vaccine design. *Proc Natl Acad Sci U S A* **2008**, *105* (22), 7857-7862.

[54] Samarakoon, U.; Regier, A.; Tan, A.; Desany, B. A.; Collins, B.; Tan, J. C.; Emrich, S. J.; Ferdig, M. T., High-throughput 454 resequencing for allele discovery and recombination mapping in Plasmodium falciparum. *BMC Genomics* **2011**, *12*, 116.

[55] Voight, B. F.; Kudaravalli, S.; Wen, X.; Pritchard, J. K., A map of recent positive selection in the human genome. *PLoS Biol* **2006**, *4* (3), e72.

[56] Sabeti, P. C.; Varilly, P.; Fry, B.; Lohmueller, J.; Hostetter, E.; Cotsapas, C.; Xie, X.; Byrne, E. H.; McCarroll, S. A.; Gaudet, R.; Schaffner, S. F.; Lander, E. S., Genome-wide detection and characterization of positive selection in human populations. *Nature* **2007**, *449* (7164), 913-918.

[57] Hurst, L. D., Fundamental concepts in genetics: Genetics and the understanding of selection. *Nat Rev Genet* **2009**, 10(2), 83-93

[58] Wootton, J. C.; Feng, X.; Ferdig, M. T.; Cooper, R. A.; Mu, J.; Baruch, D. I.; Magill, A. J.; Su, X. Z., Genetic diversity and chloroquine selective sweeps in Plasmodium falciparum. *Nature* **2002**, *418* (6895), 320-323.

[59] Nash, D.; Nair, S.; Mayxay, M.; Newton, P. N.; Guthmann, J. P.; Nosten, F.; Anderson, T. J., Selection strength and hitchhiking around two anti-malarial resistance genes. *Proc Biol Sci* **2005**, *272* (1568), 1153-1161.

[60] Mu, J.; Awadalla, P.; Duan, J.; McGee, K. M.; Keebler, J.; Seydel, K.; McVean, G. A.; Su, X. Z., Genome-wide variation and identification of vaccine targets in the Plasmodium falciparum genome. *Nat Genet* **2007**, *39* (1), 126-130.

[61] Van Tyne, D.; Park, D. J.; Schaffner, S. F.; Neafsey, D. E.; Angelino, E.; Cortese, J. F.; Barnes, K. G.; Rosen, D. M.; Lukens, A. K.; Daniels, R. F.; Milner, D. A., Jr.; Johnson, C. A.; Shlyakhter, I.; Grossman, S. R.; Becker, J. S.; Yamins, D.; Karlsson, E. K.; Ndiaye, D.; Sarr, O.; Mboup, S.; Happi, C.; Furlotte, N. A.; Eskin, E.; Kang, H. M.; Hartl, D. L.; Birren, B. W.; Wiegand, R. C.; Lander, E. S.; Wirth, D. F.; Volkman, S. K.; Sabeti, P. C., Identification and functional validation of the novel antimalarial resistance locus PF10_0355 in Plasmodium falciparum. *PLoS Genet* **2011**, *7* (4), e1001383.

[62] Cheeseman, I. H.; Miller, B. A.; Nair, S.; Nkhoma, S.; Tan, A.; Tan, J. C.; Al Saai, S.; Phyo, A. P.; Moo, C. L.; Lwin, K. M.; McGready, R.; Ashley, E.; Imwong, M.; Stepniewska, K.; Yi, P.; Dondorp, A. M.; Mayxay, M.; Newton, P. N.; White, N. J.; Nosten, F.; Ferdig, M. T.; Anderson, T. J., A major genome region underlying artemisinin resistance in malaria. *Science* **2012**, *336* (6077), 79-82.

[63] Henrichsen, C. N.; Chaignat, E.; Reymond, A., Copy number variants, diseases and gene expression. *Hum Mol Genet* **2009**, *18* (R1), R1-8.

[64] Anderson, T. J.; Patel, J.; Ferdig, M. T., Gene copy number and malaria biology. *Trends Parasitol* **2009**, *25* (7), 336-343.

[65] Price, R. N.; Uhlemann, A. C.; Brockman, A.; McGready, R.; Ashley, E.; Phaipun, L.; Patel, R.; Laing, K.; Looareesuwan, S.; White, N. J.; Nosten, F.; Krishna, S., Mefloquine resistance in Plasmodium falciparum and increased pfmdr1 gene copy number. *Lancet* **2004**, *364* (9432), 438-447.

[66] Nair, S.; Miller, B.; Barends, M.; Jaidee, A.; Patel, J.; Mayxay, M.; Newton, P.; Nosten, F.; Ferdig, M. T.; Anderson, T. J., Adaptive copy number evolution in malaria parasites. *PLoS Genet* **2008**, *4* (10), e1000243.

[67] Jensen, K.; Plichta, D.; Panagiotou, G.; Kouskoumvekaki, I., Mapping the genome of Plasmodium falciparum on the drug-like chemical space reveals novel anti-malarial targets and potential drug leads. *Mol Biosyst* **2012**, 8(6), 1678-1685.

[68] Gamo, F. J.; Sanz, L. M.; Vidal, J.; de Cozar, C.; Alvarez, E.; Lavandera, J. L.; Vanderwall, D. E.; Green, D. V.; Kumar, V.; Hasan, S.; Brown, J. R.; Peishoff, C. E.; Cardon, L. R.; Garcia-Bustos, J. F., Thousands of chemical starting points for antimalarial lead identification. *Nature* **2010**, *465* (7296), 305-310.

[69] Guiguemde, W. A.; Shelat, A. A.; Bouck, D.; Duffy, S.; Crowther, G. J.; Davis, P. H.; Smithson, D. C.; Connelly, M.; Clark, J.; Zhu, F.; Jimenez-Diaz, M. B.; Martinez, M. S.; Wilson, E. B.; Tripathi, A. K.; Gut, J.; Sharlow, E. R.; Bathurst, I.; El Mazouni, F.; Fowble, J. W.; Forquer, I.; McGinley, P. L.; Castro, S.; Angulo-Barturen, I.; Ferrer, S.; Rosenthal, P. J.; Derisi, J. L.; Sullivan, D. J.; Lazo, J. S.; Roos, D. S.; Riscoe, M. K.; Phillips, M. A.; Rathod, P. K.; Van Voorhis, W. C.; Avery, V. M.; Guy, R. K., Chemical genetics of Plasmodium falciparum. *Nature* **2010**, *465* (7296), 311-315.

[70] Guiguemde, W. A.; Shelat, A. A.; Garcia-Bustos, J. F.; Diagana, T. T.; Gamo, F. J.; Guy, R. K., Global phenotypic screening for antimalarials. *Chem Biol* **2012**, *19* (1), 116-129.

[71] Garcia-Bustos, J. F.; Gamo, F. J., Phenotypic screens, chemical genomics, and antimalarial lead discovery. *PLoS Pathog* **2011**, *7* (8), e1002156.

[72] Chakrabarti, K.; Pearson, M.; Grate, L.; Sterne-Weiler, T.; Deans, J.; Donohue, J. P.; Ares, M., Jr., Structural RNAs of known and unknown function identified in malaria parasites by comparative genomics and RNA analysis. *Rna* **2007**, *13* (11), 1923-1939.

[73] Mourier, T.; Carret, C.; Kyes, S.; Christodoulou, Z.; Gardner, P. P.; Jeffares, D. C.; Pinches, R.; Barrell, B.; Berriman, M.; Griffiths-Jones, S.; Ivens, A.; Newbold, C.; Pain, A., Genome-wide discovery and verification of novel structured RNAs in Plasmodium falciparum. *Genome Res* **2008**, *18* (2), 281-292.

[74] Bright, A. T.; Winzeler, E. A., Noncoding RNA, antigenic variation, and the virulence genes of Plasmodium falciparum. *BMC Biol* **2011**, *9*, 50.

[75] Broadbent, K. M.; Park, D.; Wolf, A. R.; Van Tyne, D.; Sims, J. S.; Ribacke, U.; Volkman, S.; Duraisingh, M.; Wirth, D.; Sabeti, P. C.; Rinn, J. L., A global transcriptional analysis of Plasmodium falciparum malaria reveals a novel family of telomere-associated lncRNAs. *Genome Biol* **2011**, *12* (6), R56.

[76] Hertel, J.; Stadler, P. F., Hairpins in a Haystack: recognizing microRNA precursors in comparative genomics data. *Bioinformatics* **2006**, *22* (14), e197-202.

[77] Iyer, J.; Gruner, A. C.; Renia, L.; Snounou, G.; Preiser, P. R., Invasion of host cells by malaria parasites: a tale of two protein families. *Mol Microbiol* **2007**, *65* (2), 231-249.

[78] Cowman, A. F.; Crabb, B. S., Invasion of red blood cells by malaria parasites. *Cell* **2006**, *124* (4), 755-766.

[79] Gaur, D.; Mayer, D. C.; Miller, L. H., Parasite ligand-host receptor interactions during invasion of erythrocytes by Plasmodium merozoites. *Int J Parasitol* **2004**, *34* (13-14), 1413-1429.

[80] Singh, S.; Plassmeyer, M.; Gaur, D.; Miller, L. H., Mononeme: a new secretory organelle in Plasmodium falciparum merozoites identified by localization of rhomboid-1 protease. *Proc Natl Acad Sci U S A* **2007**, *104* (50), 20043-20048.

[81] Yeoh, S.; O'Donnell, R. A.; Koussis, K.; Dluzewski, A. R.; Ansell, K. H.; Osborne, S. A.; Hackett, F.; Withers-Martinez, C.; Mitchell, G. H.; Bannister, L. H.; Bryans, J. S.; Kettleborough, C. A.; Blackman, M. J., Subcellular discharge of a serine protease mediates release of invasive malaria parasites from host erythrocytes. *Cell* **2007**, *131* (6), 1072-1083.

[82] Blackman, M. J., Malarial proteases and host cell egress: an 'emerging' cascade. *Cell Microbiol* **2008**, *10* (10), 1925-1934.

[83] Balaji, S.; Babu, M. M.; Iyer, L. M.; Aravind, L., Discovery of the principal specific transcription factors of Apicomplexa and their implication for the evolution of the AP2-integrase DNA binding domains. *Nucleic Acids Res* **2005**, *33* (13), 3994-4006.

[84] De Silva, E. K.; Gehrke, A. R.; Olszewski, K.; Leon, I.; Chahal, J. S.; Bulyk, M. L.; Llinas, M., Specific DNA-binding by apicomplexan AP2 transcription factors. *Proc Natl Acad Sci U S A* **2008**, *105* (24), 8393-8398.

[85] Aurrecoechea, C.; Brestelli, J.; Brunk, B. P.; Dommer, J.; Fischer, S.; Gajria, B.; Gao, X.; Gingle, A.; Grant, G.; Harb, O. S.; Heiges, M.; Innamorato, F.; Iodice, J.; Kissinger, J. C.; Kraemer, E.; Li, W.; Miller, J. A.; Nayak, V.; Pennington, C.; Pinney, D. F.; Roos, D. S.; Ross, C.; Stoeckert, C. J., Jr.; Treatman, C.; Wang, H., PlasmoDB: a functional genomic database for malaria parasites. *Nucleic Acids Res*, **2009**, 37, D539-543.

[86] Painter, H. J.; Campbell, T. L.; Llinas, M., The Apicomplexan AP2 family: integral factors regulating Plasmodium development. *Mol Biochem Parasitol* **2011**, *176* (1), 1-7.

[87] Campbell, T. L.; De Silva, E. K.; Olszewski, K. L.; Elemento, O.; Llinas, M., Identification and genome-wide prediction of DNA binding specificities for the ApiAP2 family of regulators from the malaria parasite. *PLoS Pathog* **2010**, *6* (10), e1001165.

[88] Flueck, C.; Bartfai, R.; Niederwieser, I.; Witmer, K.; Alako, B. T.; Moes, S.; Bozdech, Z.; Jenoe, P.; Stunnenberg, H. G.; Voss, T. S., A major role for the Plasmodium falciparum ApiAP2 protein PfSIP2 in chromosome end biology. *PLoS Pathog* **2010**, *6* (2), e1000784.

[89] Young, J. A.; Johnson, J. R.; Benner, C.; Yan, S. F.; Chen, K.; Le Roch, K. G.; Zhou, Y.; Winzeler, E. A., *In silico* discovery of transcription regulatory elements in Plasmodium falciparum. *BMC Genomics* **2008**, *9*, 70.

[90] Hiller, N. L.; Bhattacharjee, S.; van Ooij, C.; Liolios, K.; Harrison, T.; Lopez-Estrano, C.; Haldar, K., A host-targeting signal in virulence proteins reveals a secretome in malarial infection. *Science* **2004**, *306* (5703), 1934-1937.

[91] Marti, M.; Good, R. T.; Rug, M.; Knuepfer, E.; Cowman, A. F., Targeting malaria virulence and remodeling proteins to the host erythrocyte. *Science* **2004**, *306* (5703), 1930-1933.

[92] Maier, A. G.; Rug, M.; O'Neill, M. T.; Brown, M.; Chakravorty, S.; Szestak, T.; Chesson, J.; Wu, Y.; Hughes, K.; Coppel, R. L.; Newbold, C.; Beeson, J. G.; Craig, A.; Crabb, B. S.; Cowman, A. F., Exported proteins required for virulence and rigidity of Plasmodium falciparum-infected human erythrocytes. *Cell* **2008**, *134* (1), 48-61.

[93] Maier, A. G.; Cooke, B. M.; Cowman, A. F.; Tilley, L., Malaria parasite proteins that remodel the host erythrocyte. *Nat Rev Microbiol* **2009**, *7* (5), 341-354.

[94] van Ooij, C.; Tamez, P.; Bhattacharjee, S.; Hiller, N. L.; Harrison, T.; Liolios, K.; Kooij, T.; Ramesar, J.; Balu, B.; Adams, J.; Waters, A. P.; Janse, C. J.; Haldar, K., The malaria secretome: from algorithms to essential function in blood stage infection. *PLoS Pathog* **2008**, *4* (6), e1000084.

[95] de Koning-Ward, T. F.; Gilson, P. R.; Boddey, J. A.; Rug, M.; Smith, B. J.; Papenfuss, A. T.; Sanders, P. R.; Lundie, R. J.; Maier, A. G.; Cowman, A. F.; Crabb, B. S., A newly discovered protein export machine in malaria parasites. *Nature* **2009**, *459* (7249), 945-949.

[96] Le Roch, K. G.; Zhou, Y.; Blair, P. L.; Grainger, M.; Moch, J. K.; Haynes, J. D.; De La Vega, P.; Holder, A. A.; Batalov, S.; Carucci, D. J.; Winzeler, E. A., Discovery of gene function by expression profiling of the malaria parasite life cycle. *Science* **2003**, *301* (5639), 1503-1508.

[97] Zhou, Y.; Young, J. A.; Santrosyan, A.; Chen, K.; Yan, S. F.; Winzeler, E. A., *In silico* gene function prediction using ontology-based pattern identification. *Bioinformatics* **2005**, *21* (7), 1237-1245.

[98] Hu, G.; Cabrera, A.; Kono, M.; Mok, S.; Chaal, B. K.; Haase, S.; Engelberg, K.; Cheemadan, S.; Spielmann, T.; Preiser, P. R.; Gilberger, T. W.; Bozdech, Z., Transcriptional profiling of growth perturbations of the human malaria parasite Plasmodium falciparum. *Nat Biotechnol* **2010**, *28* (1), 91-98.

[99] Zhou, Y.; Ramachandran, V.; Kumar, K. A.; Westenberger, S.; Refour, P.; Zhou, B.; Li, F.; Young, J. A.; Chen, K.; Plouffe, D.; Henson, K.; Nussenzweig, V.; Carlton, J.; Vinetz, J. M.; Duraisingh, M. T.; Winzeler, E. A., Evidence-Based Annotation of the Malaria Parasite's Genome Using Comparative Expression Profiling. *PLoS ONE* **2008**, *3* (2),e1570.

[100] Bozdech, Z.; Llinas, M.; Pulliam, B. L.; Wong, E. D.; Zhu, J.; DeRisi, J. L., The transcriptome of the intraerythrocytic developmental cycle of Plasmodium falciparum. *PLoS Biol* **2003**, *1* (1), E5.

[101] Llinas, M.; Bozdech, Z.; Wong, E. D.; Adai, A. T.; DeRisi, J. L., Comparative whole genome transcriptome analysis of three Plasmodium falciparum strains. *Nucleic Acids Res* **2006**, *34* (4), 1166-1173.

[102] Daily, J. P.; Scanfeld, D.; Pochet, N.; Le Roch, K.; Plouffe, D.; Kamal, M.; Sarr, O.; Mboup, S.; Ndir, O.; Wypij, D.; Levasseur, K.; Thomas, E.; Tamayo, P.; Dong, C.; Zhou, Y.; Lander, E. S.; Ndiaye, D.; Wirth, D.; Winzeler, E. A.; Mesirov, J. P.; Regev, A., Distinct physiological states of Plasmodium falciparum in malaria-infected patients. *Nature* **2007**, *450* (7172), 1091-1095.

[103] Gonzales, J. M.; Patel, J. J.; Ponmee, N.; Jiang, L.; Tan, A.; Maher, S. P.; Wuchty, S.; Rathod, P. K.; Ferdig, M. T., Regulatory hotspots in the malaria parasite genome dictate transcriptional variation. *PLoS Biol* **2008**, *6* (9), e238.

[104] Huang, Y.; Wuchty, S.; Ferdig, M. T.; Przytycka, T. M., Graph theoretical approach to study eQTL: a case study of Plasmodium falciparum. *Bioinformatics* **2009**, *25* (12), i15-20.

[105] Lemieux, J. E.; Gomez-Escobar, N.; Feller, A.; Carret, C.; Amambua-Ngwa, A.; Pinches, R.; Day, F.; Kyes, S. A.; Conway, D. J.; Holmes, C. C.; Newbold, C. I., Statistical estimation of cell-cycle progression and lineage commitment in Plasmodium falciparum reveals a homogeneous pattern of transcription in *ex vivo* culture. *Proc Natl Acad Sci U S A* **2009**, *106* (18), 7559-7564.

[106] Jiang, H.; Patel, J. J.; Yi, M.; Mu, J.; Ding, J.; Stephens, R.; Cooper, R. A.; Ferdig, M. T.; Su, X. Z., Genome-wide compensatory changes accompany drug- selected mutations in the Plasmodium falciparum crt gene. *PLoS ONE* **2008**, *3* (6), e2484.

[107] Stubbs, J.; Simpson, K. M.; Triglia, T.; Plouffe, D.; Tonkin, C. J.; Duraisingh, M. T.; Maier, A. G.; Winzeler, E. A.; Cowman, A. F., Molecular mechanism for switching of P. falciparum invasion pathways into human erythrocytes. *Science* **2005**, *309* (5739), 1384-1387.

[108] Gaur, D.; Furuya, T.; Mu, J.; Jiang, L. B.; Su, X. Z.; Miller, L. H., Upregulation of expression of the reticulocyte homology gene 4 in the Plasmodium falciparum clone Dd2 is associated with a switch in the erythrocyte invasion pathway. *Mol Biochem Parasitol* **2006**, *145* (2), 205-215.

[109] Eksi, S.; Haile, Y.; Furuya, T.; Ma, L.; Su, X.; Williamson, K. C., Identification of a subtelomeric gene family expressed during the asexual-sexual stage transition in Plasmodium falciparum. *Mol Biochem Parasitol* **2005**, *143* (1), 90-99.

[110] Deitsch, K. W., A mark of silence in malaria parasites. *Cell Host Microbe* **2009**, *5* (2), 112-113.

[111] Deitsch, K. W.; Lukehart, S. A.; Stringer, J. R., Common strategies for antigenic variation by bacterial, fungal and protozoan pathogens. *Nat Rev Microbiol* **2009**, *7* (7), 493-503.

[112] Scherf, A.; Lopez-Rubio, J. J.; Riviere, L., Antigenic variation in Plasmodium falciparum. *Annu Rev Microbiol* **2008**, *62*, 445-470.

[113] Jiang, L.; Lopez-Barragan, M. J.; Jiang, H.; Mu, J.; Gaur, D.; Zhao, K.; Felsenfeld, G.; Miller, L. H., Epigenetic control of the variable expression of a Plasmodium falciparum receptor protein for erythrocyte invasion. *Proc Natl Acad Sci U S A* **2010**, *107* (5), 2224-2229.

[114] Rovira-Graells, N.; Gupta, A. P.; Planet, E.; Crowley, V. M.; Mok, S.; Ribas de Pouplana, L.; Preiser, P. R.; Bozdech, Z.; Cortes, A., Transcriptional variation in the malaria parasite Plasmodium falciparum. *Genome Res* **2012**, 22(5),925-938.

[115] Cui, L.; Fan, Q.; Miao, J., Histone lysine methyltransferases and demethylases in Plasmodium falciparum. *Int J Parasitol* **2008**, *38* (10), 1083-1097.

[116] Cui, L.; Miao, J.; Furuya, T.; Fan, Q.; Li, X.; Rathod, P. K.; Su, X. Z., Histone acetyltransferase inhibitor anacardic acid causes changes in global gene expression during *in vitro* Plasmodium falciparum development. *Eukaryot Cell* **2008**, *7* (7), 1200-1210.

[117] Fan, Q.; Miao, J.; Cui, L., Characterization of protein arginine methyltransferase I from Plasmodium falciparum. *Biochem J* **2009**, 421(1), 107-18.

[118] Merrick, C. J.; Duraisingh, M. T., Epigenetics in Plasmodium: what do we really know? *Eukaryot Cell* **2010**, *9* (8), 1150-1158.

[119] Cui, L.; Miao, J., Chromatin-mediated epigenetic regulation in the malaria parasite Plasmodium falciparum. *Eukaryot Cell* **2010**, *9* (8), 1138-1149.

[120] Volz, J. C.; Bartfai, R.; Petter, M.; Langer, C.; Josling, G. A.; Tsuboi, T.; Schwach, F.; Baum, J.; Rayner, J. C.; Stunnenberg, H. G.; Duffy, M. F.; Cowman, A. F., PfSET10, a Plasmodium falciparum methyltransferase, maintains the active var gene in a poised state during parasite division. *Cell Host Microbe* **2012**, *11* (1), 7-18.

[121] Salcedo-Amaya, A. M.; Hoeijmakers, W. A.; Bartfai, R.; Stunnenberg, H. G., Malaria: could its unusual epigenome be the weak spot? *Int J Biochem Cell Biol* **2010**, *42* (6), 781-784.

[122] Su, X. Z.; Heatwole, V. M.; Wertheimer, S. P.; Guinet, F.; Herrfeldt, J. A.; Peterson, D. S.; Ravetch, J. A.; Wellems, T. E., The large diverse gene family var encodes proteins involved in cytoadherence and antigenic variation of Plasmodium falciparum-infected erythrocytes. *Cell* **1995**, *82* (1), 89-100.

[123] Chen, Q.; Fernandez, V.; Sundstrom, A.; Schlichtherle, M.; Datta, S.; Hagblom, P.; Wahlgren, M., Developmental selection of var gene expression in Plasmodium falciparum. *Nature* **1998**, *394* (6691), 392-395.

[124] Chookajorn, T.; Dzikowski, R.; Frank, M.; Li, F.; Jiwani, A. Z.; Hartl, D. L.; Deitsch, K. W., Epigenetic memory at malaria virulence genes. *Proc Natl Acad Sci U S A* **2007**, *104* (3), 899-902.

[125] Duraisingh, M. T.; Voss, T. S.; Marty, A. J.; Duffy, M. F.; Good, R. T.; Thompson, J. K.; Freitas-Junior, L. H.; Scherf, A.; Crabb, B. S.; Cowman, A. F., Heterochromatin silencing and locus repositioning linked to regulation of virulence genes in Plasmodium falciparum. *Cell* **2005**, *121* (1), 13-24.

[126] Freitas-Junior, L. H.; Hernandez-Rivas, R.; Ralph, S. A.; Montiel-Condado, D.; Ruvalcaba-Salazar, O. K.; Rojas-Meza, A. P.; Mancio-Silva, L.; Leal-Silvestre, R. J.; Gontijo, A. M.; Shorte, S.; Scherf, A., Telomeric heterochromatin propagation and histone acetylation control mutually exclusive expression of antigenic variation genes in malaria parasites. *Cell* **2005**, *121* (1), 25-36.

[127] Deitsch, K. W.; Calderwood, M. S.; Wellems, T. E., Malaria. Cooperative silencing elements in var genes. *Nature* **2001**, *412* (6850), 875-876.

[128] Gannoun-Zaki, L.; Jost, A.; Mu, J.; Deitsch, K. W.; Wellems, T. E., A silenced Plasmodium falciparum var promoter can be activated *in vivo* through spontaneous deletion of a silencing element in the intron. *Eukaryot Cell* **2005**, *4* (2), 490-492.

[129] Lopez-Rubio, J. J.; Gontijo, A. M.; Nunes, M. C.; Issar, N.; Hernandez Rivas, R.; Scherf, A., 5' flanking region of var genes nucleate histone modification patterns linked to phenotypic inheritance of virulence traits in malaria parasites. *Mol Microbiol* **2007**, *66* (6), 1296-1305.

[130] Comeaux, C. A.; Coleman, B. I.; Bei, A. K.; Whitehurst, N.; Duraisingh, M. T., Functional analysis of epigenetic regulation of tandem RhopH1/clag genes reveals a role in Plasmodium falciparum growth. *Mol Microbiol* **2011**, *80* (2), 378-390.

[131] Crowley, V. M.; Rovira-Graells, N.; Ribas de Pouplana, L.; Cortes, A., Heterochromatin formation in bistable chromatin domains controls the epigenetic repression of clonally variant Plasmodium falciparum genes linked to erythrocyte invasion. *Mol Microbiol* **2011**, *80* (2), 391-406.

[132] Lopez-Rubio, J. J.; Mancio-Silva, L.; Scherf, A., Genome-wide analysis of heterochromatin associates clonally variant gene regulation with perinuclear repressive centers in malaria parasites. *Cell Host Microbe* **2009**, *5* (2), 179-190.

[133] Salcedo-Amaya, A. M.; van Driel, M. A.; Alako, B. T.; Trelle, M. B.; van den Elzen, A. M.; Cohen, A. M.; Janssen-Megens, E. M.; van de Vegte-Bolmer, M.; Selzer, R. R.; Iniguez, A. L.; Green, R. D.; Sauerwein, R. W.; Jensen, O. N.; Stunnenberg, H. G., Dynamic histone H3 epigenome marking during the intraerythrocytic cycle of Plasmodium falciparum. *Proc Natl Acad Sci U S A* **2009**, *106* (24), 9655-9660.

[134] Chaal, B. K.; Gupta, A. P.; Wastuwidyaningtyas, B. D.; Luah, Y. H.; Bozdech, Z., Histone deacetylases play a major role in the transcriptional regulation of the Plasmodium falciparum life cycle. *PLoS Pathog 6* (1), e1000737.

[135] Florens, L.; Washburn, M. P.; Raine, J. D.; Anthony, R. M.; Grainger, M.; Haynes, J. D.; Moch, J. K.; Muster, N.; Sacci, J. B.; Tabb, D. L.; Witney, A. A.; Wolters, D.; Wu, Y.; Gardner, M. J.; Holder, A. A.; Sinden, R. E.; Yates, J. R.; Carucci, D. J., A proteomic view of the Plasmodium falciparum life cycle. *Nature* **2002**, *419* (6906), 520-526.

[136] Lasonder, E.; Ishihama, Y.; Andersen, J. S.; Vermunt, A. M.; Pain, A.; Sauerwein, R. W.; Eling, W. M.; Hall, N.; Waters, A. P.; Stunnenberg, H. G.; Mann, M., Analysis of the Plasmodium falciparum proteome by high-accuracy mass spectrometry. *Nature* **2002**, *419* (6906), 537-542.

[137] Cooper, R. A.; Carucci, D. J., Proteomic approaches to studying drug targets and resistance in Plasmodium. *Curr Drug Targets Infect Disord* **2004**, *4* (1), 41-51.

[138] Prieto, J. H.; Koncarevic, S.; Park, S. K.; Yates, J., 3rd; Becker, K., Large-scale differential proteome analysis in Plasmodium falciparum under drug treatment. *PLoS ONE* **2008**, *3* (12), e4098.

[139] Belli, S. I.; Walker, R. A.; Flowers, S. A., Global protein expression analysis in apicomplexan parasites: current status. *Proteomics* **2005**, *5* (4), 918-924.

[140] Lal, K.; Prieto, J. H.; Bromley, E.; Sanderson, S. J.; Yates, J. R., 3rd; Wastling, J. M.; Tomley, F. M.; Sinden, R. E., Characterisation of Plasmodium invasive organelles; an ookinete microneme proteome. *Proteomics* **2009**, 9(5),1142-1151.

[141] Tarun, A. S.; Peng, X.; Dumpit, R. F.; Ogata, Y.; Silva-Rivera, H.; Camargo, N.; Daly, T. M.; Bergman, L. W.; Kappe, S. H., A combined transcriptome and proteome survey of malaria parasite liver stages. *Proc Natl Acad Sci U S A* **2008**, *105* (1), 305-310.

[142] Lasonder, E.; Janse, C. J.; van Gemert, G. J.; Mair, G. R.; Vermunt, A. M.; Douradinha, B. G.; van Noort, V.; Huynen, M. A.; Luty, A. J.; Kroeze, H.; Khan, S. M.; Sauerwein, R. W.; Waters, A. P.; Mann, M.; Stunnenberg, H. G., Proteomic profiling of Plasmodium sporozoite maturation identifies new proteins essential for parasite development and infectivity. *PLoS Pathog* **2008**, *4* (10), e1000195.

[143] Khan, S. M.; Franke-Fayard, B.; Mair, G. R.; Lasonder, E.; Janse, C. J.; Mann, M.; Waters, A. P., Proteome analysis of separated male and female gametocytes reveals novel sex-specific Plasmodium biology. *Cell* **2005**, *121* (5), 675-687.

[144] Trelle, M. B.; Salcedo-Amaya, A. M.; Cohen, A. M.; Stunnenberg, H. G.; Jensen, O. N., Global Histone Analysis by Mass Spectrometry Reveals a High Content of Acetylated Lysine Residues in the Malaria Parasite Plasmodium falciparum. *J Proteome Res* **2009**, 8(7),3439-3450.

[145] Solyakov, L.; Halbert, J.; Alam, M. M.; Semblat, J. P.; Dorin-Semblat, D.; Reininger, L.; Bottrill, A. R.; Mistry, S.; Abdi, A.; Fennell, C.; Holland, Z.; Demarta, C.; Bouza, Y.; Sicard, A.; Nivez, M. P.; Eschenlauer, S.; Lama, T.; Thomas, D. C.; Sharma, P.; Agarwal, S.; Kern, S.; Pradel, G.; Graciotti, M.; Tobin, A. B.; Doerig, C., Global kinomic and phospho-proteomic analyses of the human malaria parasite Plasmodium falciparum. *Nat Commun* **2011**, *2*, 565.

[146] Trieu, A.; Kayala, M. A.; Burk, C.; Molina, D. M.; Freilich, D. A.; Richie, T. L.; Baldi, P.; Felgner, P. L.; Doolan, D. L., Sterile protective immunity to malaria is associated with a panel of novel P. falciparum antigens. *Mol Cell Proteomics* **2011**, *10* (9), M111 007948.

[147] Crompton, P. D.; Kayala, M. A.; Traore, B.; Kayentao, K.; Ongoiba, A.; Weiss, G. E.; Molina, D. M.; Burk, C. R.; Waisberg, M.; Jasinskas, A.; Tan, X.; Doumbo, S.; Doumtabe, D.; Kone, Y.; Narum, D. L.; Liang, X.; Doumbo, O. K.; Miller, L. H.; Doolan, D. L.; Baldi, P.; Felgner, P. L.; Pierce, S. K., A prospective analysis of the Ab response to Plasmodium falciparum before and after a malaria season by protein microarray. *Proc Natl Acad Sci U S A* **2010**, *107* (15), 6958-6963.

[148] Di Girolamo, F.; Raggi, C.; Birago, C.; Pizzi, E.; Lalle, M.; Picci, L.; Pace, T.; Bachi, A.; de Jong, J.; Janse, C. J.; Waters, A. P.; Sargiacomo, M.; Ponzi, M., Plasmodium lipid rafts

contain proteins implicated in vesicular trafficking and signalling as well as members of the PIR superfamily, potentially implicated in host immune system interactions. *Proteomics* **2008**, *8* (12), 2500-2513.

[149] Singh, M.; Mukherjee, P.; Narayanasamy, K.; Arora, R.; Gupta, S.; Natarajan, K.; Malhotra, P., Proteome analysis of plasmodium falciparum extracellular secretory antigens at asexual blood stages reveals a cohort of proteins with possible roles in immune modulation and signaling. *Mol Cell Proteomics* **2009**, 8(9),2102-2018.

[150] Olszewski, K. L.; Morrisey, J. M.; Wilinski, D.; Burns, J. M.; Vaidya, A. B.; Rabinowitz, J. D.; Llinas, M., Host-parasite interactions revealed by Plasmodium falciparum metabolomics. *Cell Host Microbe* **2009**, *5* (2), 191-199.

[151] Wass, M. N.; Stanway, R.; Blagborough, A. M.; Lal, K.; Prieto, J. H.; Raine, D.; Sternberg, M. J.; Talman, A. M.; Tomley, F.; Yates, J.; Sinden, R. E., Proteomic analysis of Plasmodium in the mosquito: progress and pitfalls. *Parasitology* **2012**, 1-15.

[152] Tyagi, N.; Swapna, L. S.; Mohanty, S.; Agarwal, G.; Gowri, V. S.; Anamika, K.; Priya, M. L.; Krishnadev, O.; Srinivasan, N., Evolutionary divergence of Plasmodium falciparum: sequences, protein-protein interactions, pathways and processes. *Infect Disord Drug Targets* **2009**, *9* (3), 257-271.

[153] Bongfen, S. E.; Laroque, A.; Berghout, J.; Gros, P., Genetic and genomic analyses of host-pathogen interactions in malaria. *Trends Parasitol* **2009**, *25* (9), 417-422.

[154] Kafsack, B. F.; Llinas, M., Eating at the table of another: metabolomics of host-parasite interactions. *Cell Host Microbe 7* (2), 90-99.

[155] Balu, B.; Shoue, D. A.; Fraser, M. J., Jr.; Adams, J. H., High-efficiency transformation of Plasmodium falciparum by the lepidopteran transposable element piggyBac. *Proc Natl Acad Sci U S A* **2005**, *102* (45), 16391-16396.

[156] Janse, C. J.; Ramesar, J.; Waters, A. P., High-efficiency transfection and drug selection of genetically transformed blood stages of the rodent malaria parasite Plasmodium berghei. *Nat Protoc* **2006**, *1* (1), 346-356.

# INDEX

## A



## B



Christian Neri (Ed)

## C

**D**



**E**

**H**

# I



# K



# L



# M

**N**

**O**



**P**

**Q**



**R**

## S

**T**



**V**



**Y**

Made in United States
North Haven, CT
04 November 2026

21727474R00190